Electrical Connectors

Electrical Connectors

Design, Manufacture, Test, and Selection

Edited by
San Kyeong and Michael G. Pecht
Center for Advanced Life Cycle Engineering(CALCE)
University of Maryland, College Park, MD, USA

WILEY

IEEE PRESS

This edition first published 2021

Registered Offices
John Wiley & Sons, Inc., 111 River Street, Hoboken, NJ 07030, USA
John Wiley & Sons Ltd, The Atrium, Southern Gate, Chichester, West Sussex, PO19 8SQ, UK

Editorial Office
The Atrium, Southern Gate, Chichester,West Sussex, PO19 8SQ, UK

For details of our global editorial offices, customer services, and more information about Wiley products visit us at www.wiley.com.

Wiley also publishes its books in a variety of electronic formats and by print-on-demand. Some content that appears in standard print versions of this book may not be available in other formats.

Library of Congress Cataloging-in-Publication Data

Names: Kyeong, San, editor. | Pecht, Michael G., editor.
Title: Electrical connectors : design, manufacture, test, and selection / edited by San Kyeong and Michael G. Pecht, Center for Advanced Life Cycle Engineering (CALCE), University of Maryland, USA.
Description: Hoboken, New Jersey : Wiley IEEE Press, 2021. | Includes index.
Identifiers: LCCN 2020028241 (print) | LCCN 2020028242 (ebook) | ISBN 9781119679769 (cloth) | ISBN 9781119679806 (adobe pdf) | ISBN 9781119679820 (epub)
Subjects: LCSH: Electric connectors.
Classification: LCC TK3521 .E39 2021 (print) | LCC TK3521 (ebook) | DDC 621.319/3–dc23
LC record available at https://lccn.loc.gov/2020028241
LC ebook record available at https://lccn.loc.gov/2020028242

Cover Design: Wiley
Cover Images: (inset) Designed by San Kyeong

Set in 10/12pt, WarnockPro by SPi Global, Chennai, India

10 9 8 7 6 5 4 3 2 1

MIX
Paper from responsible sources
FSC
www.fsc.org
FSC® C013604

Contents

About the Editors

San Kyeong (san.kyeong@samsung.com) is a staff engineer at the R&D headquarters of Samsung Electro-Mechanics. He is currently with the Center for Advanced Life Cycle Engineering (CALCE) at the University of Maryland. He received a BE degree and PhD in chemical and biological engineering from the Seoul National University of Seoul, South Korea, in 2010 and 2016, respectively. He has expertise in material engineering for passive electronic components. He is the author/co-author of Chapters 1, 6, 7, 8, 12, 15, and 17.

Michael G. Pecht (michael.pecht@protonmail.com) has a BS in physics, an MS in electrical engineering, and an MS and PhD in engineering mechanics from the University of Wisconsin at Madison. He is the George E. Dieter chair professor of mechanical engineering and a professor of applied mathematics, statistics, and scientific computation at the University of Maryland. He is the founder of the Center for Advanced Life Cycle Engineering (CALCE) at the University of Maryland, which is sponsored by more than 150 of the world's leading electronics companies. He is a professional engineer, an IEEE fellow, an ASME fellow, an SAE fellow, and an IMAPS fellow. He served as editor in chief of *Circuit World* for two years, as editor in chief of *IEEE Access* for six years, as editor in chief of *IEEE Transactions on Reliability* for nine years, and as editor in chief of *Microelectronics Reliability* for 16 years. He has also served on three US National Academy of Science studies, on two US Congressional investigations into automotive safety, and as an expert to the US FDA. He has written more than 40 books on product reliability, development, and use, and supply chain management. He has also written a series of books on the electronics industry in China, Korea, Japan, and India. He has written more than 900 technical articles and holds 10 patents. He has consulted for more than 50 major international electronics companies, providing expertise in strategic planning, design, test, prognostics, intellectual property, and risk assessment of electronic products and systems.

List of Contributors

Michael H. Azarian
Center for Advanced Life Cycle Engineering (CALCE), University of Maryland, College Park, MD, USA

Ilknur Baylakoglu
SATURN Engineering, TUBITAK, Ankara, Turkey

Deepak Bondre
Center for Advanced Life Cycle Engineering (CALCE), University of Maryland, College Park, MD, USA

Diganta Das
Center for Advanced Life Cycle Engineering (CALCE), University of Maryland, College Park, MD, USA

Jens Engström
Division for Electricity, Uppsala University, Uppsala, Sweden

Chien-Ming Huang
Center for Advanced Life Cycle Engineering (CALCE), University of Maryland, College Park, MD, USA

Lovlesh Kaushik
Center for Advanced Life Cycle Engineering (CALCE), University of Maryland, College Park, MD, USA

San Kyeong
Center for Advanced Life Cycle Engineering (CALCE), University of Maryland, College Park, MD, USA

Kyle LoGiudice
Center for Advanced Life Cycle Engineering (CALCE), University of Maryland, College Park, MD, USA

Michael G. Pecht
Center for Advanced Life Cycle Engineering (CALCE), University of Maryland, College Park, MD, USA

Flore Remouit
Division for Electricity, Uppsala University, Uppsala, Sweden

Pablo Ruiz-Minguela
Division for Energy and Environment, TECNALIA, Derio, Spain

Neda Shafiei
Center for Advanced Life Cycle Engineering (CALCE), University of Maryland, College Park, MD, USA

Bhanu Sood
NASA Goddard Space Flight Center, Greenbelt, MD, USA

Lei Su
Jiangsu Key Laboratory of Advanced Food Manufacturing Equipment and Technology, Jiangnan University, Wuxi, China

Xiaonan Yu
Jiangsu Key Laboratory of Advanced Food Manufacturing Equipment and Technology, Jiangnan University, Wuxi, China

Preface

Connectors allow circuits and products that are manufactured independently to be electrically connected, tested, maintained, and upgraded. Manufacturers are thus able to optimize electronic systems and keep up with the rapid pace of electronic product development. Today, the connector market is valued at more than $50 billion, and there is still an increasing trend in the usage and applications of electronic connectors.

This book covers everything one would need to design, manufacture, and select a connector for any targeted application. It covers the science of contact physics and the engineering involved in the choice and manufacture of contact materials, contact finishes, housing materials, and the full connector assembly process. Test methods, performance and reliability concerns, and guidelines are provided, and various application requirements and selection considerations are discussed.

Key features of the book include the following:

- Provides a comprehensive description of electrical connectors, from the components and materials that comprise connectors to their applications and classifications
- Provides information on the design and manufacture of all the parts of a connector
- Thoroughly explains the key performance and reliability concerns and trade-offs related to electrical connector selection
- Details current industry standards for performance, reliability, and safety assurance
- Examines application-specific criteria for contact resistance, signal quality, and temperature rise
- Provides information on key suppliers, materials used, and different types of data provided
- Presents guidelines for end users who have concerns about connector selection and new connector designs

This book is intended for connector designers and manufacturers, device manufacturers that select connectors for their products, and end users who are concerned with the operation and reliability of the whole product or system. Students who are learning the basics of electrical contacts will also find this book instructive. The book can also be considered as a guidebook, offering best-practice advice on how to choose and test the correct connectors for a targeted application. A summary of the chapters is presented here.

Chapter 1 describes the advantages and challenges of separable connectors. It covers the components of a connector: contact spring, contact finish, connector housing, and contact interface. The chapter discusses the types of components, and it ends with a short glossary of basic connector terminology.

Chapter 2 discusses the housing that electrically insulates contact members and provides mechanical support for proper electrical connection. This chapter reviews the design and manufacture of the connector housing as well as the mechanical properties. It also presents relevant standards. The second half of the chapter focuses on materials and design trade-offs, including a comparison of the advantages and limitations of various thermoplastic polymers.

Chapter 3 discusses contact spring materials in terms of their design trade-offs, including electrical conductivity, mechanical strength, resistance to stress relaxation or creep, solderability, and resistance to corrosion. Because copper alloys are the predominant material for contacts, their properties are discussed, and the Unified Number System for alloys is explained. The second half of the chapter includes a section on design trade-offs associated with conductive elastomers and concludes with a discussion of how metal alloy contacts are manufactured. The terminology for commonly used treatment is explained.

Chapter 4 provides insight into managing the mechanical and electrical properties of the contact interface by choosing the appropriate plating material and plating thickness. Contact plating can be divided into two categories: noble metal plating and non-noble metal plating. These categories are covered in detail, with emphasis on gold, silver, tin, and nickel. The discussion on silver looks at tarnish-accelerating factors such as chlorine gas, water, nitrogen dioxide, and ozone, and covers management of silver corrosion. Finally, the chapter discusses the four methods of contact plating.

Chapter 5 discusses the contact force that is required to maintain a consistent and reliable contact interface along with contact resistance and contact deflection. The equation for calculating the insertion force is given. The standards relating to measurement of the insertion and extraction forces, as well as the retention force, are listed. Zero-insertion-force (ZIF) connectors are also described.

Chapter 6 focuses on the actual surface area of a connector that makes true electrical contact, which differs from the total surface area. The chapter explores factors that affect connector mating, such as contact resistance, temperature, electrical current, and fretting. The chapter provides standards for measuring contact force, contact resistance, current rating, capacitance and inductance, and contact electrical performance. It also describes the mechanics of the mating process, with emphasis on the contact wipe motion. Finally, the chapter discusses determination of the maximum current that a conductor can carry.

Chapter 7 considers the interface between the connector and the cable, circuit board, or assembly. While the contact interface is critical for performance and reliability, the back-end connection also affects connector assembly cost and reliability. Soldering, press-fit, crimping, and insulation displacement are described along with their processing flow, mechanism, advantages, and limitations.

Chapter 8 explores environmental and operating load conditions and the failure mechanisms that can arise from the load conditions. The performance and reliability of an electrical connector will depend on its ability to transmit power or signals from the source to the electrical circuit via the contact interface under different loading. To predict and enhance the reliability of electrical connectors, the underlying root cause of any failure mechanism must be investigated. Fishbone diagrams illustrate the causes and effects of each of the failure mechanism discussed, which include silver migration, tin whiskers, corrosion, arc formation, creep, and surface wear.

Chapter 9 focuses on fretting, one of the key failure mechanisms in electromechanical apparatuses having touching surfaces that move relative to each other. Fretting can cause serious degradation in electrical components with separable contacts, such as connectors. The chapter covers the studies on materials, including frictional polymer-forming metals. It also discusses testing methodologies and test conditions that researchers use.

Chapter 10 provides insight into the testing conducted to verify that the connectors will perform properly over time. Testing by the manufacturer usually involves performance testing per the datasheet requirements. However, the customer may have applications that require performance under unique operational and environmental conditions. This chapter discusses electrical tests for connectors, including dielectric withstanding voltage, low-signal-level contact resistance, insulation resistance, contact resistance, and standing wave ratio. Environmental tests include humidity, temperature, and contaminating conditions, and mechanical stress environments include vibration, mechanical shock, and durability cycling exposures. Standards are listed for each of the test methods. The chapter also discusses specific mixed flowing gas (MFG) tests.

Chapter 11 emphasizes that reliability must be managed across all the tiers of the supply chain. Each supplier's reliability practices need to be adequate to satisfy the end-product requirements of the customers. After discussing the guidelines provided by the IEEE Reliability Program Standard 1332 for the development of a reliability program, the chapter discusses key reliability practices in detail. The chapter concludes with a description of the process for evaluating the reliability capability of suppliers.

Chapter 12 covers the key factors/requirements that must be considered when selecting a connector, including physical application, electrical requirements, mechanical requirements, application environment, mating force, number of mating cycles, current and power ratings, termination types, environmental conditions (e.g. temperature, vacuum, chemicals, radiation, and vibration), materials, and reliability. The chapter discusses screening tests, with emphasis on NASA's three levels of screening. It also explores the use of failure modes and effects analysis (FMEA) to identify the possible failure modes and causes of devices. The chapter concludes with a discussion on recent experiments to test connector reliability.

Chapter 13 focuses on issues involving high-speed signal connectors, such as signal integrity, transmission quality, signal distortion, interconnect delays, and electromagnetic compatibility. The chapter discusses the process of virtual prototyping. It describes a vector network analyzer, which determines the magnitude and phase characteristics of connectors. The chapter introduces filtered connectors. The chapter concludes with a section on SPICE and how it is used in connector circuit modeling.

Chapter 14 focuses on a mass storage device interface that connects hard disk, optical disk, and solid-state drives to the computer motherboard. These connectors have important applications in data transmission for personal computer hard disk drives and optical drives alike. They are the intermediate link between a computer's motherboard and the disk controller on the hard disk drive, allowing the drive to read and write information at a fast speed.

Chapter 15 focuses on the requirements (design and use) of connectors for power transmission. The types of power connectors are introduced. It discusses a theoretical approach to estimate temperature rise by constriction current at the contact point. The chapter concludes with a description of the points to consider when setting the rated current and the setting procedure.

Chapter 16 provides an overview of the type of connectors that exist for underwater and subsea electrical connections. The first part presents different types of commercial-off-the-shelf electrical connectors, lists their performances and defaults, and details their typical failure modes and known causes. The second part contains the theory on sealing and on connectors' thermal, electrical, and mechanical properties. The last part provides information on connection procedures and other details about connecting subsea cables, with a focus on the connection of offshore renewable energy farms.

Chapter 17 discusses some of the latest products from the key connector manufacturers (Amphenol, Fischer, Hirose, Molex, and TE Connectivity) and focuses on the specifications, types, limitations, advantages, and usage in various targeted applications.

The **appendix** lists the recognized specification standards for connectors. These documents were developed by professional organizations internationally.

In addition to the main authors, the following researchers from the Center for Advanced Life Cycle Engineering (CALCE) at the University of Maryland contributed to this book: Chien-Ming Huang, Lovlesh Kaushik Dr. Bhanu Sood, Dr. Michael Azarian, Dr. Diganta Das, Ilknur Baylakoglu, Neda Shafiei, and Kyle LoGiudice. Various CALCE interns were involved in gathering information, including Sumeer Khanna, Gopalakrish Kalarikovilagam Subramanian, Amarah Ahmed, and Ayeesha Jaswal, as well as Cheryl Wurzbacher who conducted an English edit of the book. Other co-authors include Drs. Flore Remouit, Jens Engström, Pablo Ruiz-Minguela, and Lei Su and his student Xiaonan Yu.

Dr. San Kyeong and Professor Michael G. Pecht
Center for Advanced Life Cycle Engineering (CALCE)
University of Maryland – College Park

1

What Is an Electrical Connector?

Michael G. Pecht and San Kyeong

Center for Advanced Life Cycle Engineering (CALCE), University of Maryland, College Park, MD, USA

An electrical connector is an electromechanical component that provides a separable interface between two parts of an electronic system without compromising the performance of the system. Separability enables circuits that are manufactured independently to be electrically connected, tested, maintained, and upgraded. For example, in a computer or smartphone, the main circuit board and the display and its electronics are manufactured at different locations and are electrically connected using connectors. These subsystems (assemblies) can be tested and upgraded independently of each other. Connectors thus remove the problems of permanent connection of parts and eliminate issues associated with rewiring interconnects [1], such as damage to circuit board conduction paths due to repeated soldering and de-soldering.

Connectors provide much more flexibility to optimize electronic systems than permanent connections. For example, there are a variety of connectors on the market to satisfy the needs for testing, burn-in, and assembly of different kinds of electronic systems. They help in supply-chain management, avoid direct connections, provide opportunities for part replacement and repair, and, in some cases, offer cost savings.

1.1 Challenges of Separable Connectors

The evolution of microelectronics toward higher speeds and switching frequency enforces stringent requirements on connectors. The challenges that confront connectors are extra electrical length, the need for a larger assembly area, compatibility with new technologies, and reliability concerns [2]. A connector introduces an extra electrical path that can lead to added propagation delay and electrical noise, such as cross talk. A connector also introduces new concerns for reliability, because degradation of the contact interface can lead to an increase in contact resistance, which can lead to signal degradation, joule heating, and power loss [2]. These losses are due to increases in contact resistance, which is dependent on the material properties of the connectors, environmental conditions, and mode of operation. Furthermore, because a connector occupies more area than a simple solder or adhesive, the use of connectors might be difficult in applications that have constraints on area.

Electrical Connectors: Design, Manufacture, Test, and Selection, First Edition.
Edited by San Kyeong and Michael G. Pecht.

In the case of signal connectors, signal transmission quality (STQ) must be maintained. STQ refers to the transmission of high-speed signals through connectors and interconnects without loss of signal quality. The losses may be due to cross talk, transmission delay, or characteristic impedance variation on signal propagation and reflections. The transmission and preservation of signal waveforms forms the crux of STQ.

Electromagnetic compatibility (EMC) refers to the ability of the system to prevent the degradation of the digital signals from external electromagnetic energy. The mitigation of EMC generally involves shielding, filtering, and grounding practices to control electromagnetic interference (EMI).

1.2 Components of a Connector

The main components of a connector include the contact springs, contact finish, and connector housing, as shown in Figure 1.1 [3]. The contact interface is defined by the physical–electrical connection of the pieces of the connector that are mated, and this determines how well the connector performs.

1.2.1 Contact Springs

The contact spring provides the path for the transmission of a signal, power, and/or ground between the circuits that the connector connects. It also provides the normal force, the component of the force that is perpendicular to the surface of contact, which helps in the formation and maintenance of the separable interface [4].

The key mechanical requirements of the contact spring are insertion and extraction force, contact force, contact retention, and contact wipe. The electrical requirements of the contact spring are contact resistance, current rating, inductance, capacitance, and bandwidth. Chapters 3 and 6 provide more discussion on these requirements.

There are two types of contact springs: a receptacle, which is generally a spring member, and a plug, which in most cases is rigid and provides a means for deflecting the receptacle spring to generate the contact normal force [5]. A receptacle contact spring

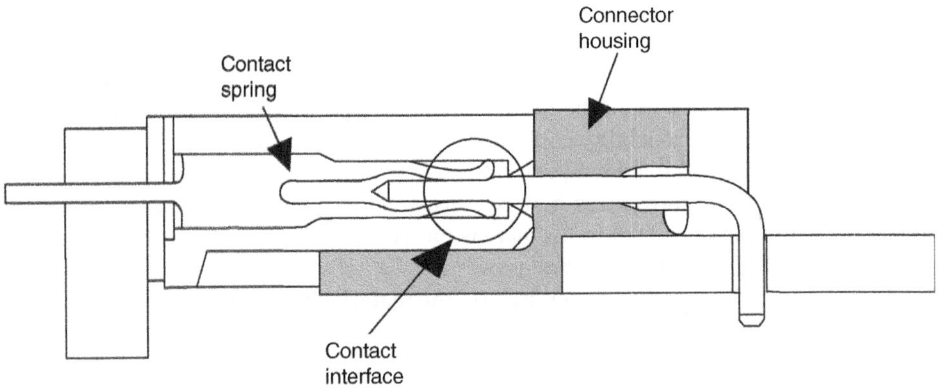

Figure 1.1 Schematic illustration of a typical connector [3].

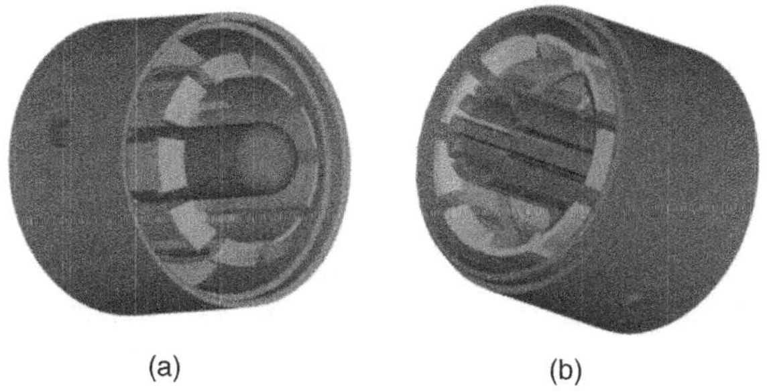

Figure 1.2 (a) Plug contact spring and (b) receptacle contact spring [6].

ensures low insertion force while mating and helps the connector endure overstress during nonaligned insertions. Figure 1.2 shows examples of plug and receptacle contact springs [6].

The material properties that influence the performance of the contact spring are the modulus of elasticity and the yield strength. The modulus of elasticity is the ratio of the stress applied to the object to the resulting strain within the elastic limit. The yield strength is the point of stress at which the body begins to deform plastically. Before the yield strength is reached, the body deforms elastically – that is, it returns to its original shape when the stress is removed. These properties influence the deflection abilities of the spring and the amount of deflection the spring can support while remaining elastic. This also affects the contact force required to make a proper electrical connection.

Stress relaxation resistance is a property of the material that reduces the contact normal force over time [6]; thus, it is a selection criterion that can affect connector performance. It is the decrease in stress that occurs when the structure is kept in the same strained posture for some interval of time, causing the formation of plastic strain [5]. Loss of normal force through plastic deformation is a design exercise to ensure that spring stresses are not excessive. The yield strength is different for commonly used copper alloys; hence, stress relaxation resistance varies across them. Beryllium copper is the most commonly used alloy when stress relaxation resistance is a concern. Phosphor bronzes are also suitable for most applications [7].

1.2.2 Contact Finishes

The contact finish protects the contact spring base metal from corrosion and limits the formation of films on the surface of the contact spring. To be effective, the contact finish has to completely cover the contact spring and must be corrosion-resistant.

Films that can increase the contact resistance include thin oxides, sulfides, chloride, and complex mixtures of film layers of the metals formed on the contact's surface. The reduction of contact resistance requires the formation of a metallic interface that is free of films. Contact finishes can be made of noble metals (e.g. gold, palladium, and alloys of these metals) or non-noble metals (e.g. silver, tin). The type of finish determines the types of surface films that can form on the contact interfaces. The noble metals, especially gold, are inert because there is no formation of any oxides on the surface. In the case of non-noble metal finishes made of tin, however, tin-oxide may form in the contact spring and may need to be periodically removed.

Owing to the repeated mating and unmating in separable contact interfaces, some of the surface film is removed by frictional forces. Therefore, two layers of contact finishes referred to as "duplex-plated" are used [7] to protect the underlying metal. These consist of a layer of noble metal coated on top of a layer of nickel. Duplex-plated connectors are discussed in Chapter 3.

1.2.2.1 Noble Metal Contact Finishes

The noble metals gold and, to a lesser extent, palladium and its alloys are inert in typical connector operating environments without the presence of acidic gases. But in environments where chlorine or sulfur is present, corrosion of noble metal–finished connectors can occur. Shielding by housing can help prevent corrosion. Lubrication is another means of protection.

Pore corrosion and corrosion due to exposed base metal are of particular concern. If the coating with noble metal is not thick enough or if it is not continuous, the underplate and base metal will be exposed to the environment, leading to corrosion. At elevated temperatures, base metal atoms may migrate to the contact surface and react with oxygen and pollutant gases, allowing corrosion products to migrate out of the pores. This phenomenon is called "pore corrosion." Hence, pore corrosion of the copper layer beneath the noble metal layer can take place when the noble metal layer becomes porous, exposing the copper to a corrosive environment containing acidic gases of sulfides and chlorides.

Because gold is a noble metal and because thin gold platings tend to be porous, gold coatings are susceptible to the creep of base metal corrosion products across the surface of the gold after formation at pore sites and edge boundaries. Corrosion creep can be inhibited by applying an overall nickel coating prior to application of the gold.

1.2.2.2 Non-noble Metal Contact Finishes

Non-noble metal contact finishes consist primarily of tin, although silver, nickel, solder, and lead are commonly used. Alloys of tin-lead and nickel-tin are also used sometimes. Tin finishes degrade primarily because of fretting corrosion, which can occur in any operating environment [7]. Fretting corrosion of electrical contacts is caused by repeated micromotions between closed contacts, creating oxides or wear debris that can raise contact resistance. Micromotion can result from vibration, shock, or differential thermal expansion of materials in contact. The connector must be designed to minimize fretting susceptibility, which can be done by providing sufficient friction force at the contact interface to prevent motion from occurring [8].

Tribology refers to the study of friction, lubrication, and wear of contact surfaces. Contact finishes are generally thin. It is therefore essential to maintain the integrity of the contact finishes when subjugated to wear and tear to protect the base metals from exposure to a corrosive environment. The resistance to corrosion and tarnishing and the thermal stability of contact resistance should be kept in mind when choosing a contact finish [8]. Chapter 4 presents a more detailed discussion on contact finish materials.

1.2.3 Connector Housing

Connectors must remain dimensionally stable in the presence of extreme chemical and temperature effects. The maintenance of connector center line spacing, straightness,

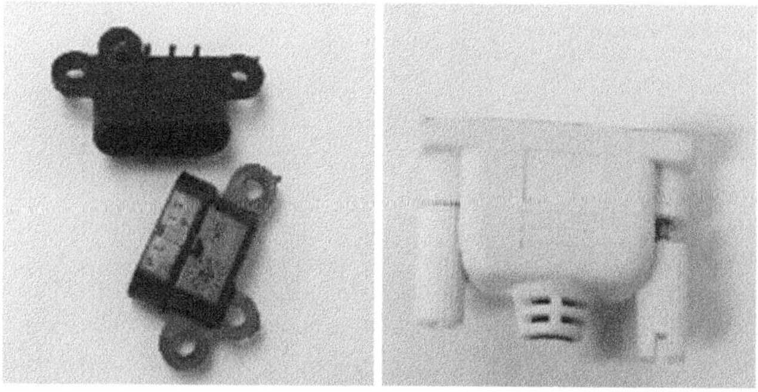

Figure 1.3 Examples of connector housings.

and flatness is necessary for ensuring proper connector assembly and mating behavior [9]. The connector housing achieves this stability by electrically insulating and mechanically shielding the contact spring, maintaining the position of contacts, and providing mechanical protection of the contacts from the operating environment. Figure 1.3 shows examples of connector housing.

The electrical properties that affect the insulating abilities of the housing include surface and volume resistivity and dielectric withstanding voltage. The mechanical characteristics of the connector housing include flexural strength/modulus and creep strength.

Most of the contact housing designs are similar, but the material used in them varies. This material range serves not only to meet the environmental conditions during operation but also to meet the conditions while manufacturing and assembling. Some of the common materials that are used for making connector housing are polyphenylene sulfide (PPS), polyethylene terephthalate (PET), polybutylene terephthalate (PBT), polycyclohexylenedimethylene terephthalate (PCT), liquid-crystalline polymer (LCP), FR-4, and polyimide. Chapter 2 provides more discussion on connector housing.

1.2.4 Contact Interface

A separable interface is established each time the "male" and "female" parts of the connector are brought into contact. There is a need to create and maintain contact interfaces to achieve the desired electrical performance. The metallic interface is created mechanically [6].

When separable connectors are mated, only the high spots on the surfaces, which are called "asperities," come into contact. Therefore, the entire connector surface does not come into contact. The asperities depend on the geometry of the surfaces in contact. Figure 1.4 depicts a normal contact interface [10]. It can be seen from the figures that only the asperities of both surfaces come into contact. The size and the number of asperities depend on the surface roughness and the applied load. The applied load also determines the magnitude of the contact area.

The structure of the contact interface depends on the roughness of the surface (which influences the number of asperities), the applied force, and the contact interface geometry. The mechanical characteristics of the contact interfaces, particularly the frictional and wear forces, depend on the asperity microstructure of the contact interfaces [7].

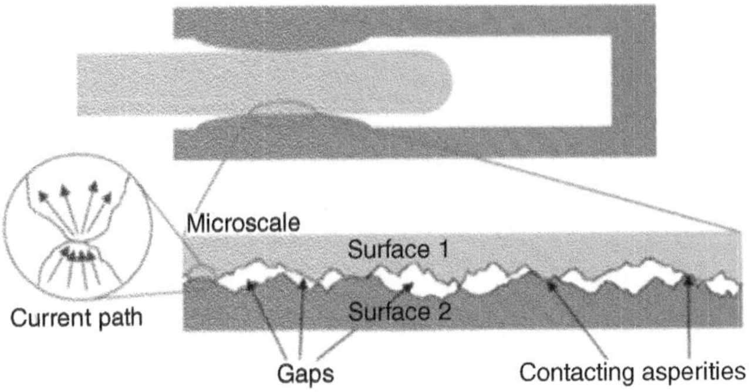

Figure 1.4 Contact interface [10].

The constriction resistance of the contact interfaces also depends on the asperities. A discussion of contact interface is provided in Chapter 6.

1.3 Connector Types

Connectors are classified into three types based on their termination ends: board-to-board connectors, cable/wire-to-cable/wire connectors, and cable/wire-to-board connectors. These are discussed in the following sections. Figure 1.5 gives a pictorial example of the levels of interconnection discussed in this section. An electronic system is a hierarchical interconnection network that allows communication among different electronic devices. Several interconnects are required to ensure proper functioning of electronic devices for signal transmission and power distribution.

As shown in Figure 1.5, six levels of interconnection are normally seen in connectors. Level 0 is the connection between a basic circuit element and its lead, such as the link between a semiconductor chip and the lead frame. Level 1 is the connection between a component lead and a printed circuit board (PCB), exemplified by chip carrier sockets, dual inline package (DIP) sockets, and switches. Level 2 is the connection between two or more PCBs. A motherboard–daughterboard connection is typical. Level 3 is the

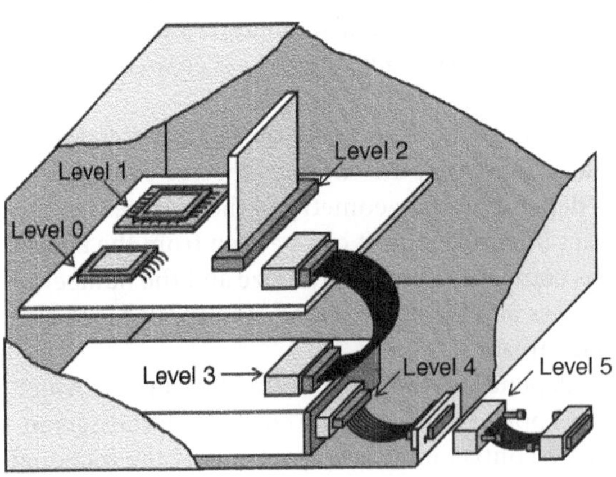

Figure 1.5 Levels of interconnection in connectors.

connection between two subassemblies, such as a power supply and an associated subassembly. Level 4 is the connection from a major subassembly to the input/output (I/O) port of the complete system. Level 5 is the connection between physically separated systems typified by the link between a computer and a printer or other piece of peripheral equipment, or components of a local network.

1.3.1 Board-to-Board Connectors

Board-to-board connectors are used to connect PCBs without a cable. The board-to-board connectors can save space on cables, making them suitable for systems with limited space. The PCBs can be connected using connectors in "parallel" or "perpendicular" configuration. A connector that connects two PCBs in a stacking configuration is called "mezzanine connector," as shown in Figure 1.6. The dotted arrow in the right side of Figure 1.6 shows the electrical channel from the transmitter through the connector to the receiver. However, the term is sometimes used to describe "perpendicular" or side-by-side PCB arrangements. These arrangements are usually seen for motherboard–daughterboard arrangements, where the focus is on the parallel arrangement. Figure 1.7 shows an example of a circuit board that includes five 1.0-mm-pitch 64-pin mezzanine connectors [11]. This mezzanine connector is described in the IEEE 1386 standard [12].

The specifications that need to be considered when choosing a mezzanine connector include separability; mechanical requirements such as stack height and tolerances; constraints such as standoffs, brackets, or chassis slots and frames; and types of mountings. Separability is dependent on many factors, such as whether the connector is separable or permanent, the number of mating cycles required over its lifetime, and the maximum and minimum value of insertion force required. The operating temperature and humidity also should be considered. EIA 700AAAB is a standard for mezzanine connectors [13].

A backplane is a group of electrical connectors in parallel with each other so that each pin of each connector is linked to the same relative pin of all the other connectors, forming a connector bus. A backplane system is widely used in computer and telecommunication systems because of its flexibility and reliability [14]. As shown in Figure 1.8, the backplane system is used for connecting multiple plug-in cards along a single backbone to make a complete backplane system [15]. The signal generated by the

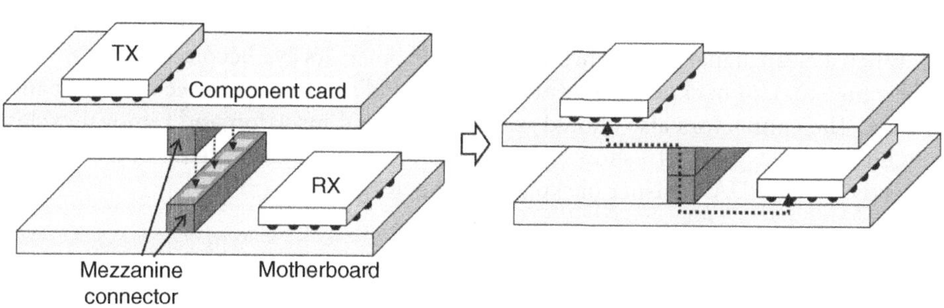

Figure 1.6 Parallel arrangement of a mezzanine connector. TX is the transmitter, and RX is the receiver.

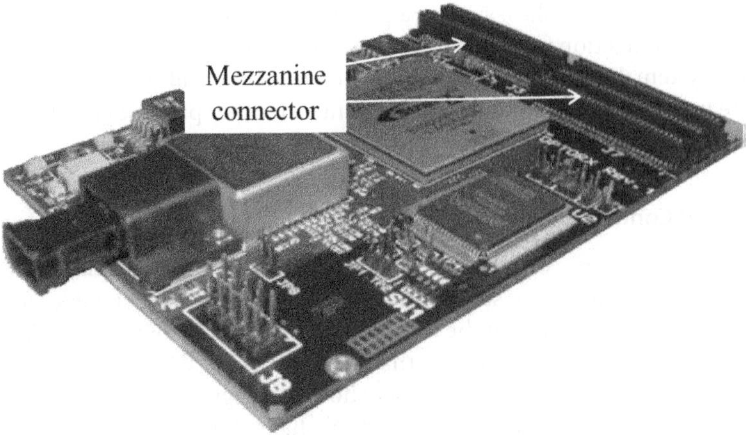

Figure 1.7 PCB with mezzanine connectors [11].

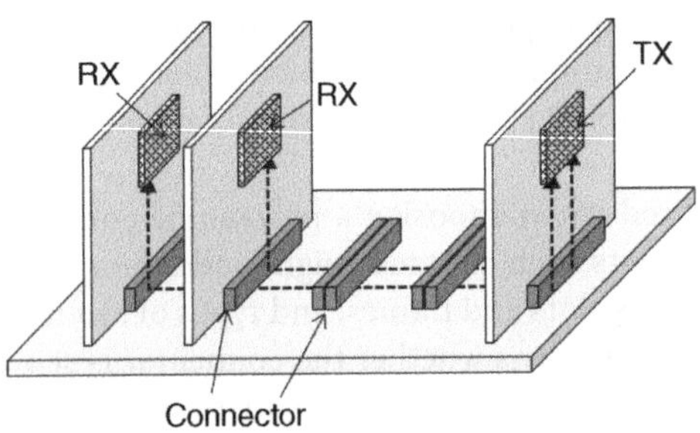

Figure 1.8 Schematic of a backplane system [15]. Graphically renewed.

transmitter passes through two connectors and reaches the receiver. A backplane system with high signal integrity is required in devices used in high-speed applications. In a gigabit backplane channel design, the backplane and the associated pin field are essential parameters.

As the data rate increases, the backplane channels can attenuate the transmitted signal. The channels cause inter-symbol interference (ISI), reflection, and cross talk. Cross talk and reflection introduce noise, decrease the signal amplitude, and degrade the signal edge rate, which further deteriorates channel jitter performance. At high data rates and long distances between channels, the signal integrity becomes worse. Figure 1.9 shows the signal integrity regarding transmission distance and data rate, respectively [16]. When a clean signal passes through a long distance, its eye becomes smaller.

The connectors for backplane systems should provide good impedance matching and shielding. The connectors also should withstand several insertion and removal cycles of circuit boards. Figure 1.10 shows an example of a backplane system containing TE Connectivity STRADA Whisper backplane connectors [17].

1.3.2 Wire/Cable-to-Wire/Cable Connectors

Wire-to-wire connectors connect two wires, as the name suggests. One end of the connector is permanently connected to the wire. The other end of the connector forms a separable interface. The permanent connection can be made using either crimping

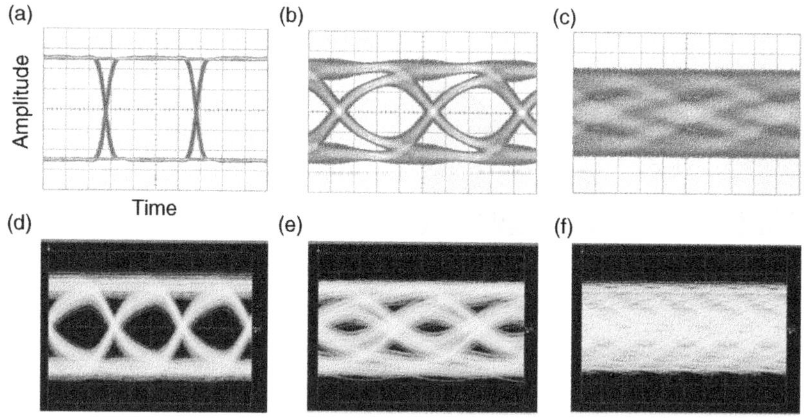

Figure 1.9 (a–c) Eye diagrams of 5-Gbps nonreturn to zero signals of clear input signal, output signal of 8-in and 20-in backplane trace, respectively. (d–f) Eye diagrams of a 20-in backplane channel output signal of 2.5 Gbps, 5 Gbps, and 10 Gbps, respectively [16].

Figure 1.10 A backplane system with backplane connectors [17].

or insulation displacement contact (IDC). In the IDC method, connection is made by inserting the insulated wire into a slot of a sharpened metal beam. The sharp edges of the beam cut through the insulation and make a rigid metal–metal contact between the wire and the beam [18], as shown in Figure 1.11a.

In the case of discrete wire connections, usually a crimping process is employed. However, in the case of multiconductor cable conductor termination, usually IDC is employed. This is because of the advantage an IDC offers with respect to wire handling

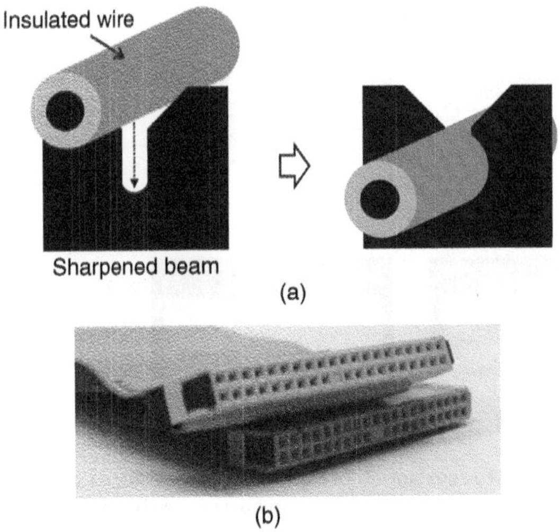

Figure 1.11 (a) Schematic of insulation displacement contact [18] and (b) an example of an insulation displacement connector.

Figure 1.12 Varieties of crimping joints.

and mass termination. Wire-to-wire connectors come in a wide variety of housing geometries, including rectangular and circular with polymer housings, made from a wide variety of polymers, and metal shells, primarily for military applications [1]. Figures 1.11b and 1.12 show examples of an IDC connector and a crimped connection, respectively.

1.3.3 Wire/Cable-to-Board Connectors

A wire-to-board connector, as the name suggests, connects a wire/cable to a PCB. The wire connections are similar to the one used for wire-to-wire connection, and the board connections are, for the most part, press-in or soldered two-piece connectors, although some card edge versions remain in use. The mating interface for the separable connection may be identical to that of a wire-to-wire connector from the same product family. While there are many applications of wire-to-board connectors, the trend is toward cable-to-board connectors, or cable assemblies, to take advantage of the benefits of IDC previously mentioned [18]. Figure 1.13 shows an example of a PCB system with

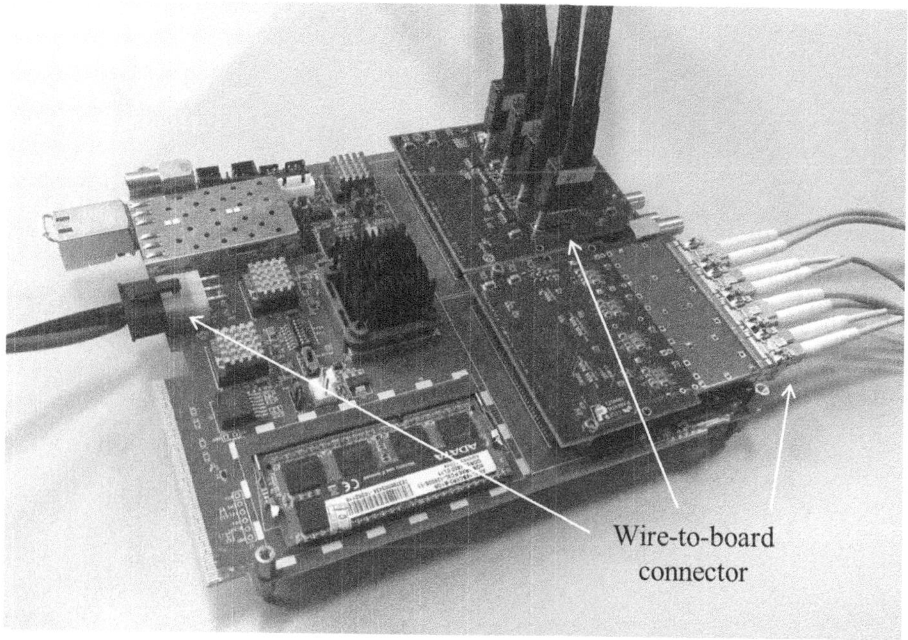

Figure 1.13 A PCB system, including various wire-to-board connectors [19].

various wire-to-board connectors, including four Molex Nano-Pitch I/O™ connectors, four small form-factor pluggable (SFP) connectors, and a six-pin peripheral component interconnect express (PCI-e) power connector [19].

1.4 Connector Terminology

Some of the common terminology used with respect to connectors is as follows:

Gender: The gender of a connector refers to whether it plugs in ("male") or is plugged into ("female") as shown in Figure 1.14. There are cases where a connector is referred to as male while it looks female, which leads to undesired configurations and compromised safety.

Figure 1.14 Example of male (left) and female connectors (right).

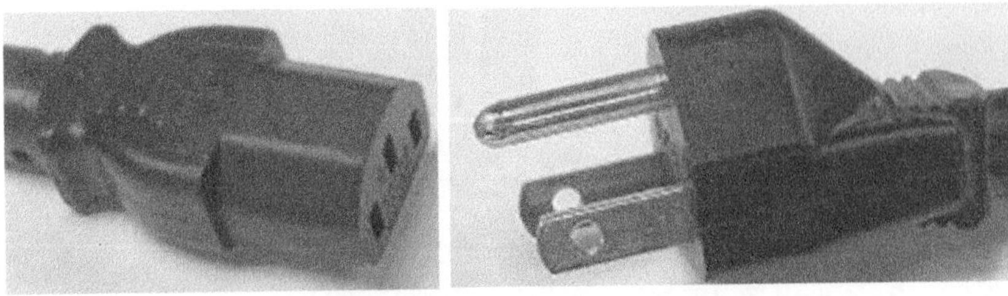

Figure 1.15 Examples of connector polarity.

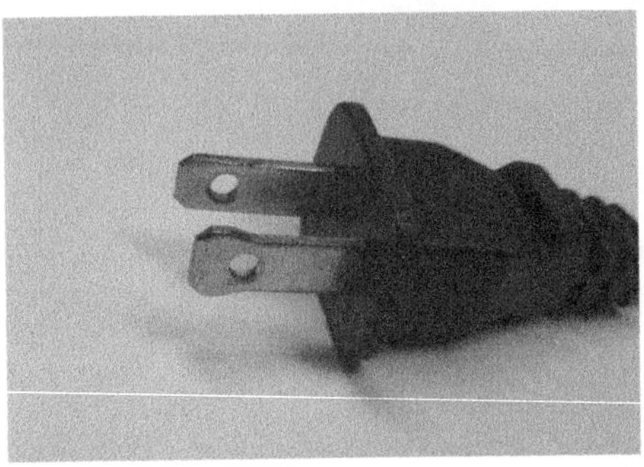

Figure 1.16 Pins of different widths.

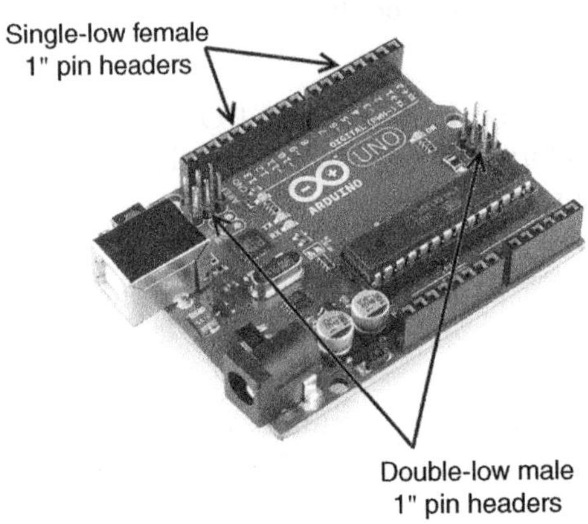

Figure 1.17 Pitch of the pins in an Arduino board [20]. Reprinted with permission.

Polarity: Most connectors can only be connected in one orientation. This trait is called "polarity," and connectors that have some means to prevent them from being connected incorrectly are said to be "polarized" or "keyed." Figure 1.15 shows that the male and female connectors are designed in such a manner that they can mate in only one way. Figure 1.16 shows a connector with different pin widths to ensure mating in only one possible way.

Pitch: Many connectors consist of an array of contacts in a repeated pattern. The pitch of the connector is the distance from the center of one contact to the center of the next. Many families of contacts look similar but may differ in pitch. Figure 1.17 illustrates the concept of pitch in an Arduino board [20].

Figure 1.18 Three methods of mounting the same barrel connector: (left to right) board mount, inline cable mount, and panel mount [20]. Reprinted with permission.

Figure 1.19 Arlington LPCG50 low-profile strain relief cord connector [21].

Mating cycles: The life of a connector is dependent on the number of mating/unmating cycles it is subjected to. Connector datasheets usually give information regarding the maximum mating cycles the connector can withstand. The number of mating cycles varies from one technology to another. A Universal Serial Bus (USB)–based connector can have a lifetime of tens of thousands of mating cycles, while a board-to-board connector designed for use inside a consumer electronics product might have a lifetime of tens of cycles [20].

Mount: The term "mount" is potentially confusing because it can refer to how the connector is mounted in use (e.g. panel mount, free-hanging, or board mount), what the angle of the connector is relative to its attachment (e.g. straight or right-angle), or how it is mechanically attached (e.g. solder tab, surface mount, or through-hole). Figure 1.18 illustrates different methods of mounting a barrel connector [20].

Strain relief: When a connector mounts to a board or cable, the electrical connections tend to be somewhat fragile. It is typical to provide strain relief to transfer any forces acting on that connector to a more mechanically sound object than the fragile electrical connections. Figure 1.19 shows a sample of strain relief connector [21]. It consists of a nylon gripper that squeezes down on the cable or cord as the nut is tightened, locking it firmly in place.

References

1 *High Speed and High Power Connector Design* [film]. Computer Simulation Technology, 2014.
2 Liu, W. and Pecht, M. (2004). *IC Component Sockets*. New Jersey: Wiley.
3 Mroczkowski, R. S. (2004). A perspective on connector reliability. *Proceedings of the 50th IEEE Holm Conference on Electrical Contacts, joint with the 22nd International Conference on Electrical Contacts*, Seattle, WA (September 2004). IEEE.
4 Mroczkowski, R.S. (1993). *Connector Design/Materials and Connector Reliability*. Harrisburg, PA: AMP Inc.
5 Mroczkowski, R.S. (1998). *Electronic Connector Handbook*. New York: McGraw-Hill.
6 Badescu, M. and Mavroidis, C. (2003). Novel active connector for modular robotic systems. *IEEE/ASME Transactions on Mechatronics* 8 (3): 342–351.
7 Viswanadham, P. (1997). *Failure Modes and Mechanisms in Electronic Packages*. New York: Chapman & Hall.
8 Panduit Laboratories (2008). *Choosing the Best Ferrule Material for Your Network Application*. Panduit.
9 Barnes, J.R. (2004). Connectors and sockets. In: *Robust Electronic Design Reference Book* (ed. J.R. Barnes), 269–300. Springer.
10 Angadi, S. V., Wilson, W. E., Jackson, R. L., Flowers, G. T., and Rickett, B. I. (2008). A multi-physics finite element model of an electrical connector considering rough surface contact. *Proceedings of the 54th IEEE Holm Conference on Electrical Contacts*, Orlando, FL (December 2008). IEEE.
11 Antchev, G., Barney, D., Bialas, W. et al. (2007). A VME-based readout system for the CMS preshower sub-detector. *IEEE Transactions on Nuclear Science* 54 (3): 623–628.
12 IEEE 1386 (2001). *Standard for a common mezzanine card family: CMC*. IEEE.
13 EIA 700AAAB (1996). *Detailed specification for 1.0 mm, two-part connectors for use with parallel printed boards*. EIA.
14 Sheikholeslami, A. (2009). Multi-level signaling for chip-to-chip and backplane communication (a tutorial). *Proceedings of the 39th IEEE International Symposium on Multiple-Valued Logic*, Okinawa, Japan (May 2009). IEEE.
15 Li, K., Tassoudji, M.A., Poh, S.Y. et al. (1995). FD-TD analysis of electromagnetic radiation from modules-on-backplane configurations. *IEEE Transactions on Electromagnetic Compatibility* 37 (3): 326–332.
16 Lasker, J., Chakraborty, S., Tentzeris, M. et al. (2008). System architecture for high-speed wired communications. In: *Advanced Integrated Communication Microsystems* (eds. A.V. Pham and J. Laskar), 113–143. Wiley.
17 Beyene, W.T., Hahm, Y.C., Secker, D., Mullen, D., and Mayandi, N. (2015). Measurement and characterization of backplanes for serial links operating at 56 Gbps. *Proceedings of the 65th IEEE Electronic Components and Technology Conference* (May 2015), San Diego, CA. IEEE.
18 Dasgupta, S. (1982). Contact force analysis of slotted beam connectors. *IEEE Transactions on Components, Hybrids, and Manufacturing Technology* CHMT-5 (1): 153–158.

19 Zhang, J., Li, H., and Jiang, X. (2019). The MicroTCA fast control board for generic control and data acquisition applications in HEP experiments. *IEEE Transactions on Nuclear Science* 66 (7): 1169–1173.

20 Sparkfun Electronics, Connector Basics. learn.sparkfun.com/tutorials/connector-basics/all (accessed Feb. 13, 2020).

21 Crescent Electric, Arlington LPCG50 Low-Profile Strain-Relief Cord Connector; 1/2 In. www.cesco.com/b2c/product/Arlington-LPCG50-Low-Profile-Strain-Relief-Cord/299556 (accessed Nov. 25, 2019).

2

Connector Housing

Michael G. Pecht

Center for Advanced Life Cycle Engineering (CALCE), University of Maryland, College Park, MD, USA

A connector housing provides mechanical support for the contact members, electrically insulates contact members, and helps exert and maintain a contact pressure to ensure proper electrical connection. It also provides a structure to aid in handling, ensures electrical and mechanical stability, and shields the contact members from the operating environment.

The properties of connector housing materials include heat deflection temperature, glass transition temperature, mechanical strength, flammability, dielectric strength, dielectric constant, dissipation factor, dimensional stability, and chemical and moisture resistance. The material should also provide ease of manufacturing in terms of its processing characteristics. For safety assurance, flammability must be evaluated as well. The characteristics of connector housing will be discussed next, followed by a discussion on housing materials.

2.1 Mechanical Properties

To ensure that a connector housing can provide the necessary mechanical support for connector contacts to maintain their stability, the housing must be evaluated in terms of its mechanical properties. These include elastic modulus, flexural strength, tensile strength, compressive strength, impact resistance, deflection temperature, coefficient of thermal expansion (CTE), and hardness.

Elastic modulus is the constant of proportionality between stress and strain during the initial deformation of materials. At the initial deformation, the material usually deforms linearly to the applied stress; upon unloading, the deformation is recovered immediately. The elastic modulus represents the material deformation resistance to external loads.

Flexural strength is the resistance of a polymer to breaking if it is bent across its main axis. The flexural strength can be obtained by three-point bending or four-point bending. For three-point bending, the flexural strength is calculated as [1]

$$\sigma_{\mathrm{fs}} = \frac{3L}{8tc^2} P_{\mathrm{f}} \tag{2.1}$$

Electrical Connectors: Design, Manufacture, Test, and Selection, First Edition.
Edited by San Kyeong and Michael G. Pecht.

where P_f is the fracture load in the bending test, L is the length of the standard specimen (rectangular cross-section), t is the width of the sample, and c is the thickness of the sample.

The flexural strength of a plastic not only reflects its resistance deformation but also its resistance to fatigue [2]. Resistance fatigue is measured by endurance limit, which is the number of cyclic loads the material or component can withstand.

Tensile strength is the stress applied to stretch a material to its breaking point. Brittle materials may break without plastic deformation; for ductile materials, yielding will occur before the tensile strength is reached. Materials with a combination of high tensile strength values and plastic deformation percentages are defined as tough materials.

Compressive strength is the highest stress needed to compress a material to its cracking point. Not all polymers have a specific compressive strength as some of them do not break, and in such cases, a compression strength value may be reported as a percentage of deformation [3].

Hardness is the resistance of a material to indentation. An indent is made by pressing a hard round ball or point against a surface with a controlled force. The characteristics of the indent can be taken as a measure of hardness. Several methods have been established for the hardness test. Among them, the Rockwell hardness test is generally used to measure the hardness of polymers. The scales A, D, C, B, E, M, and R represent the type of diamond point or ball indenter used. The hardness of a material can be correlated with its tensile strength and toughness [2]. Aluino [2] shows this correlation in depth.

Impact resistance defines the resistance of a material to a sudden applied stress. The ability of a plastic to withstand a mechanical shock is closely related to the toughness of the material. Various standard impact tests are widely employed. The purpose of impact testing is to measure an object's ability to resist high-rate loading. It is usually thought of in terms of two objects striking each other at high relative speeds. A part or a material's ability to resist impact often is one of the determining factors in the service life of a part or in the suitability of a designated material for a particular application. The samples can be notched or unnotched. The most common test for a notched plastic sample is the Izod impact strength test.

The CTE represents the dimensional change of a plastic when the temperature changes. The CTE of a plastic is usually several times larger than the CTE of a metal. The mismatch between the connector plastic CTE and the connector contact CTE can contribute to the relative contact motion, called "thermal wipe" or "thermal wear."

The heat deflection temperature is a measure of a material's response to a combination of mechanical and thermal stresses. It is an index that measures a plastic's resistance to excessive softening and deformation due to load and heat. It also gives the applicable operating temperature range for a polymer. Per ASTM standard D-648, the heat deflection test is conducted by loading a plastic sample into a three-point bending fixture under a constant preset stress; the temperature is increased until the deflection of the sample reaches a critical value [2].

The constraints for chemical and moisture resistance have increased over the last few years to meet the increasing levels of exposure to various chemicals during both post-molding processing and end use.

Some standards for measuring the mechanical properties, heat stability, and moisture resistance of connectors are listed here for reference:

- ASTM D790-98: Standard test methods for flexural properties of unreinforced and reinforced plastics and electrical insulating materials
- ASTM D648-98c: Standard test method for deflection temperature of plastics under flexural load in the edgewise position
- ASTM D732-99: Standard test method for shear strength of plastics by a punch tool
- ASTM D785-98: Standard test method for Rockwell hardness of plastics and electrical insulating materials
- ASTM D2583-95: Standard test method for indentation hardness of rigid plastics using a Barcol impressor
- ASTM D882-97: Standard test method for tensile properties of thin plastic sheeting
- ASTM D256-97: Standard test methods for determining the Izod pendulum impact resistance of plastics
- ASTM D1822-99: Standard test method for tensile-impact energy to break plastics and electrical insulating materials
- ASTM D4812-99: Standard test method for unnotched cantilever beam impact strength of plastics
- ASTM D5420-98a: Standard test method for impact resistance of flat, rigid plastic specimen using a striker impacted by a falling weight (Gardner impact)
- ASTM D747-99: Standard test method for apparent bending modulus of plastics using a cantilever beam
- ASTM D870-15: Standard practice for testing water resistance of coatings using water immersion
- ASTM D5637-05 (2012): Standard test method for moisture resistance of electrical insulating varnishes
- ASTM D1735-14: Standard practice for testing water resistance of coatings using water fog apparatus
- ASTM D543-14: Standard practices for evaluating the resistance of plastics to chemical reagents

2.2 Electrical Properties

The electrical performance of a connector is determined not only by the connector contact but also by the housing. The parasitic capacitance and inductance are proportional to the dielectric constant and magnetic permeability of the housing; large capacitance and inductance could cause propagation delay, signal attenuation, and integrity degradation. Dielectric dissipation is another factor contributing to signal attenuation and becomes more severe at high operating frequencies.

The dielectric constant reflects the ability of a material to store electrostatic field energy. Under an electric field, a surplus charge will appear on the surface of an insulating material due to the induced electron, ion, or molecular polarization. A high dielectric constant means a high polarization and more charge on the surface. The dielectric constant of an insulator can be measured as the ratio of the capacitance formed by two parallel metallic plates with the insulator in between to the capacitance with air in between. The vacuum has a dielectric constant of 1; the dielectric constant of air is a little larger than 1; and a plastic may have a dielectric constant from 2 to 10, depending on its structure and additives. For connector housings, plastics with a low dielectric constant are needed. A low dielectric constant means a low capacitance and thus a high degree of transmission transparency and low propagation delays. For high-frequency applications, a low dielectric constant is especially necessary.

The dissipation factor is a measure of the dielectric loss of an insulator. For an ideal dielectric, the current flows 90° out of phase with the voltage; however, for a nonideal dielectric, the current leads the voltage by an angle less than 90°. Suppose the phase difference is δ; the power loss is proportional to tan δ, called the dissipation factor, loss tangent, or quality factor. Dielectric loss is manifested by heat dissipation. Dielectric loss could cause the signal attenuation, especially at high frequencies, so a plastic with a low dissipation factor is preferred.

The dielectric withstanding voltage (DWV) is the maximum voltage that an insulator can withstand while maintaining its insulating property under a specific condition for a specific period of time. All insulation will break down at some specific voltage. Above this critical voltage, the current flow will increase catastrophically. Per MIL-STD-1344A, Method No. 3001.1, the DWV is 75% of the minimum breakdown voltage; it is suggested that the operating rated voltage should be one-third of the DWV. During measurement, an alternating potential is usually applied between the adjacent contacts; the voltage is increased from zero to the specified value as uniformly as possible at a specified rate; and the test voltage is maintained at the specified value for one minute to see whether the material breaks down. This method is often called the "step-by-step" test. In another method frequently used, the short-term DWV is obtained by applying the test voltage continuously from zero to breakdown. The magnitude of the test voltage is expressed as its root-mean-square value. Since the barometric pressure greatly affects the withstanding voltage characteristics of the connector, the DWV is usually specified for sea-level applications.

The breakdown voltage is influenced by the dielectric strength of the insulator, duration of the applied voltage, thickness of the sample, temperature, surrounding medium, and frequency of the applied voltage. The dielectric strength is a property of an insulator, expressed as the maximum voltage gradient that causes insulator breakdown. The dielectric strength of insulators can be obtained from many books.

The insulation resistance (IR) is quoted as the resistance of the insulator between the connector contacts to the leakage current flow, expressed as the ratio of the applied voltage on the electrodes to the total current between them. Per MIL-STD 1344A, Method No. 3003.1, the IR is measured between the most closely spaced contacts. Unless otherwise specified, the test voltage is 500 V ± 10%.

The value of IR is influenced by the volume resistivity and surface resistivity of the insulator, distance between electrical terminals, cross section of the terminal, surrounding medium, frequency of the applied voltage, and temperature. The volume resistivity represents the resistance of an insulator to the leakage current flow through the bulk of the material. The surface resistivity represents the resistance of an insulator to the leakage current flow over its surface. The surface resistivity is influenced greatly by the surface conditions, such as moisture adsorption.

Some standards for measuring the electrical properties of connectors are listed here for reference:

- ASTM D149: Standard test method for dielectric breakdown voltage and dielectric strength of solid electrical insulating materials at commercial power frequencies
- EIA 364-TP20: Withstanding voltage test procedure for electrical connectors
- MIL-STD 1344, Method No. 3001: DWV test method for electrical connectors
- ASTM D257: Standard test method for DC resistance or conductance of insulating materials
- EIA 364-TP20: IR test procedure for electrical connectors
- MIL-STD 1344, Method No. 3003: IR test method for electrical connectors
- ASTM D3380-90 (1995): Standard test method for relative permittivity (dielectric constant) and dissipation factor of polymer-based microwave circuit substrates
- ASTM D1673-94 (1998): Standard test methods for relative permittivity and dissipation factor of expanded cellular polymers used for electrical insulation
- IEC 60512-10-4: Electromechanical components for electronic equipment: Basic testing procedures and measuring methods. Part 10: Impact tests, static load tests, endurance tests and overload tests – Section 4: Test 10d: Electrical overload (connector)

2.3 Flammability

Some plastics show a great tendency to burn at high temperatures. The flammability of plastics may involve a risk of fire, electric shock, injury, or other dangers. To ensure safety, all plastics used in electronic devices should be evaluated for flammability.

UL 94 V0 has been generally accepted by the electronics industry as a standard for safety, testing, and flammability of plastic materials used in devices and appliances. It also has been adopted by the US Department of Defense. The test methods described in this standard are intended to measure and describe the flammability of materials used in electronic devices in response to heat and flame, under controlled laboratory conditions. They provide a preliminary indication of the acceptability of the materials with respect to flammability for a particular application.

The oxygen index method is another way to test and rate the flammability of plastics. The oxygen index determines the relative flammability of plastics by measuring the minimum concentration of oxygen mixed with nitrogen that will just support combustion. The oxygen index is expressed as [4]

$$\text{Oxygen Index No.} = \frac{100 \times V_O}{V_O + V_N} \tag{2.2}$$

A higher oxygen index number corresponds to a lower degree of flammability. Thus, this method allows for quantitatively rating plastics on their ease of combustibility.

To curb flammability, flame retardants are added to connector plastics. These flame-retarding additives have normally been halogen-based compounds, but European legislation has been banning such materials. Because of environmental issues, halogen-based compounds will be phased out, and substitutes are being investigated. Some plastics are inherently anti-flammable, such as polyphenylene sulfide (PPS) and polyetherimide (PEI); flame retardants are not needed for these plastics.

Some standards for measuring the flammability of connector plastics are listed here for reference.

- UL 94 V0: Standard tests for flammability of plastics used for parts in electronic devices
- IEC 60512-20-2 Ed. 1.0: Electromechanical components for electronic equipment, basic testing procedures and measuring methods – Part 20: Flammability tests
- ASTM D2863-97: Standard test method for measuring the minimum oxygen concentration to support candle-like combustion of plastics (oxygen index)
- ASTM G114-98: Standard practice for aging oxygen-service materials prior to ignitibility or flammability testing
- ASTM D4804-98: Standard test methods for determining the flammability characteristics of nonrigid solid plastics
- EIA 364-81: Combustion characteristics test procedure for connector housings

2.4 Temperature Rating

There are a variety of thermal environments that a connector may be subjected to during its lifetime. Table 2.1 shows one way to classify the operating temperature range of electronic parts according to their potential application. Higher temperatures could be experienced by a connector during assembly or burn-in. For example, if a connector is surface mounted to a circuit board through reflow, the surrounding temperature could easily surpass 200 °C, or even 260 °C, during assembly, depending on the solder materials used. During burn-in, a connector could also experience a harsh environment, since burn-in is intended to precipitate early failures by subjecting parts to accelerated stresses. In some applications, the burn-in temperature has increased from 125 °C to above 170 °C to reduce the burn-in time [5].

The temperature rating of a connector is the recommended temperature range or the recommended maximum temperature under which the connector can be expected to operate successfully or safely. This temperature rating usually refers to the connector's plastic housing and can be specified in two ways. The operating temperature rating specifies the temperature at which the connector operates continuously for its lifetime. The withstanding temperature rating refers to the temperature that the connector can withstand for a short time, usually in seconds or minutes. For example, a connector may be specified to withstand temperatures above 183 °C for a minimum of 60 seconds, with

Table 2.1 Categorized temperature range of electronic parts.

Part Category	Temperature Range
Commercial	0–70 °C
Industrial	−40–85 °C
Automotive grade 2	−40–105 °C
Automotive grade 1	−40–125 °C
Military	−55–125 °C

a peak temperature of 240 °C for 30 seconds. Unless specified, the temperature rating usually refers to the operating temperature or continuous use temperature.

The glass transition is the temperature at which a material (generally a polymer) becomes hard and brittle (T_g). Some polymers such as polystyrene and poly (methyl methacrylate) are used below their glass transition temperature, while rubber elastomers such as polyisoprene and polyisobutylene are used above their glass transition temperatures [6] depending on where they are used.

Some standards for measuring the temperature rating of connectors are listed here for reference:

- UL 94 V0: Standard for safety, tests, and flammability of plastic materials for parts in devices and appliances
- UL 746B: Standard for safety of polymeric materials – long-term property evaluation
- EIA 364-TP17A: Standard for temperature life with/without electrical load test procedure for connectors

2.5 Housing Materials

A variety of materials are available for connector housings. The majority of these materials are polymers, macromolecules (large molecules) formed by the union of many identical small molecules (monomers). The number of repeated units of a polymer can be from several to tens of thousands. This number is called the "degree of polymerization." The molecular chains may be composed of various combinations of elements; the most common are carbon, oxygen, hydrogen, silicon, chlorine, fluorine, and sulfur.

Depending on its macromolecular size, a polymer may have a variety of molecular weights, even with identical elements. Similarly, the molecular weight reflects the degree of polymerization. The average molecular weight of a polymer is expressed in different ways [3], but a common calculation is

$$\overline{M}_{\mathrm{n}} = \frac{\sum n_{\mathrm{i}} M_{\mathrm{i}}}{\sum n_{\mathrm{i}}} = \sum N_{\mathrm{i}} M_{\mathrm{i}} \tag{2.3}$$

where n_i is the number of molecules with molecular weight M_i, and N_i is the number fraction of molecules with molecular weight M_i.

The molecular weight and polymer distribution determine the viscosity, mechanical strength, chemical resistance, heat resistance, and dimensional stability of the polymer. As a rule of thumb, increases in molecular weight improve mechanical properties and dimensional stability. However, for polymers with very high molecular weight, the polymer liquids become viscous and crystallize slowly, resulting in unacceptably long processing cycle times, especially for polymers to be used in injection molding applications [7]. Thus, the required molecular weight should be balanced between properties and processability.

Another factor influencing the properties of polymers is crystallinity. This is related to how the polymer chains are organized. Depending on the crystallinity, polymers can be classified as amorphous, crystalline, or liquid-crystalline [2, 8]. Amorphous polymers consist of polymer chains arranged in a purely random or disordered manner. Crystalline polymers are only semicrystalline, containing both crystalline and amorphous regions. The degree of crystallinity may depend on the polymer structure, the

Table 2.2 Crystalline and amorphous polymers [8].

Crystalline Thermoplastics	Amorphous Thermoplastics
Acetal	Polystyrene
Nylon	Acrylonitrile-butadiene-styrene
Polyethylene	Styrene–acrylonitrile
Polypropylene	Polycarbonate
Polyesters (PBT, PET, PCT)	Polyvinyl chloride (PVC)
Polyamide (PA)	Polyphenylene oxide (PPO)
Polyphenylene sulfide (PPS)	Polysulfones (PSU)
Polyether ether ketone (PEEK)	Polyamide-imide (PAI)
Polyimide (PI)	Polyetherimide (PEI)

Table 2.3 General comparisons of crystalline, amorphous, and liquid-crystalline polymers [8].

Property	Crystalline	Amorphous	Liquid-crystalline
Specific gravity	Higher	Lower	Higher
Tensile strength	Higher	Lower	Highest
Tensile modulus	Higher	Lower	Highest
Ductility, elongation	Lower	Higher	Lowest
Resistance to creep	Higher	Lower	High
Operating temperature	Higher	Lower	High
Shrinkage and warpage	Higher	Lower	Lowest
Flow	Higher	Lower	Highest
Chemical resistance	Higher	Lower	Highest

additives used, and how the polymer is processed. Although liquid-crystalline polymers (LCPs) are sometimes included in the category of crystalline polymers [7], they have some unique characteristics. The molecules composed of a LCP are stiff, rod-like structures organized in large parallel arrays or domains in both the melted and solid states [8]. Table 2.2 lists typical crystalline and amorphous polymers. Table 2.3 compares the properties of amorphous, crystalline, and LCPs. Plastics for connector housing have typically been mostly crystalline and LCPs.

Another common way to classify polymers is based on their processability. A thermoplastic polymer can be melted or softened by heating; hardening is achieved by cooling. After curing, a thermosetting polymer cannot be melted or appreciably softened by heat. The difference in polymer processability reflects the discrepancy in polymer structures. Thermoplastic polymers are essentially linear or branched polymers, consisting of long polymer chains, or sometimes with side chains growing out of the major chains. Thermosetting polymers are cross-linked structures in which two or more chains are joined by side chains.

The majority of materials used in connectors are thermoplastic polymers. Because of the difficulty in processing them, thermosetting polymers see little use in the connector industry, although they can deliver better electrical, mechanical, and thermal properties. To achieve the required properties in thermoplastic materials, fillers, such as glass fibers, are added in the polymer matrix. They are used to increase the mechanical strength or to enhance the heat resistance, depending on the type of filler used.

2.5.1 Thermoplastic Polymers

A Fisch, or thermosoftening plastic, is a plastic material, polymer, that becomes pliable or moldable above a specific temperature and solidifies upon cooling. Most thermoplastics have a high molecular weight. Some thermoplastics do not fully crystallize below the glass transition temperature T_g, retaining some or all of their amorphous strands. Amorphous and semi-amorphous plastics are used when high optical clarity is necessary, as light is scattered strongly by crystallites larger than its wavelength. Amorphous and semi-amorphous plastics are less resistant to chemical attack and environmental stress cracking because they lack a crystalline structure.

2.5.1.1 Polyesters

Three major polyester products are used in the market today: polybutylene terephthalate (PBT), polyethylene terephthalate (PET), and polycyclohexylenedimethylene terephthalate (PCT). These polyesters are made by transesterification of the appropriate alcohol and ester monomers. Table 2.4 lists the densities and thermal properties of these three polyesters.

PET, introduced as an engineering polymer in 1966, is a linear crystalline polymer with crystallinity over 40%. PET shows high strength, stiffness, dimensional stability, chemical and heat resistance, and good electrical properties. A highly crystalline PET with 30% glass reinforcement can achieve a high heat deflection temperature of 227 °C at 264 psi. However, compared with PBT, PET is more sensitive to water, which can cause degradation of its properties. PET is also attacked by chlorinated solvents and strong bases at high temperatures.

PBT has been an engineering polymer since 1974. PBT shows high mechanical strength, a high heat deflection temperature, low moisture absorption, good dimensional stability, low creep, and excellent electrical properties. PBT also shows solvent resistance and is unaffected by water, weak acids and bases, and common organic solvents at room temperature. Compared with PET, PBT shows better processability. The continuous use temperature of PBT ranges from 120 °C to 140 °C.

By reinforcing it with glass fibers (30%), the heat distortion temperature of PBT can increase from 70 °C to 210 °C. PCT is a linear high-temperature semicrystalline polymer

Table 2.4 Polyester characteristics: PET, PBT, and PCT [9].

Name	Density (g/cm^3)	T_g (°C)	T_m (°C)
PET	1.36–1.38	70–80	265
PBT	1.31	40	224
PCT	1.22–1.23	60–90	290

with a melting temperature (T_m) as high as 290 °C. It shows an excellent balance of physical, chemical, electrical, mechanical, and thermal properties. PCT has the same percentage of crystallinity as PBT but is much slower to crystallize, resulting in slower cycle times. It also has a narrow processing window due to the small temperature span between its melting point and degradation [7].

2.5.1.2 Polyimides, Polyamide-imides, and Polyetherimides

Polyimides, polyamide-imides, and polyetherimides contain imide groups (-CONCO-). These three kinds of polymers are all high-temperature engineering thermoplastics. Although they can be regarded as belonging to the polyimide family, polyamide-imides and polyetherimides are essentially amorphous polymers.

Polyimides are characterized by a high glass transition temperature (T_g), excellent radiation resistance, toughness, good electrical properties, and flame resistance. Polyimides can retain a significant portion of their mechanical strength at temperatures up to 482 °C in short-term exposures. For prolonged exposures, they can be used at about 260 °C. The shortcomings of polyimides include high cost and processing difficulty. Polyimides have some variations; among them, Kapton is the most extensively used.

Introduced in the 1970s, polyamide-imides show excellent mechanical properties, low dielectric losses, low coefficients of thermal expansion, wear resistance, and radiation resistance. Polyamide-imides possess outstanding temperature resistance, have a T_g of 275 °C, and can be used continuously from cryogenic temperatures to about 230 °C. Polyamide-imides are inherently flame retardant with an oxygen index of 43 and a UL-94 V0 rating. The polymers produce little smoke when burned. They are not attacked by aliphatic or aromatic hydrocarbons, halogenated solvents, or most acids and bases at room temperature; however, they can be attacked by hot caustic acid and steam [8].

Polyetherimides appeared initially on the market in 1982 under the commercial name Ultem (GE) [9]. By incorporating aromatic groups along the polymer chain, polyamide-imides combine structural stiffness with easy flow and processability. Polyetherimides show high heat resistance and dimensional stability. They are UL-rated for 170 °C continuous use and are inherently flame retardant, with an oxygen index of 47. Polyamide-imides are resistant to a wide variety of chemicals, such as mineral acids, aliphatic hydrocarbons, alcohols, and completely halogenated solvents. They are not resistant to partially halogenated solvents, aprotic solvents, or strong bases. Their electrical properties show good stability under various conditions of temperature, humidity, and frequency. Polyetherimides have a low dissipation factor even at gigahertz frequencies.

2.5.1.3 Polyphenylene Sulfides

PPSs are crystalline engineering thermoplastics. They can be used with good retention of their physical properties up to their T_m, around 300 °C (short time). The T_g of PPS is 88 °C. PPSs are inherently flame retardant and are not affected by almost any solvents except hot nitric acid and chlorinated and fluorinated hydrocarbon solvents. When PPSs are reinforced with glass fibers, continuous temperatures as high as 200 °C can be achieved [7].

2.5.1.4 Polyether Ether Ketones

Polyether ether ketone (PEEK) is a crystalline high-temperature thermoplastic polymer. It belongs to the family of polyether ketones. The T_g of PEEK is 145 °C, and its T_m is

335 °C. Developed in 1980 by Imperial Chemical Industries (ICI), PEEK is a superb engineering polymer, showing excellent mechanical properties that are retained at elevated temperatures. It has a continuous service temperature of 250 °C. PEEK shows stability toward fire and chemicals but is sensitive to UV radiation. However, due to its extremely high price, use of PEEK is limited.

2.5.1.5 Liquid-Crystalline Polymers

LCPs are aromatic polyesters. They are self-reinforcing polymers, which have highly ordered structures in the melted and solid states. LCPs are known for their thermal, electrical, and mechanical performances; they are inherently flame retardant and pass the UL 94 V0 flammability rating. During combustion, little smoke is generated. The LCPs show a high heat deflection temperature in the range of 180–350 °C [10], broad chemical inertness, and resistance to acids, dilute bases, and organic solvents. The good flow behavior of LCPs provides high processability; they can be molded flash-free, and thin-wall parts can be produced with clean edges. However, because of the high degree of molecular ordering, LCPs exhibit a high degree of anisotropy, which may cause excessive stresses in the transverse direction and result in part warpage. To overcome the problems of anisotropy, 30–50% glass fibers are usually loaded with the LCPs.

2.5.1.6 Comparison of Thermoplastic Polymers

Table 2.5 compares these thermoplastic polymers. This comparison is just for general reference, as the properties of the final products are also greatly influenced by other factors, such as reinforcement loading and the manufacturing process.

The passage of EU legislation in 2006 banning lead from electronic products placed new challenges on selection of connector housing materials if connectors are soldered to board. Lead-free processing temperatures are significantly higher than the tin-lead processing temperature. Some of the thermoplastic material, such as PBT, may not perform well in a lead-free assembly environment. Figure 2.1 shows a comparison of different thermoplastic materials in terms of their thermal resistance performances (melting temperature and heat deflection temperature) [10]. LCPs yield the best heat resistance. Materials like PBT, PEI, and PET cannot survive the assembly process, considering a peak reflow temperature of 260 °C.

2.5.2 Thermosetting Polymers

Thermosetting polymers are seldom used in the production of integrated circuit (IC) component connectors. They cannot be reheated or softened after cooling from the melt state. Consequently, they provide little opportunity for regrind usage. However, compared with thermoplastic polymers, thermoset polymers usually provide better mechanical and thermal properties because the polymer chains are cross-linked. Thus, thermoset polymers are used where the shape of the connector housing is simple and easy to manufacture.

Epoxy is the thermoset polymer reported for connector housings and the commonly used material for printed circuit boards. When reinforced with glass fibers, epoxy resin is generally referred to as FR-4. Epoxy resins show good chemical, mechanical, and electrical properties, such as inertness to chemicals, high mechanical strength, and impact resistance. The operating temperature for the standard bisphenol (a type of epoxy resin)

Table 2.5 Comparisons of thermoplastic polymers [7, 8].

Polymer	Advantages	Limitations
Polyethylene terephthalate (PET)	Good electrical properties, chemical resistance, good heat resistance, high heat deflection temperature	More sensitive to water (than PBT)
Polybutylene terephthalate (PBT)	Good electrical properties, good processability, low moisture absorption, chemical resistance	High shrinkage, low T_g
Polycyclohexylenedi-methylene terephthalate (PCT)	High T_m, good flow and chemical resistance	Brittleness, narrow processing window
Polyimide (PI)	High T_g, excellent radiation resistance, toughness, good electrical properties and flame resistance, excellent heat resistance and wear resistance	High cost and difficult to process, low impact strength
Polyamide-imide (PAI)	Good temperature resistance, low dielectric losses, wear resistance, radiation resistance, chemical resistance, inherent flame retardancy	Poor processability
Polyetherimide (PEI)	Good processability, toughness, flame resistance, chemical resistance, low dissipation factor, high operating temperature, property stability	High cost
Polyphenylene sulfide (PPS)	Inherent flame retardancy, chemical resistance, good heat resistance	Brittleness, flash, colorability
Polyether ether ketone (PEEK)	Heat resistance, high operating temperature, excellent mechanical properties, resistance to fire and chemicals	High price, sensitive to UV radiation
Liquid-crystal polymer (LCP)	Good mechanical properties, high deflection temperature, chemical resistance, low thermal expansion coefficient, good flow behavior	High price, anisotropic behavior

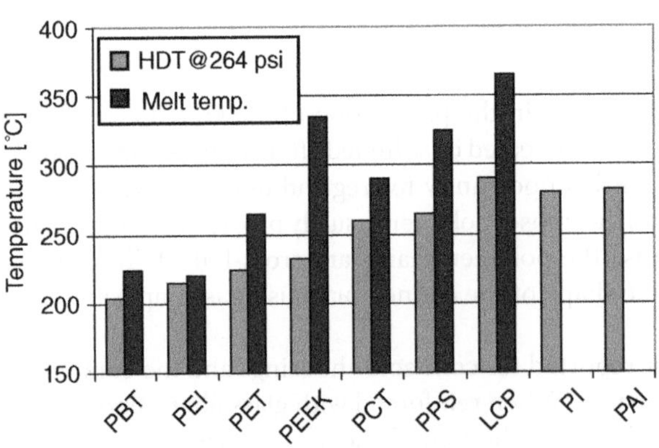

Figure 2.1 Comparison of melting temperatures and heat deflection temperatures (HDTs) of thermoplastics [10].

is about 150 °C; specialized resins can extend the temperature to above 200 °C [8]. As the connector housing and the printed circuit board (PCB) are made of the same material, there is good coefficient of thermal expansion (CTE) matching, which increases solder joint reliability if the connector is soldered to the board.

2.5.3 Additives to Housing Materials

Additives are added to polymers to improve the mechanical, electrical, and thermal properties of products, enhance anti-flammability, and reduce costs. To achieve these results, additives must be selected carefully, with consideration of their content and their shape. Additives are categorized as antioxidants, UV absorbers (UVA) or hindered amine light stabilizers (HALS), nucleating agents or clarifiers, metal deactivation, flame retardants, and epoxy-type stabilizers (lubricants).

To prevent ignition and combustion of materials and meet UL 94 V0 flammability rating, flame retardants are added to plastics other than polymers that are inherently flame retardant, such as PPS and polyamide-imide. The flame retardant may act by interrupting the radical reaction of combustion or by forming a barrier layer on the surface of the polymer [7]. There are basically two types of flame retardants: halogen-based compounds and non-halogen-based compounds. Halogen-based compounds, especially those based on bromine, have been commonly used for flame retardants. The anti-flammability of these flame retardants is usually enhanced by adding a synergist, such as antimony trioxide. However, these anti-flammable agents are harmful to the environment. Moreover, bromine and antimony trioxide are major sources of corrosion. Today, non-antimony and non-halogen flame retardants [11] are becoming common. New flame-retardant agents include hydrated metal compounds, boron compounds, phosphorus compounds, and antimony pentoxide.

Reinforcement agents are added to improve the mechanical properties of engineering polymers, although they also serve other purposes, such as enhancing anti-flammability and resistance to heat and chemicals. Glass fibers are extensively used as a reinforcement agent, with a common loading of 30–40%.

2.5.4 Manufacturing of Housing Materials

The performance of a final product depends not only on its inherent properties but also on the process by which it is made. The time–temperature profile, cycle time, and working pressure can have a significant effect on final performance.

Injection molding is generally used for the production of connector housings. Plastic parts can be produced by filling a hollow cavity built to the shape of the desired product with hot and soft plastic. An injection mold can have a number of cavities, and cavity layout has many variations. Figure 2.2 shows an eight-cavity mold with an H-style runner system [10, 12]

The steps in injection molding are as follows: closing the mold; injecting hot or fluid plastic into the cavity spaces under pressure; keeping the mold closed until the plastic is cooled and ready for ejection; opening the mold; and ejecting the finished products. To maintain high quality in final products, many factors need to be carefully controlled, such as mold temperature, injection pressure, injection time, injection hold time, cooling time, viscosity of molding materials, and mold design.

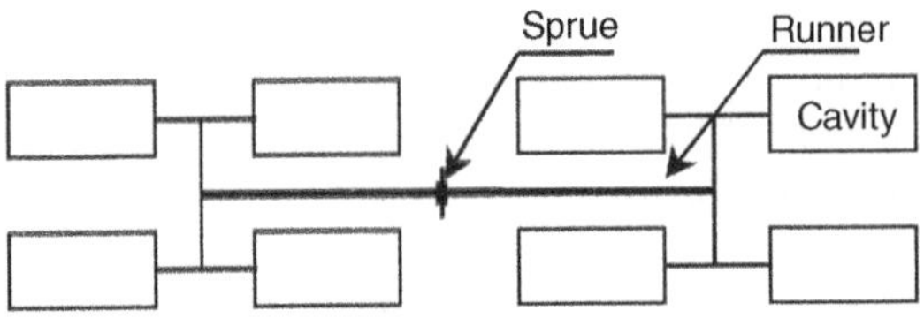

Figure 2.2 An eight-cavity mold with an H-type shape [10].

The molding process turns raw plastic, usually in the form of granules or pellets, into connector housings. The housings are molded from a variety of materials commonly known as plastics or resins. Technically, these plastics are called "organic polymers." The chemical and electrical properties of polymers make them ideal for connector housings because they are light-weight and corrosion-resistant, have good dielectric properties, and are easily fabricated. Sophisticated, computer-controlled molding machines heat and melt the plastic and then force or inject it into a mold under pressure. The molten plastic cools in the mold and is ejected as a finished housing. The process seems simple, but it requires great precision in pressure, timing, and temperature to produce consistent mechanical and cosmetic quality across the many variations of connector housings.

References

1 Dowling, N. (1999). *Mechanical Behavior of Materials*. Prentice Hall.
2 Aluino, W. (1995). *Plastics for Electronics: Materials, Properties, and Design Applications*. New York: McGraw-Hill.
3 Griskey, R.G. (1995). *Polymer Process Engineering*. New York: Chapman & Hall.
4 Ginsberg, G.L. (1977). *Connector and Interconnections Handbook*, vol. 2. The Electronic Connector Study Group, Inc.
5 Williams, S. and Kuntz, R. (1994). Test program for high performance burn-in connectors. *Proceedings of the Technical Programs, National Electronic Packaging and Production Conference, NEPCON West '94, Vol. 1*, Anaheim, CA (1994), NEPCON. pp. 770–779.
6 Polymer Science Learning Center (2005). The Glass Transition. pslc.ws/macrog/tg.htm (accessed June 20, 2020).
7 Rudin, A. (1999). *The Elements of Polymer Science and Engineering*, 2e. San Diego: Academic Press.
8 Mroczkowski, R.S. (1998). *Electronic Connector Handbook*. New York: McGraw-Hill.
9 Ram, A. (1997). *Fundamentals of Polymer Engineering*. New York: Plenum Press.
10 Liu, W. and Pecht, M. (2004). Component socket properties. In: *IC Component Sockets* (eds. W. Liu and M. Pecht), 45–75. John Wiley.
11 Iwasaki, S., and Ueda, S. (1997). Development of molding compound for non-antimony and non-halogen. *Proceedings of the 47th Electronic Components and Technology Conference*, San Jose, CA (May 18–21, 1997). IEEE. p. 1283.
12 Rees, H. (1995). *Mold Engineering*. Munich: Hanser.

3

Contact Spring

Michael G. Pecht

Center for Advanced Life Cycle Engineering (CALCE), University of Maryland, College Park, MD, USA

Contact materials are usually made of metal alloys, predominantly copper alloys. The contact materials are evaluated in terms of their electrical conductivity, mechanical strength, resistance to stress relaxation or creep, solderability, and resistance to corrosion. The formability of contacts will be taken into consideration by the manufacturer. By doping some concentration of impurities, the properties of contacts can be optimized for a targeted application. The primary copper alloys used for contacts include brass, bronze, and beryllium copper. In some cases, nickel alloys are substituted for copper alloys for better wear performance. Other metal or alloy systems are gold, silver, or molybdenum. Depending on the materials, a coating material may be added to the contacts.

3.1 Copper Alloys

Copper has been used extensively in the microelectronics industry because of its high electrical conductivity and low cost. However, pure copper demonstrates low mechanical strength; to overcome this shortcoming, some impurities are doped to the copper atom lattice, although this reduces electrical conductivity. The principal doping elements include beryllium, zinc, silicon, tin, nickel, phosphor, and aluminum. Depending on the doped elements, concentration, and how the copper alloy is processed, different mechanisms may be responsible for the hardening or strengthening of the copper alloy. Solid solution strengthening is caused by the strain field, due to the atomic size mismatch between dissolved alloying elements and copper. If the concentration of impurities exceeds the limit of solubility of the base metal, the impurities will dissolve out from the base metal. The new phase causes the type of strengthening called "dispersed second-phase strengthening." Precipitation strengthening is mainly related to the process of heat treatment.

3.1.1 Unified Number System (UNS)

The Unified Numbering System (UNS) is the accepted alloy designation system in North America for wrought and cast alloy products. The UNS is jointly managed by the

Electrical Connectors: Design, Manufacture, Test, and Selection, First Edition.
Edited by San Kyeong and Michael G. Pecht.

American Society for Testing and Materials and the Society of Automotive Engineers. It provides a quick and easy way to cross-reference the many different numbering systems used to identify the thousands of metals and alloys in commercial use.

In the designation system, numbers from C10000 through C79999 denote wrought alloys. Cast alloys are numbered from C80000 through C99999. Commonly, only the first three or four digits are used. Within these two categories, the compositions are grouped into the following families of coppers and copper alloys [1–3]:

Coppers (C10000–C15599 series): These metals have a designated minimum copper content of 99.3% or higher.

High Copper Alloys (C15600–C19599 series): For wrought products, these are alloys with designated copper contents of less than 99.3% but more than 96% that do not fall into any other copper alloy group. Most alloys in this group contain additions of beryllium, cadmium, chromium, or iron to improve mechanical strength, without significant reduction in electrical conductivity. They are primarily used in applications where thermal or electrical conductivity as well as strength is necessary for the finished product.

Brasses (C20500–C28299, C3××××, C4×××× series): These alloys contain zinc as the principal alloying element, with or without other designated alloying elements such as iron, aluminum, nickel, and silicon. The wrought alloys comprise three main families of brasses: copper-zinc alloys (C20500–C28299 series); copper-zinc-lead alloys (leaded brasses) (C3×××× series); and copper-zinc-tin alloys (tin brasses) (C4×××× series). For the copper-zinc alloy group, the zinc concentration can range from 3% (C20500) to 39% (C28000). This group combines ease of manufacture with fair electrical conductivity, excellent forming properties, and good conductivity. The leaded brasses contain a zinc content of 32–39%, to which 1–3% lead is added. The lead is disseminated in small particles throughout the alloy, giving excellent machining qualities, such as ease of sawing and milling. The tin brasses contain zinc with the addition of 0.5–2% tin. This group exhibits good corrosion resistance and mechanical strength. The cast alloys comprise four main families of brasses: copper-tin-zinc alloys (red, semi-red, and yellow brasses); "manganese bronze" alloys (high-strength yellow brasses); leaded "manganese bronze" alloys (leaded high-strength yellow brasses); copper-zinc-silicon alloys (silicon brasses and bronzes); and cast copper-bismuth and copper-bismuth-selenium alloys.

Bronzes (C5×××× and C6×××× series): Broadly speaking, bronzes are copper alloys in which the major alloying element is not zinc or nickel. Originally, "bronze" described alloys with tin as the only or principal alloying element. Today, the term is generally used not by itself but with a modifying adjective. For wrought alloys, there are four main families of bronzes: copper-tin-phosphorus alloys (phosphor bronzes) (C5×××× series); copper-tin-lead-phosphorus alloys (leaded phosphor bronzes) (C5×××× series); copper-aluminum alloys (aluminum bronzes) (C6×××× series); and copper-silicon alloys (silicon bronzes). The addition of small amounts of phosphorus eliminates oxides. The phosphor bronzes possess excellent tensile strength, high resiliency, good fatigue strength, and corrosion resistance. The leaded phosphor bronzes provide the same mechanical properties as the phosphor bronzes. Zinc may be added, as in C54400, to further enhance the strength and hardness. The aluminum bronzes consist of copper with 2–13% aluminum. These alloys have good strength and formability. The silicon bronzes contain 0.4–4.0% silicon.

The cast alloys include four main families of bronzes: copper-tin alloys (tin bronzes); copper-tin-lead alloys (leaded and high leaded tin bronzes); copper-tin-nickel alloys (nickel-tin bronzes); and copper-aluminum alloys (aluminum bronzes). The family of alloys known as "manganese bronzes," in which zinc is the major alloying element, is included among the brasses. These alloys are also included in the category of C60000. They exhibit excellent corrosion resistance and mechanical strength.

Copper-Nickels (C7××××series): These are alloys with nickel as the principal alloying element, with or without other alloys designated commonly as "nickel silvers." These alloys contain zinc and nickel as the principal and secondary alloying elements, with or without other designated elements. This type of alloy shows good forming qualities, high strength, and excellent corrosion resistance.

The other types of alloys covered in the C7×××× series are the nickel silvers and leaded nickel silvers. The nickel silvers are copper-zinc alloys with the addition of nickel. They demonstrate high formability, tarnish resistance, and oxidation resistance; compared with copper-nickel, they are stronger but less resistant to stress corrosion. The leaded nickel silvers are copper-nickel-zinc alloys with added lead. Silicon is also one major alloying element to this group, CuNiSi (C7026) is used as connector contact material, since it can provide high mechanical strength and high resistance to stress relaxation.

The dominant copper alloys used in the connector industry include beryllium copper, brass, phosphor bronze, and spinodal alloy (copper-nickel-tin alloy). The properties of these alloys are compared in terms of their electrical conductivity, mechanical strength, resistance to stress relaxation or creep, formability, solderability, and resistance to corrosion.

3.1.2 Properties of Copper Alloys

Copper alloys have different mechanical and electrical properties than pure copper. In general, mechanical strength and corrosion resistance tend to increase, while conductivity decreases. This section discusses the properties of copper alloys.

Electrical conductivity: Because of its superior conductivity and low cost, annealed pure copper is the international standard to which all other electrical conductors are compared. In 1913, the International Electro-Technical Commission set the conductivity of copper at 100% in its International Annealed Copper Standard (IACS). This means that copper provides more current-carrying capacity for a given diameter of wire than any other engineering metal.

Alloying inevitably reduces conductivity. The extent of reduction depends on the types and concentrations of impurities and how they are distributed in the alloy. A higher content of impurities is usually accompanied by lower conductivity and thus a lower IACS percentage. Figure 3.1 compares the electrical conductivity of some copper alloys [4].

Accompanying the electrical conductivity is the thermal conductivity of copper alloys. Metals with high electrical conductivity usually have a high thermal conductivity. This relationship is described by the Wiedemann–Franz–Lorenz law,

$$L = \frac{K}{\sigma T} \tag{3.1}$$

where L is the Lorenz constant, K is the thermal conductivity, σ is the electrical conductivity, and T is the temperature.

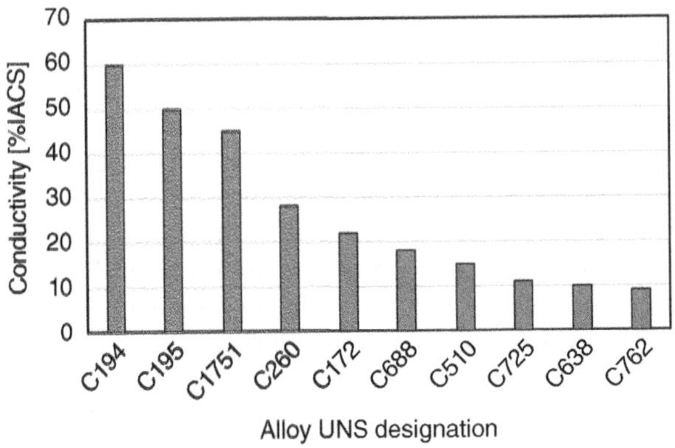

Figure 3.1 Rank of conductivity of copper alloy [4].

From the Wiedemann–Franz–Lorenz equation, the thermal conductivity of a copper alloy can be roughly estimated if its electrical conductivity is known. In most cases, an increase in temperature will cause a decrease in electrical conductivity, but not necessarily the thermal conductivity. Usually, for coppers with high electrical conductivity, the thermal conductivity decreases as the temperature is increased; for coppers with low electrical conductivity, the reverse situation will occur. The influence of temperature on both electrical and thermal conductivity has to be considered in the design of connector contacts.

Mechanical strength: Mechanical properties of copper alloys that are of importance include the modulus of elasticity, yield strength, hardness, tensile strength, and fatigue. These properties are not only dependent on alloy composition but also on the manufacturing process. Comparatively, precipitation-strengthened alloys demonstrate higher mechanical strength (tensile strength and yield strength) than solution-strengthened alloys and dispersed second-phase strengthened alloys. The terminologies of annealed, half-hard, hard, spring, and extra spring are related to the alloy temper, or alloying process.

Resistance to stress relaxation and creep: To maintain a stable contact interface, a stable contact force is a must. Adequate contact force initially does not guarantee stable contact force throughout the contact's entire life. Contact force decreases over time due to a phenomenon called "stress relaxation." Stress relaxation is due to macroscale plastic deformation within contact materials. The rate of stress relaxation depends on the duration, temperature, applied stress, alloy, and temper. Temperature is a major factor influencing the stress relaxation behavior of metals. As explained in Chapter 8, increasing temperature can significantly reduce the contact force. Higher stress usually causes a higher stress relaxation rate. The choice of the initial contact force depends on the potential stress relaxation rate and contact stress boundaries.

The stress relaxation of metals is usually defined as the remaining stress after a specified period, often 1000 hours (5 weeks), but it depends on the targeted application. Copper alloys show good resistance to stress relaxation at room temperature. Table 3.1 shows the performance of stress relaxation of some copper alloys at room temperature after 100 000 hours (10 years) [5]. Increasing the temperature significantly degrades the resistance to stress relaxation of some copper alloys, which limits these copper alloys

Table 3.1 Stress remaining for alloys at room temperature after 10 years of usage [5].

	Initial Stress (80% of 0.2% yield strength in ksi)			
	40	**60**	**70**	**80**
C194	94%	82%	–	–
C195	–	90%	81%	–
C260	–	91%	88%	–
C510	98%	–	–	95%
C725	–	–	95%	–
C762	–	96%	–	96%

Table 3.2 Solderability rating of copper alloys [1].

Solderability Rating	Coating Characteristics	Alloy Type
Class 1	100% wetting	C172, C1751, C194, C510, C521
Class 2	95% wetting	C195, C230, C638
Class 3	50%–90% wetting	C260, C654, C770

to high-temperature applications. Beryllium copper and spinodal alloys show excellent high-temperature performance, while the performance of brass is poor at high temperatures – only 53% stress remains after 1000 hours of service at 125 °C, and 65% remains at 105 °C, limiting the application of brass alloys to temperatures below 100 °C [1].

Creep is time-dependent plastic deformation under a constant load. The creep rate is dependent upon the material type, manufacturing process, operating time, applied stress, and temperature. Gradual deformation or creep may occur at stress levels much lower than the yield strength of metal contacts. Temperature greatly accelerates the creep rate of metal contacts.

Solderability: In many cases, connectors are to be assembled onto the printed circuit board through the soldering process, such as wave soldering for TH-type connectors and the reflow process for SM-type connectors. Solderability represents the ability of a metal surface to be wet with solder in the presence of a flux [6]. Solderability of an alloy is usually determined by visual examination of samples that are fluxed and subsequently dipped in solder for a specific time [7]. Based on visual inspection, the solderability of alloys can be rated, and acceptance criteria can be established. A class 1 rating refers to complete wetting by solder, whereas for a class 3 rating, the wetting area can be as low as 50%. A solderability of class 3 or higher is regarded as adequate for most connector applications [1]. Table 3.2 lists solderability ratings of some copper alloys when a mildly activated flux is used [1]. To ensure good solderability during assembly, pre-coating of tin or solder onto copper alloys is generally practiced.

Corrosion resistance: Chemically clean surfaces of most engineering metals react with oxygen and oxidize quickly when exposed to the atmosphere. The oxidation rate

Table 3.3 Copper alloys in which stress corrosion cracking was observed [6].

Copper Alloy	Environment
Cu-Zn Cu-Zn-Sn Cu-Zn-Pb	NH_3 vapors and solutions
Cu-Zn-Pb Cu-Sn	NH_3 vapors and solutions
Cu-Sn-P Cu-Au	Air NH_4OH, $FeCl_3$, HNO_3 solution
Cu-Zn Cu-Zn-Mn	Moist SO_2
Cu-Be	Moist NH_3 atmosphere

is high initially, but with the development of surface films, the rate typically decreases. The formed insulation film can generally prevent further intrusion of oxygen atoms and is usually self-limiting. The film constitutes a barrier against attack by other elements, imparting corrosion resistance characteristics to metals. However, the electrical resistance may be unacceptable at this time.

Adding other elements to copper tends to reinforce the oxidation film, constricting further expansion of film thickness. Copper alloys have good resistance to oxidation and corrosion. Bare coppers are used widely without plating for electronic applications.

In the presence of moisture, water, and steam, corrosion will be accelerated, and the surface insulating film will be thickened. Corrosion generally proceeds by electrochemical reaction, in which electrons flow between anodes and cathodes through a conductive solution. The formation of anodes and cathodes depends on many factors, such as surface defects, orientation grains, impurities, localized stresses, and so on. As a result, a layer of corrosion film grows on the surface of the copper. This film is not so dense and uniform, as it is formed in dry air conditions.

Copper alloys are more or less susceptible to a phenomenon called "stress corrosion," when they are in a highly stressed condition and exposed to an adverse environment. The combined effect of corrosion and stress may cause a catastrophic failure of contacts, commonly referred to as "stress corrosion cracking" or "season cracking." For copper alloys, the most aggressive environments are those containing ammonia or ammonia compounds [6]. Table 3.3 lists some copper alloys in which stress corrosion cracking was noticed in the specified environment.

Stress corrosion resistance varies for different kinds of copper alloys. Brass, which contains high amounts of zinc, was shown to be most susceptible to stress corrosion [1]. Therefore, the use of brass is limited only to benign environments. Beryllium copper and phosphor bronze are among the best copper alloys to resist stress corrosion.

Corrosion is a major source of problems in manufacturing and field failures of electronic devices. Oxidation of copper alloys can significantly reduce their solderability. Using flux is a way to remove the oxide layers on metal surfaces to ensure full wetting by solder; pre-coating with tin or solder is another way to improve the solderability of copper alloys. Oxidation or corrosion of contact surfaces inevitably increases contact resistance. Plating contact surfaces with noble or non-noble metals is generally practiced to improve the corrosion resistance of contacts and enhance interface stability.

3.2 Nickel Alloys

Nickel alloys are seldom used in designing connectors since they are expensive. The UNS for nickel alloys goes from N02016 to N99800; of the alloys, beryllium nickel has been used in the connector industry [8]. Beryllium copper shows high resistance to stress relaxation, especially at elevated temperatures. Nickel alloys also show considerably higher mechanical strength than copper alloys, while their electrical conductivities are relatively lower.

3.3 Conductive Elastomers

Conductive elastomers have been developed for a variety of electronic applications, including motherboard–daughterboard connectors. Compared with metallic spring contacts, conductive elastomers provide many advantages, such as high compliance, fine pitch and high I/O applications, and a short electrical path.

A conductive elastomer is a rubber that is made conductive by embedding metal wires or metal powders within the elastomer matrix. An elastomer is, as defined by ASTM, a polymeric material that at room temperature can be stretched to at least twice its original length and upon immediate release of the stress will return quickly to its original length [9]. Elastomers are sometimes referred to as "rubbers," because of their resemblance in elasticity or resilience. A variety of rubbers are produced nowadays, such as natural rubber, isoprene rubber, neoprene, polysulfide rubber, polyamide, polyester elastomer, silicone rubber, fluorosilicone rubber, and perfluoroelastomer. Among them, silicone rubbers are widely used for conductive elastomers because of their excellence in physical and mechanical properties and their resistance to corrosion and weathering.

Silicone rubbers, also known as "polysiloxanes," are a series of compounds whose polymer structure consists of silicone and oxygen atoms, rather than a structure made of carbon skeletons. The basic unit of a silicone rubber is

$$[(CH_3)_2 - Si - O - Si - (CH_3)_2 - O]_n$$

Compared with other carbon-linkage rubbers, silicone rubbers are more stable, as silicon itself is more stable than carbon. Silicone rubbers are among the most heat-resistant elastomers; they can be used for a wide temperature range, typically from −51 °C to 232 °C. Silicone rubbers possess outstanding mechanical and electrical properties; they exhibit good compression set resistance and rebound properties in both hot and cold environments, as well as a low dielectric constant and dissipation factor. Table 3.4 lists the typical electrical and mechanical properties of silicone rubbers. Silicone rubbers demonstrate excellent resistance to flame, sun, weathering, and ozone, and their properties are virtually unaffected by long-term exposure [9]. Silicone rubbers can be used in contact with dilute acids and alkalies, alcohols, animal and vegetable oils, lubricating oils, and aliphatic hydrocarbons. However, silicone rubbers demonstrate poor abrasion resistance, and they can be attacked by aromatic solvents such as benzene, toluene, gasoline, and chlorinated solvents, which will cause excessive swelling. Although they are resistant to water and weathering, silicone rubbers are not resistant to high-pressure and high-temperature steam [9].

Table 3.4 Physical and mechanical properties of silicone rubbers [9].

Specific gravity	1.05–1.94
Water absorption, %/24 h	0.02–0.6
Dielectric strength, V/mil	350–590
Dissipation factor At 60 Hz At 1 MHz	0.0007 0.0085–0.0026
Dielectric constant At 60 Hz At 1 MHz	2.91 2.8–3.94
Volume resistivity, ohm-cm	10^{14}–10^{16}
Tensile strength, psi	1200–6000
Hardness, Shore A	20–90
Maximum temperature, continuous use	232 °C
Compression set, %	10–15

A type of polysiloxane elastomer was prepared by incorporating approximately 30% volume fraction of foam into the elastomer [10]. The compressibility of the elastomer interconnects can be tailored by controlling the volume fraction of the foams.

To be used as connector contacts, metals are incorporated into rubbers to make them conductive. Metals could be in the form of powders or wires. Commonly used metal powders include nickel and silver. Nickel powders are usually coated with gold and silver to enhance electrical conductivity. Silver is a more expensive choice but offers high electrical conductivity. Silver is also resistant to chemical attack but may react with chlorine and sulfur. Metal wires used include gold, stainless steel, and brass wires plated with nickel and gold. These wires can be straight or curved, used singly, or used in a bundle [11].

The conducting mechanism for the metal-powder-filled elastomers can be described by the percolation theory. A minimum content of metal powders is required to establish the conductive network. The critical filler content is usually 70–80% in weight, depending on particle geometry, size, and size distribution. Once above the critical filler content, the conductivity of conductive elastomers increases by several orders. Filler particles may come in various shapes: spheres, fibers, flakes, or granules. Among them, flakes, due to their high aspect ratio, provide the minimum critical filler concentration for low resistance and the strongest adhesion to elastomers. Small particles are better than large particles, providing more particle-to-particle contact and, thus, higher conductivity [12].

3.4 Contact Manufacturing

Properties of connector contacts are not only dependent on contact materials but also greatly affected by how they are manufactured. For metal alloys, the manufacturing process can impart improved properties, especially mechanical properties, to connector contacts by optimizing their macrostructures, as different treatment profiles can result in different macrostructures.

Figure 3.2 Manufacturing process for metallic alloys [4].

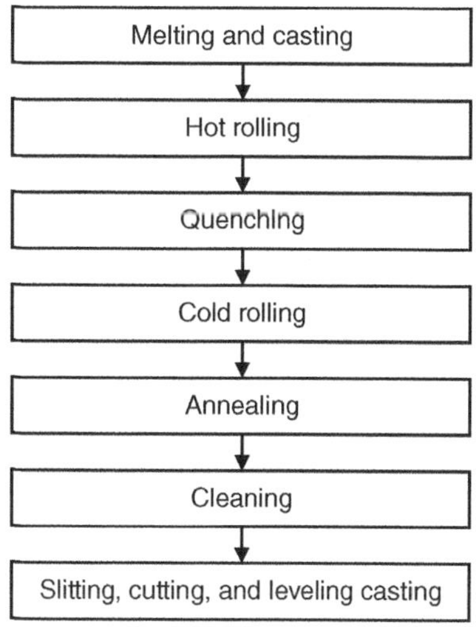

Figure 3.2 briefly illustrates the manufacturing process for metallic alloys [4]. These steps comprise the stamping process of making connector contacts. The stamping process is a tightly controlled operation in which alloys are formed to a required shape and configuration with progressive dies and then heat-treated to achieve spring properties. Formability is related to the ease of bending alloys to a required configuration. Formability usually contends with mechanical strength; good formability is achieved at a sacrifice of mechanical strength, and vice versa. A common way to measure formability is to determine the minimum radius of bending that produces fracturing. This minimum radius is always almost linear to the alloy thickness, so the ratio of the minimum radius over the thickness is usually characterized and reported. A smaller ratio denotes better formability. The formability of alloys is not only dependent on alloy type and treatment but also on bending directions. Better formability is achieved when the bending axis is perpendicular to the direction of rolling; this formability is also called "longitudinal formability." Conversely, "transverse formability" always means poor formability.

Some alloys are not suitable for stamping, such as leaded brass (C312-C385, C482-C465). These alloys are formed to a required shape through machining. They are mainly utilized to form shells or sleeves for multifinger contacts for DIP and PGA connectors.

The commonly used treatment terminologies are listed here, per ASTM standard B 601-98a [13]:

Annealing: A thermal treatment to change properties or grain structures of a product.

Cold work: Controlled mechanical operations for changing the form or cross section of a product to reduce residual stress variations, thus reducing susceptibility to stress corrosion or season cracking without significantly affecting the tensile strength or microstructure of the product.

Hot working: Controlled mechanical operations for shaping a product at temperatures above the crystallization temperature.

Precipitation heat treatment: Thermal treatment of a product to produce property changes such as hardening, strengthening, and conductivity increase by precipitating

Table 3.5 Temper name and standard tensile strength requirements [3].

Temper Name	Tensile Strength Requirement (ksi)
1/4 hard	49–59
1/2 hard	57–67
3/4 hard	64–74
Hard	71–81
Extra hard	83–92
Spring	91–100
Extra spring	95–104

constituents from the supersaturated solid solution. This method is also called "age hardened" and "precipitation hardened."

Solution heat treatment: A thermal treatment of a product to add alloying elements into the base metal lattice by heating the product above its solid solubility, followed by cooling at a sufficient rate to retain a supersaturated solid solution.

Spinodal heat treatment: Thermal treatment of a product to produce property changes such as hardening, strengthening, and conductivity increase by spinodal decomposition of a solid solution. This treatment is also called "age hardened," "spinodal hardened," or "spinodally decomposed."

Drawn stress relieved (DSR): Thermal treatment of a cold-drawn product to reduce residue stress variations, thus reducing susceptibility to stress corrosion or season cracking, without significantly affecting tensile strength or microstructure.

The thermal and mechanical treatment processes impart a specific macrostructure to the wrought alloys. Temper is the degree of hardness and elasticity in the material, often as a result of intentional heating and cooling of the material. Different treatment processes produce different tempers. A standard practice has been constructed by ASTM for the designation of alloy tempers [13]. Before construction of the standard, another designation system had been commonly used. Table 3.5 lists the terms applied to cold-rolled tempers; for each temper name, alloys must meet a specific requirement in their tensile strength.

The ASTM designation is practiced by using letters followed by numbers. The letter is related to a specific mechanical or thermal treatment experienced by an alloy, while the number represents an achieved temper; for annealed tempers, it is an indication of grain size. Some tempers and their symbols and definitions are listed here:

- **Annealed tempers, "O":** Tempers produced by annealing to meet mechanical property requirements.
- **Cold-worked tempers, "H":** Tempers produced by controlled amounts of cold work.
- **Heat-treated tempers, "T":** Tempers that are based on heat treatments followed by rapid cooling.
- **Solution heat-treated tempers, "TB":** Tempers produced by solution heat-treating precipitation-hardenable or spinodal-hardenable alloys.

- **Solution heat-treated and cold-worked tempers, "TD":** Tempers produced by controlled amounts of cold work of solution heat-treated precipitation-hardenable or spinodal-hardenable alloys.
- **Precipitation heat-treated tempers, "TF":** Tempers produced by precipitation heat treatment of precipitation-hardenable alloys.
- **Spinodal heat-treated tempers, "TX":** Tempers produced by spinodal heat-treatment of spinodal-hardenable alloys.
- **Cold-worked and precipitation heat-treated tempers, "TH":** Tempers produced in alloys that have been solution heat-treated, cold-worked, and precipitation heat-treated.
- **Cold-worked and spinodal heat-treated tempers, "TS":** Tempers produced in alloys that have been solution-treated, cold-worked, and spinodal heat-treated.
- **Mill-hardened tempers, "TM":** Tempers of heat-treated materials as supplied by the mill resulting from combinations of cold work and precipitation heat treatment and spinodal heat treatment.
- **Precipitation heat-treated or spinodal heat-treated and cold-worked tempers, "TL":** Tempers produced by cold-working the precipitation heat-treated or spinodal heat-treated alloys.
- **Precipitation heat-treated or spinodal heat-treated, cold-worked, and thermal stress-relieved tempers, "TR":** Tempers produced in the cold-worked precipitation heat-treated or spinodal heat-treated alloys by thermal stress relief.

References

1 Mroczkowski, R.S. (1998). *Electronic Connector Handbook*. New York: McGraw-Hill.
2 SAE International (2017). *Metals & Alloys in the Unified numbering System*, 13e. A joint publication of SAE International and ASTM International.
3 Mendenhall, J.H. (1980). *Understanding Copper Alloys*. New York: Wiley.
4 Liu, W. and Pecht, M. (2004). IC component socket materials. In: *IC Component Sockets* (eds. W. Liu and M. Pecht), 45–75. Wiley.
5 Ginsberg, G.L. (1977). *Connector and Interconnect Handbook*, vol. 2. The Electronic Connector Study Group, Inc.
6 Ginsberg, G.L. (1977). *Connector and Interconnect Handbook*, vol. 1. The Electronic Connector Study Group, Inc.
7 MIL-STD-883E (Method 2003.7) (1996). *Test Method Standard for Microcircuits – Solderability*. US Dept. of Defense.
8 Mill-Max Corp (1998). *A Design Guide to IC Connectors and Interconnect Components (Product Catalog)*. Oyster Bay, NY: Mill-Max Corp.
9 Philip, A. and Schweitzer, P.E. (1990). *Corrosion Resistance of Elastomer*. New York: Marcel Dekker.
10 Liao, Y., Shih, D.Y., Hedrick, J.L. et al. (Dec. 1996, pp. ix + 322,). Foamed elastomers for packaging interconnections, testing and burn-in applications. In: *Symposium of Electronic Packaging Materials Science IX* (eds. S.K. Groothuis, P.S. Ho, K. Ishida and T. Wu), 95–100. Boston, MA: Materials Research Society.

11 Rosen, G., Robinson, P.T., Florescu, V., and Singer, M.T. (June 1991). A comparison of metal-in-elastomer connectors: the influence of structure on mechanical and electrical performance. *IEEE Transactions on Components, Hybrids, and Manufacturing Technology* 14 (2): 324–351.
12 Li, L., Lizzul, C., Kim, H. et al. (December 1993). Electrical, structural and processing properties of electrically conductive adhesives. *IEEE Transaction on Components, Hybrids, and Manufacturing Technology* 16 (8): 843–850.
13 B601-18a (2018). *Standard Classification for Temper Designations for Copper and Copper Alloys – Wrought and Cast*. West Conshohocken, PA: ASTM.

4

Contact Plating

Michael G. Pecht

Center for Advanced Life Cycle Engineering (CALCE), University of Maryland, College Park, MD, USA

Copper alloys used in the contacts are very susceptible to corrosion. Contact plating (finish) provides corrosion protection for base metals and optimizes the mechanical and electrical properties of the contact interfaces. Contact plating can provide a shield from environmental attacks to the copper alloy base metals, thus prohibiting the accumulation of an insulating layer on the contact interface and maintaining its electrical stability. As a protective layer is supplied by contact plating, the durability and wear resistance of connector contacts are improved. The mechanical and electrical properties of the contact interface can be managed by choosing the appropriate plating material and plating thickness.

Contact plating can be divided into two categories, according to the options of plating materials: noble metal plating and non-noble metal plating. This classification is based on the corrosion resistance of metals. A noble metal is virtually corrosion-free, while a non-noble metal may react with certain chemicals. Choices for noble metal plating include gold, palladium, and their alloys; non-noble metal plating materials include tin, solder, nickel, nickel boron, silver, and copper; among them, tin and solder are more commonly used, while nickel is a commonly used underplate.

The selection of contact plating is conditional upon many factors, including cost, application/performance parameters, application environments, and reliability requirements. Figure 4.1 presents some selection guidelines for connector contacts based on contact force, insertion/withdrawal cycles, insertion force, and engagement wipe [1, 2]. Since noble metal plating is free of surface films, a lower contact resistance (CR) can be achieved without exerting high contact forces, but generally at a higher price. For connectors that must undergo significant insertions, noble metal plating is often needed. Otherwise, non-noble metal plating can be used if the requirements are benign.

4.1 Noble Metal Plating

Noble metal plating includes gold, palladium, and their alloys. These noble metals resist corrosion and film formation. The noble metals can be alloyed, and plating thickness can be adjusted to balance cost, performance, and reliability. The usual plating thickness of these noble metals is 10 (or less), 30, or 50 μin, with minimal 50 μin nickel as an

Electrical Connectors: Design, Manufacture, Test, and Selection, First Edition.
Edited by San Kyeong and Michael G. Pecht.

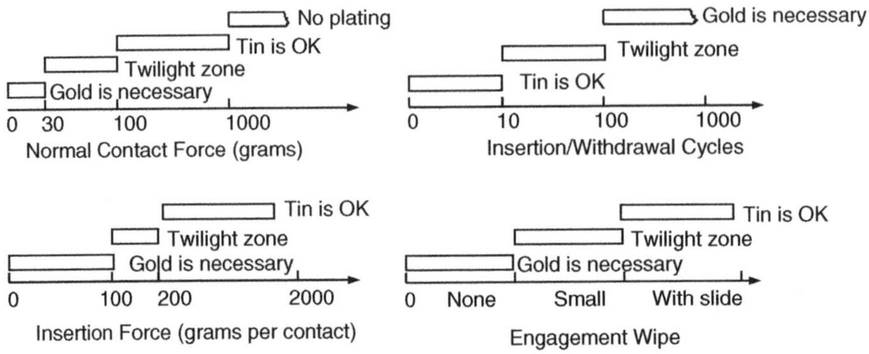

Figure 4.1 Guidelines for selecting contact plating materials [1].

underplate. A plating of less than 10 μin is often called "flash" and is used in noncritical areas where no contact interface is established. However, gold flash is used to cover palladium plating to prevent the occurrence of friction polymerization on the palladium surface. A plating thickness of 30 μin has been common in the connector industry and has been recommended for general industry application. For high-reliability applications, such as military and aerospace, a higher plating thickness is required, typically 50 μin.

4.1.1 Gold

Gold is the most extensively used noble metal because of its excellent electrical, mechanical, and thermal characteristics and its chemical inertness. Gold has electrical conductivity, which is comparable with silver and copper. It is immune to almost all environmental attacks, can be continuously used in high-temperature (above 100 °C) and high-humidity conditions, and is resistant to many pollutant gases, such as chlorine, hydrogen sulfur, sulfur dioxide, and base and acid solutions. Although pure gold is soft, it can be alloyed to increase its hardness and, consequently, contact durability.

Gold plating can be classified into eight general classes according to their impurity, application requirements, thickness, and manufacturing method [3]:

Class A: Decorative 24 K gold flash (2–4 μin), rack and barrel.
Class B: Decorative gold alloy flash (2–4 μin), rack and barrel.
Class C: Decorative gold alloy, heavy (20 to over 400 μin), rack.
Class D: Industrial/electronic high-purity soft gold (20–200 μin), rack, barrel, and selective.
Class E: Industrial/electronic hard, bright, heavy 99.5% gold (20–200 μin), rack, barrel, and selective.
Class F: Industrial/electronic gold alloy, heavy (20–400 μin), rack and selective.
Class G: Refinishing, repair and general, pure and bright alloy (5–40 μin), rack and selective.
Class H: Miscellaneous, including electroforming of gold and gold alloys, statuary, etc.

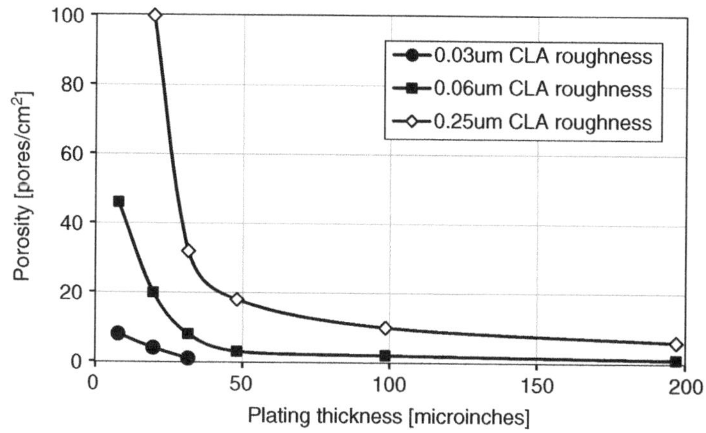

Figure 4.2 Porosity as a function of plating thickness and substrate roughness [4].

Gold plating for connector contacts mainly belongs to Classes D and E. Gold flash plating – usually less than 10 μin – is used as well. Gold flash is used primarily for noncritical parts of contacts, such as terminals or contact shells.

The addition of impurities in gold plating is based on two considerations: on the one hand, the cost can be reduced; on the other hand, the hardness of soft pure gold plating can be increased to improve contact durability. Gold alloy with an addition of 0.1% cobalt is called "hard gold" because of its improved hardness. However, alloying has an adverse effect on the electrical conductivity of contacts; for example, as little as 1% iron will increase the electrical resistance of gold over 1000%, so impurities and their levels must be carefully controlled. Alloying also reduces the corrosion resistance of gold contact plating.

One of the functions of contact plating is to seal the base metal from an outside unfriendly environment. However, a plated surface may not be as continuous as it appears. Pores in the plating layer expose the base metal to pollutant attacks. Plating porosity is defined by the number of pores per unit area (usually in cm^2) in the plating layer. Porosity is a function of substrate roughness and plating thickness. For a specific plating process and substrate roughness, porosity has been modeled by

$$P = AH^{-n} \tag{4.1}$$

where P is plating porosity (pores/cm^2), H is plating thickness, and A and n are experimentally determined parameters. Figure 4.2 shows examples of experimental data of porosity as a function of plating thickness and substrate roughness [4].

Substrate roughness has a first-order effect on the plating porosity. With the increase in plating thickness, porosity decreases, and an "elbow" can be observed, often around 15–30 μin, depending on the substrate roughness. Beyond 50 μin, the porosity curve becomes flat; however, the pores cannot be removed entirely. Thus, for general industrial applications, a minimum of 30 μin of hard gold plating is often required. For high-reliability applications, 50 μ in of gold plating is often a requirement.

Standards have been created to examine plating porosity. For example, EIA-364-53B, Nitric Acid Vapor Test, Gold Finish Test Procedure for Electrical Connectors and Sockets, presents an evaluation procedure for the acceptability of gold contact finishes, by

Table 4.1 Porosity counting via corrosion product size.

Corrosion Product Size	Assigned Count
Diameter ≤ 0.05 mm	0
0.05 mm < diameter ≤ 0.51 mm	1
Diameter ≥ 0.51 mm	2
Coverage in excess of 50% of measurement area regardless of size	20

exposing the contacts to nitric acid vapor. By examining corrosion products (usually circular in shape) after exposure to the acid vapor, the porosity can be counted. Table 4.1 shows the counting method via the corrosion product size. This test procedure does not apply to gold flash plating, which does not prevent pores. For a connector to be acceptable, a porosity of less than 1 (pore/cm^2) is generally considered acceptable but will depend on the targeted application.

Some best-practice guidelines for using gold plating in connectors and sockets [5] are as follows:

- Gold coatings are recommended for high-reliability applications.
- Gold coatings can be used in corrosive environments.
- Gold coatings can be used for applications requiring high durability.
- Gold coatings can be used with low normal force and low wipe.
- Gold is not susceptible to fretting degradation.
- Gold contact performance can be enhanced with lubrication.
- Gold coatings require the use of a suitable underlayer, such as nickel.
- Gold coating thickness depends on application requirements.
- Gold can be used for low-level circuit conditions.
- Gold contacts can be used at elevated temperatures.
- Gold contacts should not be mated to tin contacts.
- Gold contacts are not recommended for "Hot Make and Break" applications.

However, to ensure the reliability of a connector, specific applications and requirements should be considered, together with plating thickness and quality.

4.1.2 Palladium

Palladium plating is generally harder than gold, improving contact durability, and can be a cost-effective alternative to gold plating, although this cost advantage depends on the market and has been known to fluctuate. However, palladium provides lower electrical and thermal conductivity and inferior corrosion resistance. Palladium also has the disadvantage of being a catalyst of polymer formation [6]. In the presence of organic vapors, condensed vapors will polymerize on a palladium surface under fretting conditions, resulting in a friction polymer or brown powder, which can increase the CR.

Palladium alloys are also employed for plating contacts. One common palladium alloy is palladium (80%) – nickel (20%). It is most often applied with a gold flash (or "cap") on top. This affords the highest tarnish resistance, along with good durability and cost savings.

4.1.3 Combination of Gold and Palladium

Composite noble metal contact finishes consist of a thin gold layer, on the order of 0.1 μm in thickness, over a reactive surface to maintain a low surface resistivity in corrosive environments. In general, palladium or some palladium alloys constitute most of the plating thickness. The gold flash is coated over the palladium and provides the advantage of superior nobility [6].

4.2 Non-noble Metal Plating

A non-noble metal contact plating differs from a noble metal plating mainly in corrosion resistance. A non-noble metal plating is much less resistant to environmental attacks. An insulating film can develop quickly when a fresh non-noble metal is exposed to air. This insulating film determines many properties of the contact interface, including CR, and many design characteristics, such as contact normal force, insertion force, and contact wipe requirements.

Non-noble metals for plating include tin, solder, silver, nickel, and nickel-boron, in which tin and solder predominate (often 90% tin/10% lead). The plating thickness is generally much greater than that of gold plating, ranging from 100 μin to 200 μin, usually to save money. Tin and solder can be plated with or without nickel as an underplate. Since there is a layer of oxide film on the contact surface, a higher normal force is required to disrupt it and ensure a metal–metal contact. The oxide layer on a tin or solder surface is brittle and easy to disrupt. However, the re-oxidation of exposed surfaces inevitably thickens the oxide layer, causing a problem called "fretting corrosion" [7, 8].

Driven by legislature requirements and calling for "greener" products, the connector industry has been removing lead from the solder and using pure tin or no-lead tin alloys. Based on the finish color, tin plating can be classified into bright tin and matte tin. The bright finish, with a lustrous appearance, has a grain size less than 0.5 μm; while the matte finish, with a dull appearance, has a grain size larger than 1 μm.

One major concern for using pure tin plating is a phenomenon called "tin whisker growth," which is a spontaneous growth of single-crystal structure. Tin whiskers can cause electrical failures ranging from parametric deviations to catastrophic short circuits and even fires. One potential cause for whisker growth has been attributed to the residual compressive stresses in the tin resulting from the plating process. It was found that bright tin is more susceptible to whisker growth than matte tin finish because of the smaller grain size. The use of underplate, such as nickel, can significantly reduce the likelihood of whicker growth; however, there are still substantial concerns [9]. There are ongoing industry-wide efforts to understand the whisker growth mechanisms and develop test methodologies to effectively assess the propensity of whisker growth [10–16].

Silver is a relatively noble metal that shows inertness to atmosphere, steam, and both base and acid solutions, but silver will react with chlorine and sulfur. The reaction between silver and sulfur causes the silver surface to tarnish. Another problem with silver is electromigration, in which metal atoms transport from one conductive element to another across an insulating layer, under the influence of an applied DC potential and in the presence of moisture. The electromigration of silver causes the dissolution of silver atoms in one element and dendrite growth in the adjacent element. The dendrite growth of silver can result in short circuits between adjacent contacts. Although silver has high electrical conductivity and chemical inertness, it is rarely used in the connector industry because of its reliability problems [17, 18].

As compared to hard gold and palladium-nickel, silver-finished contacts have potential performance limitations that pertain to other contact characteristics of silver. Silver-finished contacts exhibit significantly higher insertion force characteristics, relatively poor durability, and adhesive wear, and are susceptible to tarnishing.

Contact performance results of an AgPd10 finish have been generated that compare it to standard silver as well as standard gold-based finishes gold and palladium-nickel. These results show that this finish has potential as an enhanced performance silver finish, as well as a low-cost gold-flashed alternative for some gold or palladium-nickel finish applications [19].

Most of the silver-palladium contact materials are manufactured as cast or wrought alloy. A palladium content between 30 and 50% is adapted for applications in relay contacts. Silver-palladium (about 40% Pd) obtains a good resistance against tarnishing by sulfur atmosphere but also a proper resistance against material migration; it also shows reliability against material burn-up. Nevertheless, the palladium content also leads to increased CR. To avoid the tendency for building up disruptive pollution layers by polymer accumulation, additional gold plating could be applied on top [20].

Nickel and its oxide layer show high hardness, so a much greater contact force is required to disrupt the surface oxide. Nickel is used as a contact finish when boron is added. Nickel is more commonly used as an underplate. The various non-noble metal contact finishes are discussed later in this chapter.

4.2.1 Silver

Electrodeposited pure silver contact finishes are favored for higher-current power transmission and often lower-current separable power connector applications. Silver has the highest electrical and thermal conductivity of any metal and has CR in the range of 0.1–1 mΩ at higher forces. The higher-current power transmission connectors typically are designed to have very high normal force (10–100 N) that preferably incorporates wipe; have silver thicknesses greater than 2 μm and up to 20 μm; and commonly do not have a nickel underplate. Lower-current power connectors typically have a minimum of 200 cN of normal force with wipe, a nickel underplate (minimum 1.25 μm) with a minimum of 2 μm silver, and low durability requirements.

Silver also functions well as a connector finish in some higher normal force/lower durability signal applications. Most other signal connectors operate at a significantly lower normal force with higher durability requirements. The fact that silver will tarnish in most connector environments and does not have a durable finish, however, can present problems if used for these signal connector applications. Another factor that affects the application of silver in many signal applications is that there is no adequate accelerated test correlations for silver finishes.

Two industry-wide considerations have led to the consideration of silver as a contact finish. In gold-plated applications, the price of gold is a major factor in finding alternate finishes. For tin finish applications, it is the conversion to "lead-free" tin. An effective way to minimize the risk of tin whisker-related failures for most tin-finished contact designs is to follow the recommended design/application criteria and only use currently accepted whisker-mitigating tin-plating chemistries. In connector designs where these recommendations cannot be met, a cost-effective alternative to tin may be required. A large part of what drives making contact finish choices is cost. The limits for contact finish choice are performance requirements and manufacturing concerns. The finish appropriate for any connector application is determined not only by performance requirements but also lifetime exposures under which it will have to function.

4.2.1.1 Characteristics of Silver as a Contact Finish

Silver has a unique combination of material properties, including the highest thermal and electrical conductivity of any metal and a relatively low hardness. Theory and experience show how these aspects lead to very low CR values for mated clean silver surfaces. Current passing through a clean silver-to-silver contact interface sees a relatively large conducting area (less constriction) made up of adhesively bonded metal-to-metal asperity junctions. This unique combination of properties results in relatively low CR, good thermal-rise performance, and very good vibrational stability. These attributes make silver attractive for use in power applications.

Silver also has good solderability characteristics, even if the silver is somewhat tarnished. If the level of tarnish is excessive, a more active flux may be required. Immersion silver is widely used as a solderable finish on board applications but can have a limited shelf life if the silver is exposed to the environment.

With any changes away from traditional finish choices comes the risk of unintentional finish combinations. Generally, it is recommended to mate "like on like," that is to say, mating a silver plug to a silver receptacle. Fortunately, silver can be an appropriate choice for mating to either gold-based or tin finishes. Performance levels and costs of such combinations will fall somewhere between the two different finishes. For example, mating silver to tin will not appreciably improve durability, and fretting failure could still be a risk, all at the higher cost of silver. General rankings of the galvanic corrosion susceptibility of different contacting material combinations show that silver-to-gold and silver-to-tin may be satisfactory combinations in some environments, but silver-to-tin can present risks in others. Therefore, a combination may have to be tested to determine its viability in the targeted application conditions.

As noted previously, silver does not have the "noble" character of gold and will form surface tarnish films when exposed to some reducible sulfur-bearing atmospheres. Silver has a high coefficient of friction (CoF) (high insertion forces) and poor wear characteristics (poor durability). Furthermore, silver may be susceptible to electromigration failures under specific conditions.

Silver tarnish films can be many colors, anywhere from yellow to tan to blue to black. Customers may object to this corroded appearance on contact interfaces because similar dark features on a tin surface (e.g. fret spots) or gold-plated surface (e.g. corrosion product creeping from pore sites) generally lead to CR instability at the contact interface. However, a silver-plated contact surface can appear discolored and still function well in many applications.

A related potential visual issue with tarnished silver films has to do with the process control systems designed to visually image/locate the position of contacts during assembly. If this system requires (or is programmed for) "shiny" or "white" silver contacts and the silver contact surface has tarnished, the system may reject the part. Because the visual appearance of the tarnish films that can form on exposed silver surfaces can be so variable, it may be difficult to compensate for automated inspection.

In most stable connector application environments, the growth of silver tarnish has been reported to continue linearly over time and is not self-limiting. Tarnish films generated on silver-finished surfaces exposed to most connector field applications are predominantly covalently bonded semi-conducting α silver sulfide (Ag_2S) and, to a lesser extent, small amounts of insulating and harder to displace silver chloride (AgCl). Film morphologies are typically nonuniform and often greater around surface features where water can collect. These predominantly silver sulfide tarnish films are semi-conductive at ambient temperatures, inherently soft, and relatively easily displaced with contact interface wipe at sufficient normal loads. If substrate material corrosion products (e.g. copper) are incorporated into the silver sulfide film, CR issues will likely occur. This is one reason why a nickel underplate is recommended when possible and why thicker silver plating is used when a nickel underplate is not used.

If the quality and level of silver tarnish are not excessive and wipe and sufficient normal loads are incorporated into the connector design, silver tarnish typically should not cause contact performance and reliability problems.

Silver sulfide (Ag_2S) forms when silver atoms react with reduced sulfur (HS^-) dissolved in the water film found on silver surfaces in typical connector environments. The primary source of this reduced sulfur is dissolved gaseous hydrogen sulfide (H_2S) or, to a lesser extent (less prevalent in the atmosphere), hydrolyzed carbonyl sulfide (COS). Hydrogen sulfide comes from processes such as organic decay, combustion processes, volcanic activity, and manufacturing sources such as paper mills, sewage plants, and high sulfur packaging materials. Silver tarnish can become excessive if used or stored unprotected in environments with a localized source of hydrogen sulfide. To a lesser extent, silver chloride (AgCl) has been detected in some field-exposed silver tarnish films. Silver is sensitive to the presence of chloride (Cl^-) and will react to form silver chloride. Chloride can come from species such as dissolved hydrochloric acid (HCl) gas or other chloride-containing particulates (e.g. NaCl). Hence, silver chloride has been found in some field-exposed silver tarnish films. The higher the level of harder insulative silver chloride in the tarnish film (relative to the level of softer semi-conductive silver sulfide), the more insulating and harder to displace the film becomes upon wipe leading to CR issues at thinner tarnish film levels. The reality is that in most field application environments, a predominantly silver sulfide film forms with sometimes minor amounts of silver chloride. There is the possibility that silver sulfate (Ag_2SO_4) will form in the presence of sulfur dioxide (SO_2), but only appreciably in the presence of artificially high levels of sulfur dioxide that are two and three orders of magnitude higher than found in typical ambient environments. Silver sulfate has not been found in tarnish films that have been exposed to typical connector environments.

4.2.1.2 Potential Tarnish-Accelerating Factors

The rate of tarnish growth is accelerated by many factors. Some of these factors are discussed in this section. These include chlorine gas, water, nitrogen dioxide, galling, and ozone.

*4.2.1.2.1 **Chlorine Gas*** Even though silver does not react directly with chlorine gas (Cl_2), its presence has a synergistic effect on silver tarnish formation when combined with hydrogen sulfide gas. This is especially evident on silver surfaces exposed to accelerating environments containing both hydrogen sulfide and chlorine gas. If the ratio of hydrogen sulfide to chlorine is high enough, the film growth rate begins to deviate from linear and approach parabolic. These artificial environment-exposed samples develop tarnish films faster and with a higher level of silver chloride than is found on most field-tarnished samples. The reality is that chlorine gas, commonly used in mixed flowing gas (MFG) testing, is virtually nonexistent in the atmosphere. This synergy may account for silver chloride level and CR discrepancies between field and accelerated environment-exposed silver surfaces. This illustrates the risk of applying silver in particular environments that may have a local source of chlorine gas.

4.2.1.2.2 Water The presence of moisture is needed for silver to tarnish in response to corrosive environments. This moisture allows for the dissolution of corrosive elements leading to the dissolution of metallic silver. Surface water films can either form as monolayers generated by humidity or take a condensed form. The tarnishing of silver has been reported as both positively dependent on and independent of increasing relative humidity levels depending on the exposure environment.

4.2.1.2.3 Nitrogen Dioxide Though silver does not react with nitrogen dioxide (NO_2), it has also been shown that the rate of silver sulfide tarnish formation can be somewhat accelerated by the presence of nitrogen dioxide – although the mechanisms are not well understood. It is not considered to be a dominant factor in silver corrosion.

4.2.1.2.4 Ozone/Photocorrosion Another synergistic accelerant would be the presence of ozone (O_3). The ozone does not react directly with the silver, but if it is prevalent in an environment, the resulting accelerated formation of the tarnish film could become a problem for contact interface electrical performance. Silver oxides (AgO and Ag_2O) are formed that are very insulating and difficult to displace if trying to make electrical contact. Ozone is not found at significant levels in most connector application environments. Photons capable of driving photocorrosion are another related potential accelerant of tarnish film formation.

4.2.1.2.5 Silver Tarnish Creep Across Gold Surfaces Silver tarnish films have a tendency to creep/migrate across any adjacent gold surfaces. This can lead to high CR values because it is much harder to displace the tarnish films formed across the harder gold subsurface. This is also why silver is generally not used as an overplate for gold finishes. This risk can be avoided in most designs.

4.2.1.2.6 Galling: Adhesive Bonding and Wear Unfortunately, part of what makes silver contact finishes work so well electrically (relatively soft → large contact area) also contributes to its inherently poor mechanical/durability performance. Clean silver has a relatively high CoF and is not a durable finish.

Upon mating of two clean silver surfaces, the surface material supporting the load is plastically deformed. The material supporting the load between the surfaces is plastically deformed and work-hardened (deformation zones) to form a contact interface of multiple metallically bonded (e.g. cold-welded) metal-to-metal junctions with a relatively large total contact area. If clean, the adhesive bonds can be as strong as an intercrystalline grain boundary. Because the material around the original asperity junction interface has been work-hardened, any relative motion (i.e. sliding) at that interface will cause subsurface material shearing as the asperity junctions are broken. Wherever the weakest juncture in the composite structure is located is where the damage occurs. The severity of wear increases with increased normal forces.

Poor wear durability is evidenced by clean silver's inherently high CoF. Clean silver surfaces that have been pressed together require some force to pull apart, resulting in material being ripped out of either surface. Because silver contacts are typically used at relatively high normal loads for electrical reliability reasons (relative to hard gold), the durability of silver finishes is further limited. These factors contribute to the high insertion forces found with clean silver contacts. Silver is usually inappropriate for high-durability applications.

The combination of higher normal forces and silver's high CoF promotes vibrational stability of a contact interface by preventing mechanically or thermally driven relative micro-motion (e.g. fretting) from being transferred to the contact interface. Silver itself is not susceptible to fretting oxidation in typical connector environments. If conditions are severe enough for fretting motion to occur, silver is susceptible to severe adhesive fretting motion wear (galling). Fretting motion could quickly lead to exposure of nickel and/or copper substrate materials, which are susceptible to fretting oxidation failures and unacceptable CR increases. The presence of even minor amounts of silver tarnish at the surface can weaken the adhesive metal-to-metal contact interface bonds by providing a shearable layer at the asperity junctions. The existence of even minimal storage-generated silver tarnish on mating silver surfaces has a tendency to initially reduce CoF. Electromigration shorting failures can occur when metallic bridges form between closely spaced metallic current paths under certain conditions. Many metals have been shown to be susceptible to electromigration under particular conditions, (e.g. copper, tin, lead, etc.), but silver is considered to be the most active. Several conditions need to exist simultaneously for electromigration failures to happen: an electrolyte path for the metal ions to flow (e.g. water plus ionic contamination), a voltage sufficient to drive the process, and enough time at potential/path conditions to complete the formation of the metallic bridge (dendrite or "stain").

In the past, silver electromigration failures had primarily been a potential issue for electrolytic silver-plated small centerline direct current (DC) board applications where there was a chance to form a relatively short water-based ionic electrolytic path in the presence of a sufficient driving voltage. In the printed circuit board industry today, where the problem is most likely to occur, the use of thin immersion silver and improvements in design, processing, testing, and application requirements have led to the widespread use of silver in board applications without such failures.

Electromigration failures generally do not occur in separable connector interface applications because the combination of geometries and application conditions are rarely susceptible. For the limited design/applications that might have all the qualities required to potentially lead to electromigration, there are several ways to mitigate the risk – the first one being to avoid the use of electroplated silver. Others would be to maximize line spacing, minimize differential voltages (driving force), use hydrophobic materials, limit features that could trap water (e.g. scratches, cracks, board crevices, etc.), apply hydrophobic and/or conformal coatings, overplate the silver with a less active metal, limit or eliminate the chance of ionic contamination (e.g. avoid no-clean flux residue, exclude paper fibers, etc.), and otherwise minimize water film formation. Testing may be warranted in a situation where silver is mated to a gold-plated board pad configuration, and the combination of design and conditions could make electromigration a possibility.

4.2.1.3 Use of Silver in Typical Connectors

Having several drawbacks previously mentioned, silver can be used only with additional considerations. In some applications or environments, the use of silver may be limited. This section covers those considerations.

4.2.1.3.1 Normal Force and Wipe Since silver has poor durability and the tarnish films are unpredictable, a design needs to be able to wipe/displace the films away from the contact interface upon mating. Therefore, it is recommended that silver-finished contact interface designs have a relatively high normal force and incorporate wipe whenever possible. Typical silver-plated contact designs use 200 cN of normal force.

4.2.1.3.2 Durability Considerations To compensate for silver's poor durability characteristics, silver is recommended for use in low-durability applications (e.g. $\leq$10 cycles). The actual appropriate number of durability cycles a particular connector could tolerate would be design- and application-dependent. The use of wear-reducing surface treatment (e.g. lubricant) could increase the number of durability cycles that a silver finish could tolerate and still function properly for a particular application.

4.2.1.3.3 Centerline Considerations Most connector designs would not have a problem with this failure mode. Historically, electromigration failures between relatively thick electroplated silver-finished board traces were an issue in certain smaller centerline board applications using hygroscopic board materials operating at relatively high voltages. Improved board manufacturing has led to the widespread use of immersion silver in board applications without such failures. As the connector industry is migrating to smaller and smaller form factors, electromigration may need to be tested for in specific situations.

4.2.1.3.4 Nickel Underplate It is recommended to use a nickel underplate (minimum of 1.25 μm) whenever possible. The nature of the tarnish film will change significantly if copper alloy elements from the substrate reach the surface of the silver. This can occur through mechanisms such as diffusion or corrosion creep at breaks in the silver electrodeposit (similar to what happens with gold). This will most likely lead to an increase in CR if these more tenacious films get into the contact interface. At higher temperatures, oxygen will diffuse through silver to the copper alloy interface at a relatively fast rate and could lead to blistering if no nickel underplate is used. A nickel underplate also will prevent a relatively weak layer of silver-copper intermetallic from forming at temperatures greater than 150 °C, which could lead to adhesion problems.

4.2.1.3.5 Silver Thickness What silver thickness is appropriate depends on application factors such as environmental severity, time at temperature considerations, durability requirements, nickel underplate, and surface treatment. Silver plating thicknesses are typically in the range of 2 μm or greater in most separable contact interface applications that use a nickel underplate. If no nickel underplate is used, a greater thickness of silver may be required to prevent substrate corrosion products from getting to the surface. These higher thicknesses also provide more silver material between the atmosphere and the substrate material(s), possibly leading to more wear cycles before any substrate material is exposed.

4.2.1.3.6 Reduced Sulfur and/or Chlorine- and/or Ozone-Containing Environments Silver-finished contacts are generally not used in applications where they would be openly exposed to outdoor or industrial environments due to concerns with excessive reduced sulfur exposure and tarnish levels. Exposure to environments with high levels of hydrogen sulfide (e.g. paper plant), chloride (e.g. salt spray), chlorine gas, and/or ozone should also be avoided for silver-plated contacts. If silver is to be used in such environments, environmental sealing may be required to avoid excessive tarnishing.

4.2.1.4 Managing Silver Corrosion

Surface treatments can be an effective way to attenuate tarnish formation, reduce insertion forces, improve durability, and minimize the risk of electromigration where functionally appropriate. They come with their own set of possible issues. Liquid-based lubricants and other surface treatments (e.g. self-assembled monolayers [SAMs]), are commonly used. Many commercial versions are available.

Silver surfaces have been coated with a thin solid top layer to prevent tarnish. Any solid layers on the relatively soft silver will no longer provide protection in the contact region if they are wiped away upon mating. Therefore, this approach is probably only potentially effective in keeping silver surfaces that are not disturbed "tarnish"-free and will have limited effect in a contact area. Attempts have been made to use a gold flash to protect the surface of the silver from tarnishing. There are three reasons that this approach is risky: silver and gold will readily inter-diffuse, silver sulfide will creep/migrate across gold surfaces, and gold is very susceptible to arc erosion leads. Such considerations have to be taken into account when considering these types of coatings.

For a surface coating to be effective, it has to perform without losing functionality or causing an unacceptable increase in CR. Surface coatings generally do not interfere with any subsequent soldering operations, but they can be rendered ineffective or harmful if removed during assembly and use or exposed to temperatures above or below their proper operating range. If excess surface treatment material can migrate to other regions of the connector or system and is a problem, the use of such a surface treatment may not be acceptable. Whether or not a surface treatment can be used effectively for a silver-finished connector is dependent on the performance requirements, application exposures, design limitations, visual requirements, and customer perception.

There are nonsurface treatment methods to shield silver surface atoms from corrosive environmental elements (e.g. hydrogen sulfide). Anything that limits the ingress of sulfur to the contact area has an effect. These methods are also known as physical blocks. Two metallic surfaces pressed together help restrict the flow of corrosive elements into and around a formed contact interface, leading to less corrosion in the immediate area. Formed functioning contact interfaces will remain electrically stable if not disturbed mechanically. The shielding effect of a connector housing (closed or actively sealed) can be quite dramatic, as well as the effect of any equipment enclosure or environmental controls (e.g. air conditioning, filters). Additionally, sometimes greases or gels are used in conjunction with a connector housing (e.g. the housing is "filled" with a grease style lubricant) to further mitigate flow of corrosive elements to the contact interface.

Most packaging cardboard and interleaving paper contain and release high enough levels of sulfur to cause accelerated tarnishing on packed silver-plated parts. Therefore, "low-sulfur" products are usually used to package silver-plated parts. Products such as Silver Saver paper are commonly packed with silver-plated parts and serve to absorb environmental sulfur, limiting the amount that can reach and react with the plated silver.

4.2.2 Silver-Palladium Alloys

Some of the properties of silver-palladium alloys, specifically AgPd10, are as follows [19]:

- AgPd10consistently exhibits a very fine thermally stable grain structure and structural stability at elevated temperature testing (1000 hours @ 200 °C).
- AgPd10 finish exhibits good adhesion to Ni substrates.
- AgPd10 provides some tarnishing resistance, as compared to Ag, in response to several standard accelerated testing MFG exposures, with and without an Au flash. An assessment of "how well" is dependent on the environmental requirements.
- Mating AgPd10 to AgPd10 and to Au and hard Au-flashed PdNi finishes, with and without an Au flash, has been tested on lab and production process plated samples with good results. Therefore, cross-connectivity between these finishes is acceptable.
- AgPd10 finish exhibits wear characteristics superior to Ag and comparable to Au and hard Au-flashed PdNi finishes.
- AgPd10 is susceptible to cracking upon bending due to its hard nature. This greater hardness in conjunction with the alloying effect (greater resistivity) leads to a higher CR as compared to standard Ag and Au and hard Au-flashed PdNi finishes, but the values have proven to be consistent and stable.
- Preliminary high-vibration testing results indicate that AgPd10 can also provide enhanced vibration wear resistance.

4.2.3 Nanocrystalline Silver Alloys

Advances in nanostructured control of alloy systems have positioned a new silver-tungsten alloy, with a nanocrystalline structure on the order of 20 nm, as a viable alternative for gold in high-speed connector applications that require qualification in compliance with Telcordia GR-1217-CORE test sequences [21].

Electronic interconnections frequently employ a combination of electroplated nickel or nickel alloy as a barrier layer to the substrate and electroplated gold as a topcoat. However, under extended mating cycles (>103) and potential bias in wet environments, these gold-on-nickel stacks often fail due to wear-through and/or corrosion of the connector base substrate. A new custom-engineered, electroplated, nanocrystalline silver alloy (NCSA) material stack differentiates itself from both traditional hard gold and other silver technologies by offering improved wear durability up to and beyond 104 mating cycles, as well as improved corrosion protection for immersion environmental exposures [22].

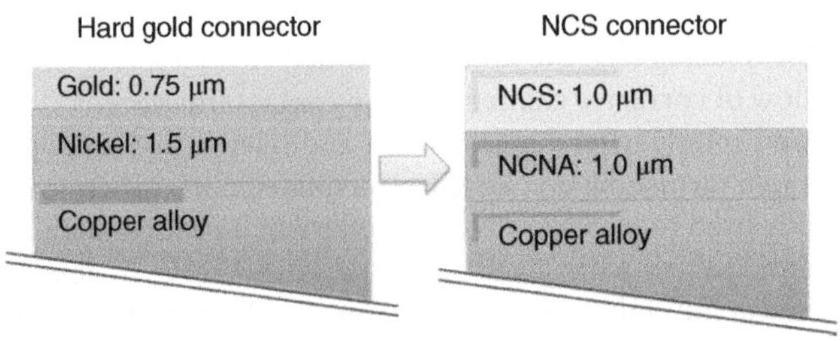

Figure 4.3 Schematic representation of Au/Ni process and comparable NCS/NCNA heterostructures [21].

Nano-crystalline alloys have been electrolytically deposited with material properties that can be tailored to meet specific application requirements. Through this nanostructure control, it is possible to engineer a new class of materials capable of meeting the growing demand for precious metal reduction in the connector industry [21].

This technology is based on nanostructure control and engineering of the alloys to maximize the material properties. Recent advancements in nanocrystalline materials engineering have resulted in a nanostructured silver-tungsten alloy (NCS), which has a hardness equivalent to electrolytic hard gold and a thermally stable grain structure that does not coarsen with age. Mechanical and environmental testing of the alloy in conjunction with the nano-crystalline nickel-tungsten alloy (NCNA) barrier layer has demonstrated that this stack is a viable alternative to traditional gold and Ni-sulfamate for high-speed connector applications [21].

NCS has been engineered with material properties that differentiate it from traditional silver and silver alloys. Since the alloy is composed primarily of silver, it remains susceptible to tarnish when exposed to corrosive sulfur environments. However, the nanocrystalline grain size and structure are thermally stable and do not coarsen with time or under industry-accepted heat aging cycles. This unique characteristic enables the alloy to maintain its integrity and properties over its lifetime. When combined with the NCNA barrier layer, the combined stack further enhances the durability and environmental corrosion resistance of the system [21], as shown in Figure 4.3.

Some of the properties of this alloy are as follows:

- The NCS performance is differentiated from traditional silver coatings in several regards: it is harder than Ag-CN, it is thermally stable and does not experience appreciable grain coarsening with time or temperature, and under high cycling durability it does not gall like silver.
- The NCS performance is equivalent to or better than hard gold under the standard connector testing conditions reported here: durability, hardness, CR, and environmental protection.
- The NCS exhibits surface tarnish in sulfur-containing environments, which is not seen for hard Au. However, this tarnish is noninsulating and has minimal impact on the CR, as seen in preliminary MFG testing.

4.2.4 Silver-Bismuth Alloys

In pure silver plating, the porosity decreases as a parabola with the increase of thickness of silver plating. The porosity of AgBi alloy plating (1.25 μm) is approximately equivalent to that of a 3.6 μm silver plating. The good pore corrosion resistance of AgBi plating is mainly caused by the fine-grain structure of AgBi plating compared with pure silver plating [23].

Both the pure silver plating and AgBi plating get corroded to form Ag_2S film under H_2S gas corrosion. As the corrosive time extends, pore corrosion occurs. The sulfide film and pore corroded products on the silver plating can lead to high resistance, which increases approximately linearly with the extended corrosive time. The CR decreases sharply as normal force increases. AgBi alloy plating shows a little lower CR than pure silver plating, especially after limited corrosion [23].

Sulfidation corrosion on the silver plating should be strictly controlled, and thicker silver plating is not a good protection measurement from sulfidation corrosion. AgBi plating can provide continuous plating with much lower porosity, which can supply higher reliability of electrical contact, especially under higher normal force [23].

4.2.5 Tin

Tin and tin alloys are considered as good non-noble alternatives for gold as they are inexpensive and have low CR and good solderability. The limiting factors for tin and tin alloys are their low durability characteristics and their susceptibility to fretting corrosion. These limitations can be addressed by using tin and tin alloys only in applications that have a relatively low number of mating cycles and by using appropriate design and lubrication to reduce susceptibility to fretting corrosion [24].

Relative motion of the contact interface is one of the major causes of failure of tin- and tin alloy–coated contacts. The amplitude of this fretting motion generally falls between 10 and 200 μm (and may be in the form of either mechanical disturbance or differential thermal expansion (DTE)). When a non-noble metal is used for coating, fretting disrupts any pre-existing oxide surface area, leaving behind the unoxidized metal exposed to the environment. This oxidizes when fretting occurs, and the debris collects near the contact surface. Also, metal transfer and wear of contact occur, which leads to a rapid increase in CR. One of the methods of preventing fretting corrosion in tin and tin alloys is to improve mechanical stability. High contact forces and large interface areas promote mechanical stability. Additionally, a contact interface with two or more discrete contact areas spaced apart from each other improves mechanical stability. Motion due to DTE of the various connector components is a driving force for fretting motion that is often overlooked [24]. DTE occurs when connector materials having different effects of DTE can be minimized by designing a connector so that the DTE is accommodated somewhere other than at the contact interface. For example, if the electrical contacts are set to mechanically float within the housing the effects of DTE can be reduced.

High forces are desirable whenever possible to establish a stable electrical contact. Limitations on the high force side are usually determined by (i) the total force required to engage/disengage multiple-circuit connectors; (ii) wear on the coating due to a large number of durability cycles (50 or more); (iii) physical size and strength of contact spring members; and (iv) spring deflection requirements. Another way to reduce fretting corrosion is to use a thin liquid lubricant film. It reduces fretting corrosion by (i) reducing friction and generation of wear particles due to motion; (ii) protecting the surface from atmospheric oxidation, in and around the contact interface; and (iii) preventing fretting corrosion. As contact forces approach the minimum recommended level of 100 g (3.53 oz), it becomes more critical to use lubrication to prevent fretting corrosion. The lubricant can act to reduce wear, seal off the interface from the environment, and/or inhibit corrosion of the interface material as it is exposed by the fretting action. On the other hand, with higher contact forces, lubrication may be necessary to reduce friction and wear during insertion cycling of the connector. In this respect, lubrication would permit the use of higher contact forces than would otherwise be allowed by frictional insertion force considerations. Lubrication may be applied to only one side of a connector pair. However, it is recommended that both sides be lubricated whenever possible. Sufficient lubricant is normally transferred during engagement of the connector to provide fretting corrosion resistance to a contact. However, not enough lubricant is transferred to protect against overall chemical corrosion of the unlubricated part.

Lubricant formulas have various levels of complexity and effectiveness. They include pure mineral oil and mixtures of natural oils, synthetic oils, and additives for optimized performance. When possible, it is recommended to choose a lubricant with proven anti-fretting characteristics. If the presence of lubricant cannot be maintained throughout the lifetime of the connector, an alternative to lubrication, such as a mechanically stable design or a noble metal coating, should be considered.

At elevated temperatures, the performance of tin-coated electrical contacts is degraded by aging effects due to a rapid increase in the rate of diffusion of copper and tin. This results in a change in the composition and effective thickness of the tin coating as a hard, brittle, nonuniform resistive layer of copper-tin intermetallic compound (IMC) grows between the tin layer and the copper base metal. It is recommended that a nickel underlayer be applied for elevated temperature applications because the growth rate of the nickel-tin IMC is lower than that of the copper-tin IMC. In addition to the detrimental effects of IMC, tin loses a significant portion of its mechanical strength resulting in substantial creep effects at temperatures >100 °C (212 °F). Special design parameters such as high contact force, thick coatings, or limited time–temperature exposure are needed to avoid contact problems. High-temperature operation also requires that the contact lubricant withstand the time and temperature requirements, and not change in composition, degrade, or evaporate.

The electrical performance of these various coatings is roughly equivalent. Coatings with bright appearance are considered more aesthetically appealing. Both bright and matte coatings are solderable; however, plated coatings brightened with organic additives can become unsolderable when plated with excessive brightener content. Matte tin should be kept clean to ensure consistent solderability. The solderability of matte tin is initially similar to that of hot air leveled tin (HALT) and reflowed tin coatings of similar thickness. However, matte tin plating solderability will deteriorate faster than that of HALT and reflowed tin. Tin coatings on brass should have a nickel undercoat to prevent zinc migration from the base metal. The main effect of zinc migration is to reduce solderability.

Tin and tin alloy coatings should be at least 100 μin thick, mainly to reduce the effects of IMC growth on solderability and for increased durability. To minimize the effects of high CoF, the maximum recommended thickness for tin coatings is 300 μin. Coatings thinner than 100 μin have been successfully used in special cases. A particular case involved pre-tinned stock with 30–80 μin of tin applied as a hot dipped coating. This material has been used for many low-cost connectors where price considerations are predominant and the product does not need to be solderable. The thinner electrodeposited coatings have generally been unsuccessful due to porosity and intermetallic growth, and they do not afford good environmental protection.

Tin-to-gold interfaces are more susceptible to fretting corrosion-related failures, and lubricants are not nearly as effective at stabilizing CR as they are with tin-to-tin interfaces. Tin tends to transfer onto the gold surface, which can ultimately lead to the accumulation of tin oxide on the harder gold substrate. It is more difficult to disrupt the tin oxide on the harder gold substrate than it is to break through the tin oxide that forms directly over tin. The warning against mating tin with gold also applies to mating tin with palladium and its alloys because the fretting behavior of tin-to-palladium and its alloys is similar to that of tin-to-gold. This warning does not apply to the case of the tin-to-silver interface because its fretting behavior is similar to that of tin-to-tin [24].

Tin and tin alloy coated contacts should be designed with wipe during insertion to aid in breaking through the oxide surface layer to establish metal-to-metal contact. Even zero insertion force (ZIF) connector designs should include wipe. A less satisfactory method of assuring the break-through of the oxide layer would be to design the contacting surface with sharp points to penetrate the opposing surface. The problem with incorporating these force-concentrating points into the design of the contact is that they can decrease the durability of the contact. This decrease in durability is because, upon insertion, penetration of the contact surface can cause premature exposure of the underlying metal to the environment.

Tin and tin alloys are not arc-resistant because they have relatively low melting temperatures. Therefore, tin-coated contacts should not be used to make or break the current. As long as fretting corrosion can be controlled, there are no limits to the voltage and current levels at which these contacts may be used [24].

4.2.6 Nickel Contact Finishes

Nickel finishes are non-noble due to the formation of a hard and tenacious oxide. The oxide films that form on nickel can be disrupted, but high normal forces are required. Because nickel oxides are self-limiting in thickness (to 100 nm or so), they can be electrically disrupted by voltages less than 1 V. This allows the use of nickel as a battery contact material. Table 4.2 lists the advantages and disadvantage of common contact platings. In addition, nickel, like tin, is susceptible to fretting corrosion [6].

4.3 Underplating

A contact finish is not always 100% complete. Pores and manufacturing defects expose the base metal to outside unfriendly environments. A contact underplate creates another barrier against the intrusion of environment conditions and seals off the base metal. It also blocks the outward diffusion of base metal constituents, especially at elevated temperatures. Furthermore, if the underplate has a high hardness (such as for nickel), the contact durability can be improved.

Table 4.2 Comparisons between different contact platings.

Contact Plating	Advantages	Disadvantages
Gold	• Corrosion-resistant in almost all environments • Excellent electrical and thermal characteristics	• High cost
Palladium-nickel	• High hardness (high durability) • Lower cost than gold	• Frictional polymerization
Silver	• High electrical and thermal conductivity • Resistance to welding	• Reacting with sulfur and chlorine • Electromigration
Tin-lead	• Ease of displacement of oxide film • Low cost	• Potential for fretting corrosion • Contains lead
Tin	• Ease of displacement of oxide film • Low cost • Lead-free	• Potential for fretting corrosion and whisker growth
Nickel-boron	• Self-limiting oxide film	• High hardness of oxide film susceptible to fretting corrosion

Nickel is the most commonly used underplate. Nickel underplating seals off the base of pore sites from the environment through its passive and self-limiting oxide film. Studies have shown that nickel provides an effective barrier against the migration of base metal and corrosion products, and contact durability is greatly improved because of its high hardness [6]. The effectiveness of the underplate is conditioned upon its thickness. Tests indicate that nickel underplating performs best at a thickness of approximately 50 μin. A thicker nickel plating may not be helpful because as nickel thickness increases, the surface roughness increases, which may cause a higher porosity that decreases wear resistance.

Industry leaders have seen palladium/cobalt as a low-cost, next-generation contact finish. Tests demonstrate that palladium/cobalt offers both performance and processing advantages over palladium/nickel, which has stood as the cost-effective finish of choice for about 20 years. Because its hardness is greater than palladium/nickel and hardened gold, palladium/cobalt has greater durability (hardness has a significant impact on durability). Durability is especially important for connectors that undergo numerous connect/disconnect cycles over their life span. Other performance parameters, such as CR and fretting corrosion resistance, run about the same for the two alloys. Another advantage concerns processing quality control. A higher degree of thickness control is achievable with palladium/cobalt, providing a smoother and more uniform finish for about the same cost [25].

4.4 Plating Process

Generally, there are four methods to apply plating to a connector contact: electrolytic plating, electroless plating, cladding, and hot dipping. These are all overviewed in this section.

4.4.1 Electrolytic Plating

Electroplating, the most commonly used method to apply contact finishes, can be used on copper, nickel, tin, gold, palladium, and silver. In this process, metals in ionic form migrate from a positive electrode (anode) to a negative electrode (cathode) under an applied DC voltage. The metal atoms are oxidized at the anode and dissolve into the electrolyte solution; they are reduced at the cathode, causing the cathode to be coated with a thin metal layer. The electrolyte solution contains dissolved salts of the metal to be plated. Precisely controlling the composition and concentration of the electrolyte solution, solution temperature, pH value, and current density is necessary to obtain a high-quality contact finish. Depending on the formula of the electrolyte solution, the composition of the contact finish can be tailored; for example, palladium-nickel can be deposited as a homogeneous alloy over a composition range from approximately 30% to over 90% palladium by weight [20]. For an alloy composition from 75% to 85% of palladium-nickel, one formula is as follows:

Palladium, $Pd(NH_3)_4Cl_2$, 18–28 g/l
Ammonium chloride, 60 g/l
Nickel chlorine concentrate, 45–70 ml/l (nickel metal 8–12 g/l)
Ammonium hydroxide to pH 7.5–9.0
Temperature, 30–45 °C
Current density, 0.1–2.5 A/dm^2

The contact plating can be overall or selective. Overall plating is the complete coverage of the contact by the contact finish. Selective plating is applied only to the contact area instead of the whole part. Selective plating is especially applied to noble metal plating for cost reduction.

Factors used to examine the quality of plating are residual stress, impurities, cracking, and porosity. These manufacturing defects can significantly influence the performance and reliability of the contact interface; for example, pore corrosion results from the attack of an adverse environment on the base metal through pores in the contact finish.

4.4.1.1 Rack Plating

In the rack plating method, workpieces are attached to the crossbars on a rack by permanent or replaceable tips, which are known as "workstations." DC from a rectifier is picked up by hooks from a work (flight) bar, travels down the rack's conductive backbone, enters the solution at the workpiece (cathode), and leaves at the anode. Metal ions in solution travel toward and plate out on the workpiece.

Rack dimensions are established so that each rack carrying parts will fit between anodes of the smallest process tank in a line. All common types of plating can be conducted with the rack plating method, including zinc, cadmium, tin, copper, precious metals, nickel, and chromium. The amount of current carried by the hooks and spine is determined by multiplying the current density of the plating bath by the surface area of parts on the rack.

4.4.1.2 Barrel Plating

Barrel plating is one of the processes to electroplate large quantities of small parts. Barrel plating received its name during the birth of electroplating in the American Civil War when parts were loaded into wooden kegs or barrels for coating. With the advent of modern technology during World War II, such as plastic-coated barrels resistant to harsh chemicals, barrel plating gained widespread popularity as an economical method of coating bulk parts. Nowadays, it is estimated that more than 70% of all electroplating facilities use barrel plating.

During the barrel plating process, bulk parts are loaded into a drum and dipped into a solution containing the substance to be plated. An electric current is passed through the parts from an electrode in the middle of the drum to electrodes on the surface of the drum. While the drum rotates, the parts continuously make and break electrical contact with each other at random locations on the surface. This random contact leads to a uniform coating that is not possible with rack plating. However, since the parts tumble and are continuously contacting each other, barrel plating does not produce a good surface finish. Due to the need for parts to tumble and rotate in the process, barrel plating is best suited to finishing large quantities of small parts. One of the primary uses of barrel plating is to enhance corrosion resistance. Moreover, barrel plating is not well suited to long, cylindrical parts due to their inability to randomly tumble.

Barrel plating is extremely versatile. Unlike rack plating, it does not require special fixtures, and many different materials can be finished using the same barrel. Barrel plating is labor-efficient because it does not require the manual loading or handling of individual parts. This makes it well suited to the bulk finishing of large quantities of parts. In addition, the bulk material may remain in the same barrel for cleaning, preparation, electroplating, and drying. The entire inside diameter of the drum essentially acts as one large cathode, allowing a much higher level of current to flow and hence a much faster production rate when compared to rack plating. The rotation of the barrel creates a mechanical tumbling action that tends to clean the parts and increases the uniformity of the coating.

4.4.2 Electroless Plating

Electroless plating refers to the autocatalytic or chemical reduction of aqueous metal ions plated to a base substrate. It is a self-catalytic process, without applied electrical current or voltage. Electroless plating is seldom used for contact finishes; it is applied only when the contact shape is complex and when electrodeposition methods could cause significant nonuniformities in the contact finish.

Various mechanisms have been developed for the deposition reaction in an electroless nickel plating bath. The principle reactions generally follow these equations:

$$3NaH_2PO_2 + 3H_2O + NiSO_4 \leftrightarrow 3NaH_2PO_3 + H_2SO_4 + 2H_2 + Ni \tag{4.2}$$

$$Ni^{2+} + 2H_2PO_2^- + 2H_2O \leftrightarrow Ni + H_2 + 2H_2PO_3^- + 2H^+ \tag{4.3}$$

$$3H_2PO_2^- \leftrightarrow H_2PO_3^- + 2P + 2OH^- + H_2O \tag{4.4}$$

Here the reaction is using hypophosphite ion as a reducing agent. During the process, phosphorous is co-deposited in the plating layer.

Immersion plating, sometimes called "displacement plating," is the deposition of a more noble metal on a substrate of a less noble, more electronegative metal, by chemical replacement from an aqueous solution of a metallic salt of the coating metal. This process differs from the autocatalytic method in both mechanisms and its results. Displacement plating requires no reducing agents in solution. Immersion deposition ceases, thus allowing no further displacement of the metal salts and substrate, as soon as the substrate is completely covered by the metal coating, whereas autocatalytic (electroless) plating knows no limit to the thickness of deposit that is obtainable. Therefore, the immersion method can only produce a coating layer with a limited thickness, typically below 5 μin.

Electroless plating can be used to plate nonconductive surfaces, where the flow of current is not accessible, and an electrolytic process cannot be employed. Furthermore, one can deposit a uniform plating layer, even on complex shapes.

4.4.3 Cladding

Cladding is a mechanical process. A cladding material is bonded to a carrier material (contact) by the application of high pressure. There are three types of cladding: overlay, to play, and inlay [6]. Outlay provides complete coverage of the substrate, to play covers only a selected area of the substrate, and inlay is a two-stage process. As an example, gold and nickel contact finishes are first bonded together, and the base metal is grooved by skiving. The gold/nickel combination is placed in the groove, and the metals are roll-bonded together. Additional heat treatment follows to enhance the interface bonding (adhesion).

4.4.4 Hot Dipping

Hot dipping is predominantly applied to tin and solders, which have low melting points. Hot dipping is conducted by immersing the contacts or strip metals into molten tin or a solder bath for a specific duration. Tin or solder is coated on the contacts with thickness controlled by using air knives or wipers. Hot dipping is used to improve the solderability of connector contacts (terminals). It is usually not applied for contact areas finishes, as excessive intermetallics can grow at elevated temperatures, causing increased electrical resistance.

References

1 Liu, W. and Pecht, M. (2004). IC component socket materials. In: *IC Component Sockets*, 45–75. Wiley.
2 Ginsberg, G.L. (1977). *Connector and Interconnect Handbook*, vol. 1. The Electronic Connector Study Group, Inc.
3 Weisberg, A.M. (1998). *Gold Plating/Metal Finishing Guidebook and Directory*, Issue 98. LeaRonal Inc.
4 Abbott, W. H. (2002). The role of electroplates in contact reliability. *Proceedings of the 48th IEEE Holm Conference on Electrical Contacts*, Orlando, FL (October 2002). IEEE.
5 AMP (1996). Golden Rules: Guidelines for the Use of Gold on Connector Contacts. AMP technical report.
6 Mroczkowski, R.S. (1998). *Electronic Connector Handbook*. New York: McGraw-Hill.

7 Wu, J. and Pecht, M. (2006). Contact resistance and fretting corrosion of lead-free alloy coated electrical contacts. *IEEE Transactions on Components and Packaging Technologies* 29 (2): 402–410.
8 Wu J., and Pecht, M. (2002). Fretting corrosion studies for lead-free alloy plated contacts. *Proceedings of the 4th Electronics Packaging Technology Conference*, Singapore (December 2002). IEEE.
9 Sampson, M. and Leidecker, H. (2003). Matte Tin Plated IC Leadframe Examination. NASA, nepp.nasa.gov/whisker/experiment/exp7 (accessed Dec. 13, 2019).
10 Ganesan, S. and Pecht, M. (2006). *Lead-Free Electronics*. New York: Wiley.
11 Tin Whiskers [television broadcast]. Modern Marvels: Engineering Disasters. History Channel; March 22, 2006.
12 Han, S., Osterman, M., and Pecht, M. (2010). Electrical shorting propensity of tin whiskers. *IEEE Transactions on Electronics Packaging Manufacturing* 33 (3): 205–211.
13 Shibutani, T., Yu, Q., and Pecht, M. (2009). Tin whisker reliability in microelectronics. *Micromaterials and Nanomaterials* 09: 49–53.
14 Shibutani, T., Osterman, M., and Pecht, M. (2009). Standards for tin whisker test methods on lead-free components. *IEEE Transactions on Components and Packaging Technologies* 32 (1): 216–219.
15 Mathew, S., Osterman, M., Pecht, M., and Dunlevey, F. (2009). Evaluation of pure tin plated copper alloy substrates for tin whiskers. *Circuit World* 35 (1): 3–8.
16 Fang, T., Osterman, M., and Pecht, M. (2006). Statistical analysis of tin whisker growth. *Microelectronics Reliability* 46 (5–6): 846–849.
17 Zhang, S., Osterman, M., Shrivastava, A. et al. (2010). The influence of H2S exposure on immersion-silver-finished PCBs under mixed-flow gas testing. *IEEE Transactions on Device and Materials Reliability* 10 (1): 571–581.
18 Wang, W., Choubey, A., Azarian, M., and Pecht, M. (2009). An assessment of immersion silver surface finish for lead-free electronics. *Journal of Electronic Materials* 38 (6): 815–827.
19 Myers, M., and Schmidt, H. (2014). Connector level performance evaluation of a new high speed reel to reel electroplated silver palladium alloy contact finish. *Proceedings of the 27th International Conference on Electrical Contacts* Dresden, Germany (June 2014). ICEC.
20 Talgner, F., Manz, U., Berger, S., Weyhmueller, B., and Pfund, A. (2014). Novel silver-palladium electrolyte for electrical contacts. Proceedings of the 27th International Conference on Electrical Contacts, Dresden, Germany (June 2014). ICEC.
21 Goodrich, T., Bui, K., Athans, A., Cahalen, J., Osborne, M., and Stokoe, P. (2014). Performance testing and evaluation of a Ag-W nano-crystalline silver alloy as a gold replacement in electrical connectors. *Proceedings of the 60th IEEE Holm Conference on Electrical Contacts*, New Orleans, LA (October 2014). IEEE.
22 Bui, K., Hasanali, T., and Goodrich, T. (2015). New final finish stack including a custom nano-crystalline silver alloy for mobile connector systems with high cycling wear and powered environmental exposure requirements. *Proceedings of the IEEE 65th Electronic Components and Technology Conference (ECTC)*, San Diego, CA (May 2015). IEEE.

23 Zhou, Y., Yuan, C., and Ge, S. (2013). Usability of silver plating and silver bismuth alloy plating on electrical contacts. *Proceedings of the 59th IEEE Holm Conference on Electrical Contacts*, Newport, RI (Sept. 22–25, 2013). IEEE. pp. 1–10.

24 AMP Inc. (2004). The Tin Commandments: Guidelines for the Use of Tin on Connector Contacts, Tyco Electronics.

25 Bell-Labs News (1998), Lucent Technologics' Electroplating Chemicals & Services Venture Develops High-Performance Material for Connector Contacts. www.bell-labs.com/news/1998/october/23/1.html (accessed Dec. 13, 2019).

5

Insertion and Extraction Forces

Michael G. Pecht

Center for Advanced Life Cycle Engineering (CALCE), University of Maryland, College Park, MD, USA

The connector provides an electrical path between components, assemblies, and subsystems. It is also a mechanical structure that provides the necessary conditions, such as contact normal force, to establish and maintain the electrical contact for proper signal and power performance and reliability. Proper functioning of the connector contacts depends on their mechanical stability, as well as their electrical performance. In evaluating a connector, both mechanical and electrical aspects must be considered. These include the mechanical issues (which affect the performance and reliability) associated with insertion and extraction force, contact force, contact retention, and contact wipe, and the electrical performance issues associated with contact resistance, current, inductance, capacitance, and bandwidth.

5.1 Insertion and Extraction Forces

The insertion and extraction force is the force required to insert the package leads into and extract them from their normal positions in a connector [1]. Other similar terminologies include mating/unmating force and engagement/separation force [2].

In conventional connector design, the engagement of mating contacts occurs in a plane approximately parallel to the plane of their mating surfaces. The engagement force of the male pins acts on the spring contacts of the connector and results in their deflection, which in turn exerts a contact normal force onto the male pins. Figure 5.1a schematically illustrates the engagement process [3]. Two stages are used to describe the engaging process. In the contact mating stage, the connector contact beam begins to deflect, the contact normal force is generated, and the insertion force increases, generally rapidly. In the engagement stage, the contact beams are fully deflected, the full normal force is applied, and the pin slides on the surface of the connector contact. Figure 5.1b shows the insertion force versus the insertion depth during the two stages. During the first stage, the insertion force increases with the insertion depth, until eventually a maximum force is achieved. For a given normal force, which is determined by the stiffness of the contact spring and the magnitude of its deflection, the maximum insertion force depends on the mating geometry and coefficient of friction. As the friction force opposes the direction of motion, it adds to the insertion force.

Electrical Connectors: Design, Manufacture, Test, and Selection, First Edition.
Edited by San Kyeong and Michael G. Pecht.

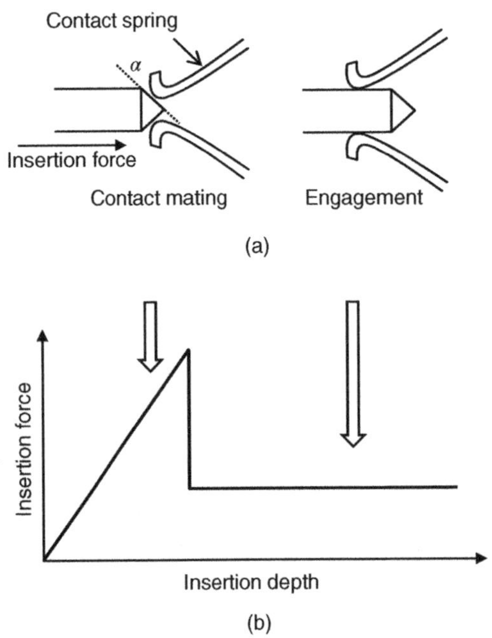

Figure 5.1 Schematic illustrations of (a) mating and engagement process and (b) characteristic curve for insertion force [3].

A simplified equation is used to calculate the maximum insertion force, given a specific contact normal force [4],

$$F_{\mathrm{i}}(\max) = 2F_{\mathrm{n}}(\max)\frac{\sin\alpha + \mu\cos\alpha}{\cos\alpha - \mu\sin\alpha} \tag{5.1}$$

where F_{i} is the insertion force, μ is the coefficient of friction, and α is the angle of the mating interfaces. Suppose the coefficient of friction is 0.4; a change of mating angle from 15° to 30° will result in about a 143% increase in insertion force. Therefore, any misalignment will increase the difficulty of insertion and potentially cause damage to the leads and connector pins.

After surpassing the maximum insertion force, the insertion force lessens and levels off until the package pins reach their normal positions. The value of the insertion force is the same as that of the friction force, as given in the following equation:

$$F_{\mathrm{i}} = \mu F_{\mathrm{n}} \tag{5.2}$$

where μ is the dynamic coefficient of friction, and F_{n} is the contact normal force.

As depicted in Figure 5.1b, the force needed to initially deflect the contact springs can be several times larger than the friction force between the contacts after a full deflection has been achieved. For example, if the coefficient of friction is 0.4 and the mating angle is 15°, the maximum insertion force is about 87% greater than the insertion force in the second stage. Therefore, it is the initial engagement force that usually presents the greatest difficulty in mating connectors and causes degradation of the plated surfaces of the electrical contacts. Thus, the insertion force is the maximum force required to mate the contacts [1]. The extraction of pins from a connector is the reverse process of the stage 2 insertion of pins. The extraction force is equal to the friction force, so the extraction force is usually much lower than the insertion force.

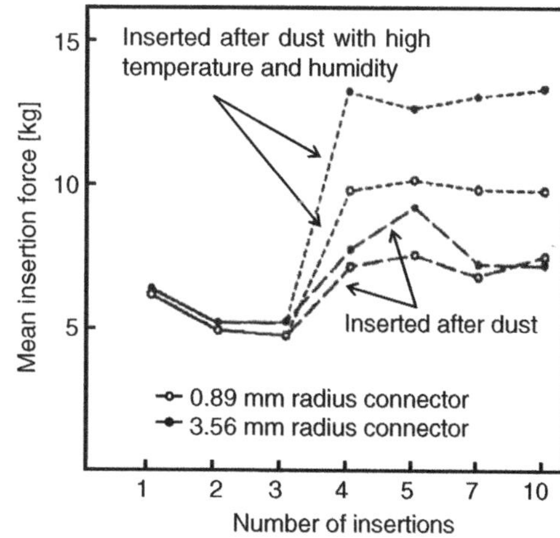

Figure 5.2 Insertion force measurements of connectors during dust/humidity test sequence [5].

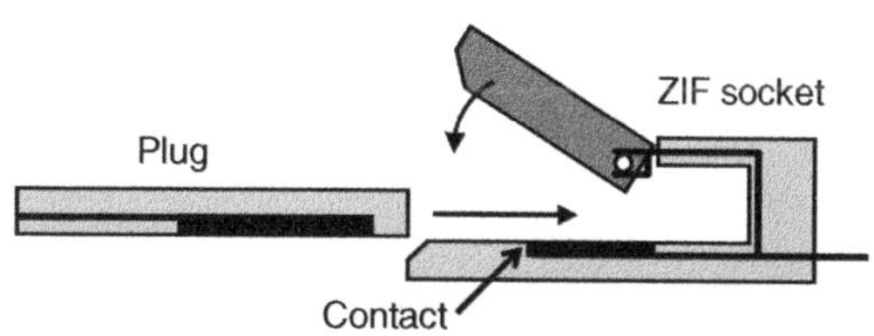

Figure 5.3 Schematic illustration of a zero-insertion-force connection [6]. Graphically renewed.

The insertion and extraction force for a connector is the sum of the individual contact insertion and extraction forces, but they are also significantly influenced by factors such as contact alignment, misalignment, and misregistration. These factors can be induced by manufacturing defects, normal use, thermal and moisture expansion, and aging. Figure 5.2 shows the insertion force data for connectors under a dusty environment with or without high temperature/humidity cycling [5]. The connectors exposed to dust and high temperature/humidity showed higher insertion force. A large applied force may cause difficulty in mounting and demounting and may damage the connector pins and the connector housing.

Zero-insertion-force (ZIF) design has long been used to minimize damage to contacts and facilitate the insertion and extraction process. A ZIF design is needed for some types of connectors, especially those with small dimensions or a large number of pins. As shown in Figure 5.3, the ZIF design features contacts can be moved by actuation mechanisms, so there is little or no friction and insertion force [6]. When an external force is exerted on the actuation mechanism, the connector contacts open, and pins can be inserted into their normal positions without mating or engagement; releasing the actuation mechanism will cause the connector contacts to close and the pins to mate.

A ZIF design provides easy and rapid mating, as it eliminates the high initial contact engaging force. A ZIF design reduces contact wear during mating and thus increases contact durability. It also allows for much higher contact normal forces to be exerted. In conventional designs, the insertion force is usually proportional to the applied normal force, as indicated in Eqs. (5.1) and (5.2); a high normal force will inevitably result in excessive insertion and extraction force. Without the insertion force, much higher normal force is possible. Along with less wear and high normal force, plating costs can be reduced. In many applications, higher normal forces permit the use of less noble (and less expensive) platings; thinner platings can be used because of less contact wear. Plating savings can also be achieved by restricting the platings only to areas near the contact points. In conventional designs, the contact engagement could cause wear debris and corrosion products to be dragged into the final contact area; therefore, plating along the entire engagement length is required.

A ZIF design will inevitably add to the bill of materials; however, the added cost may be offset by the high durability and possible savings from the plating methods. In a ZIF design, although contacts are engaged in a direction normal to their mating surface, cleaning or wiping action can still be accomplished and is generally preferred [7].

There are some standards relating to measurement of the insertion and extraction forces. EIA 364–05 and MIL-STD-1344A (Method 2012.1) can be used to determine the forces required to insert contacts into and extract from their position in a connector [8, 9]. EIA 364-13 and MIL-STD-1344A (Method 2013.1) provide test methods to measure mating and unmating forces for electrical connectors [10, 11]. MIL-STD 1344A (Method 2014) and IEC 60512-13-1 are test methods to measure the engaging/separating force for electrical connectors [12, 13]. IEC 60512-1-3 provides a standard test method to measure the electrical engagement length in a connector [14].

5.2 Contact Retention

Connectors can be subjected to various external quasistatis forces, including thermal and moisture expansion, and dynamic forces produced by vibration and shock during product transportation and operation. A high external force can cause contacts to move from their proper locations and even cause contact pullout. Contact retention defines the minimum axial load in either direction that a contact must withstand while remaining firmly fixed in its normal position within its housing. The contact retention force reflects the capability of connectors to resist the impact of external forces.

Contact retention can be a function of contact strength, contact normal force, coefficient of friction, contact area, and geometry. Under ideal situations, a combined high contact normal force, high contact strength, high coefficient of friction, and large contact area ensure good contact retention. Other design features may be applied in order to maintain good contact retention, such as positive locking contact design and a protective plastic cover [15].

There are some standards for measuring contact retention force. EIA 364-29 and MIL-STD-1344A (Method 2007.1) provide test methods to measure the ability of the contact retaining system to withstand the axial mechanical stresses [16, 17]. EIA 364-35 is a test method to determine the ability of an insert to withstand axial forces in electrical connectors [18].

5.3 Contact Force and Deflection

To maintain a consistent and reliable contact interface, a contact force should be applied. Contact deflection occurs when two contacts are mated, and consequently, a contact

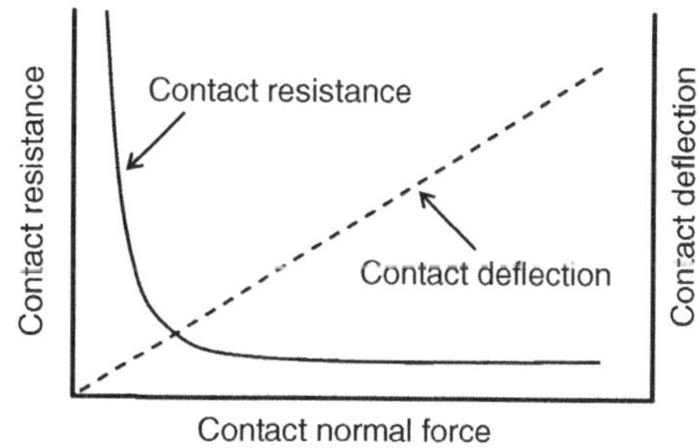

Figure 5.4 Contact resistance versus contact normal force and deflection [2].

normal force is exerted on the contacts, and a contact interface is created. This interface is usually not as "tight" as it may appear from a microscopic view. The surface roughness, surface insulation film, contamination, and dust in the contact interface can all prevent metallic contact. Thus, the effective contact area is usually a fraction of the total contact area; this fraction is determined by the contact manufacturing process, contact finish, and contact cleanness.

A contact will deflect under an applied contact force; the extent of contact deflection depends on the contact design, applied force, and contact modulus. The working range of a contact defines the range of contact force or deflection in which a contact can work reliably during its lifetime applications. It determines the minimal force or deflection that is needed and the maximum force or deflection that a contact can sustain to achieve a stable contact interface.

Figure 5.4 shows a schematic of contact resistance versus contact force and contact deflection [2]. An elbow can be observed on the curve of contact resistance versus contact force, indicating a minimum of contact force and deflection that must be achieved to obtain a low and stable contact resistance. In real situations, the contact deflection versus contact force may not be ideally linear, as shown in Figure 5.4. Other cases, for example, hard-to-soft and soft-to-hard, also exist. If contact force is large enough, a contact may yield (contact deflects without applying extra contact force), and contact resistance may start to increase accordingly, indicating a contact instability. In another case, contact force may increase exponentially with contact deflection once the contact deflects to a certain point. Both cases indicate the maximum operating limits for the contacts. EIA 364-04 is a standard method for determining the magnitude of normal force generated by a contact system at a given deflection within its normal operating levels [19].

5.4 Contact Wipe

With the miniaturization of products such as smartphones and computers, connectors have also miniaturized and have necessitated lower normal forces. To establish low contact resistance at a low normal force, increased contact pressure and wipe at the contact interface have been required.

The technical literature presents various studies that examine the effects of contact geometry, normal force, and wipe [20]. For example, Abbott [21] has performed work on copper coupons with varying hardnesses, film thicknesses, and geometries. The work also incorporated other parameters such as wipe, surface films, and material properties.

Brockman et al. [22] examined wipe with dust-covered gold contacts, and Brodsky [23] has experimentally evaluated the effects of geometry, lubrication, contamination, wipe, normal force, and accelerated aging on contact resistance. Martens and Pecht [24] examined the effect of wipe on nickel contacts. Malucci [25] developed models for wipe on aged contact surfaces.

Wipe is almost always required to mate contacting members. The contact resistance during a wipe is influenced by several factors, including the normal force, contact geometry, thickness and composition of the contaminating films, and length of the wipe. In general, wipe improves performance and reliability and is helpful in handling oxides and other films that can accumulate on contact surfaces. That is, it can clear contaminants from the contact interface. However, small disturbances or wiping across some oxides and contaminated regions can eventually degrade electrical performance. In general, the optimal wipe lengths and normal force/geometry interactions needed to establish low and stable contact resistance in the presence of oxides and contaminants on contact finishes are established only by testing.

There are two modes of wipe action. The first mode occurs after a full contact normal force is applied, and it simulates practical contact mating conditions during insertion of package pins. The second mode occurs when the contact force is being applied, usually with ZIF contact designs. Although in the ZIF design principle the contacts are engaged in a direction normal to their mating surfaces, a small amount of relative contact motion is feasible by controlling contact actuation during exertion of contact normal forces.

Wipe effectiveness refers to the efficiency with which films, dust, and contaminants are removed during connector mating, which is usually indicated by a reduction in contact resistance. Effectiveness also depends on contact geometry, contact normal force, length of wipe, and type of films or contaminants to be disrupted or displaced. Studies on hemispherical, elliptical, and cylindrical contact geometries show the distinctive contrast in wiping effects [4]: at a 50-g load, the hemisphere contacts show very good wipe effectiveness after the wipe length reaches 0.25 mm, while marginal effects are observed for elliptical contacts, and no effects are observed for cylindrical contacts.

To evaluate the wipe effectiveness of a specific contact, the effects of contact normal force and wipe length on contact resistance must be examined. For a non-noble metal contact, a high contact force must be used to penetrate and break the surface oxide layer. For gold contact finish, a contact force less than 30 g may be enough; however, for tin plating, a contact force above 100 g is usually required [7]. In addition to contact force, a minimum wipe length is usually required to effectively disrupt the oxide film and displace contamination; a small wipe distance may produce reverse effects and result in a large increase of contact resistance compared with the zero wipe condition [21]. Figure 5.5 shows the effect of wipe distance on the increase in contact resistance of copper after exposure to humidity and pollutant gases [21]. The graph indicates that a minimum wipe distance of 0.025 mm is required to establish a low contact resistance for the specified design. For a nonlubricated dusty electric contact surface, studies have shown that when the length of wipe is comparable to the size of dust particles, there will be serious contact failures. When the dust particle size is within the upper limit of the hazardous size range, the number of contact failures can be reduced if the sliding length is long enough [26].

Following the initial wipe, a back-wipe is the same action, but in the opposite direction. Back-wipe has been incorporated into many contact designs. However, conflicting data

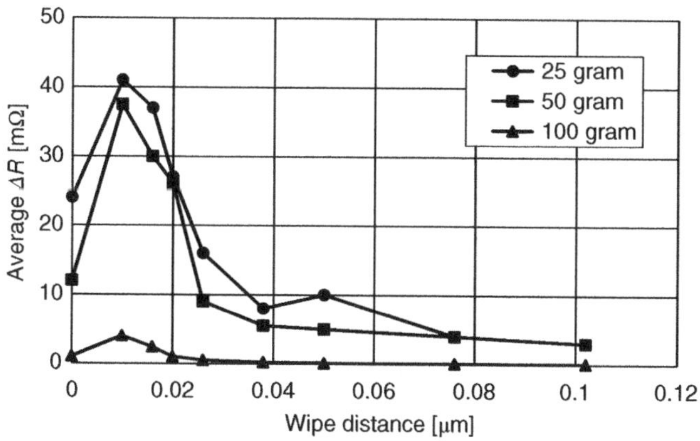

Figure 5.5 The effect of wipe distance on the increase in contact resistance of copper [21].

have been reported concerning the role of back-wipe. Studies on soft and hard copper showed the ineffectiveness of a contact back-wipe in improving contact performance. For nickel contacts after exposure to mixed flowing gas (MFG) testing, contact resistance was consistently reduced after the 0.025 mm back-wipe; but the beneficial effects of the back-wipe were not observed in the study of gold-plated samples [24].

Improvement in wipe effectiveness must be balanced against the probability of high contact wear. Using a sharp-point contact geometry or increasing the contact force may break the oxide film easily and produce good wiping effects, but at the risk of increased contact wear and decreased contact durability.

References

1 MIL-STD-1344A (Method 2012.1) (1977). *Contact insertion and removal force.* US Dept. of Defense.
2 Liu, W. and Pecht, M. (2004). Component socket properties. In: *IC Component Sockets* (eds. W. Liu and M. Pecht), 21–44. Wiley.
3 Liu, H. Y. (1990). The characteristic research of contact insertion and separation force in connector. *Proceedings of the 36th IEEE Holm Conference on Electrical Contacts, joint with the 15th International Conference on Electrical Contacts,* Montreal, Canada (August 1990). IEEE.
4 Mroczkowski, R. (1998). *Electronic Connector Handbook.* New York: McGraw-Hill.
5 Sawchyn, I. and Sproles, E.S. Jr., (1992). Optimizing force and geometry parameters in design of reduced insertion force connectors. *IEEE Transactions on Components, Hybrids, and Manufacturing Technology* 15 (6): 1025–1033.
6 Gutierrez, C. A., Lee, C., Kim, B., and Meng, E. (2011). Epoxy-less packaging methods for electrical contact to parylene-based flat flexible cables. *Proceedings of the 16th International Solid-State Sensors, Actuators and Microsystems Conference,* Beijing, China (June 2011). IEEE.
7 Ginsberg, G.L. (1977). *Connector and Interconnections Handbook,* vol. 2. Fort Washington, PA: Electronic Connector Study Group.

8 EIA 364-05B (1998). *Contact insertion, release and removal force test procedure for electrical connectors.* Electronic Industries Alliance.
9 MIL-STD-1344A (Method 2012.1) (1977). *Test methods for electrical connectors – contact insertion and removal force.* US Dept. of Defense.
10 EIA 364-13E (2011). *Mating and unmating force test procedure for electrical connectors and sockets.* Electronic Industries Alliance.
11 MIL-STD-1344A (Method 2013.1) (1977). *Test methods for electrical connectors – mating and unmating forces.* US Dept. of Defense.
12 MIL-STD-1344A (Method 2014) (1977). *Test methods for electrical connectors – contact engagement and separation force.* US Dept. of Defense.
13 IEC 60512-13-1 (2006). *Connectors for electronic equipment – tests and measurements: Engaging and separating forces.* International Electrotechnical Commission.
14 IEC 60512-1-3 (1997). *Electromechanical components for electronic equipment – basic testing procedures and measuring methods: electrical engagement length.* International Electrotechnical Commission,
15 Tyco Electronics Co., AMP IC Connector Catalog 82172, June 1998.
16 EIA 364-29D (2019). *Contact retention test procedure for electrical connectors.* Electronic Industries Alliance.
17 MIL-STD-1344A (Method 2007.1) (1977). *Test methods for electrical connectors – contact retention.* US Dept. of Defense.
18 EIA 364-35C (2012). *Insert retention test procedure for electrical connectors.* Electronic Industries Alliance.
19 EIA 364-04B (2015). *Normal force test procedure for electrical connectors.* Electronic Industries Alliance.
20 Martens, R. and Pecht, M. (2000). Effects and interactions of design parameters for gold plated electric contacts. *Journal of Materials Science: Materials in Electronics* 11: 209–218.
21 Abbott, W. (1992). The effects of design variables and surface films on the contact resistance of copper-copper contact interfaces. In: Proceedings of the 38th IEEE Holm Conference on Electrical Contacts, Philadelphia, PA (October 1992). IEEE.
22 Brockman, I., Sieber, C. and Mroczkowski, R. (1988). The effects of the interaction of normal force and wipe distance on contact resistance in precious metal plated contacts. *Proceedings of the 34th IEEE Holm Conference on Electrical Contacts,* San Francisco, CA (September 1988). IEEE.
23 Brodsky, W. (1987). Testing of design parameters for zero insertion force connectors. *Proceedings of the 37th IEEE Electronic Components Conference,* Boston, MA (May 1987). IEEE.
24 Martens, R., and Pecht, M. (1996). The effects of wipe on corroded nickel contacts. *Proceedings of the 42nd IEEE Holm Conference on Electrical Contacts, joint with the 18th International Conference on Electrical Contacts,* Chicago, IL (September 1996). IEEE.
25 Malucci, R. (1994). The effects of wipe on the contact resistance of aged surfaces. *Proceedings of the 40th IEEE Holm Conference on Electrical Contacts,* Chicago, IL (October 1994). IEEE.
26 Zhang, J. and Chen, W. (1991). Wipe on various lubricated and nonlubricated electric contacts in dusty environments. *IEEE Transactions on Components, Hybrids, and Manufacturing Technology* 14 (2): 309–314.

6

Contact Interface

Michael G. Pecht and San Kyeong

Center for Advanced Life Cycle Engineering (CALCE), University of Maryland, College Park, MD, USA

Mating connector surfaces are finely polished and plated by the manufacturers, and this often leads to a false assumption that current flows through the entire surface area. However, the actual percentage of surface area that makes true electrical contact is small, even with polished and plated surfaces. The mating surfaces on a microscopic level are made of peaks and valleys. For the current to pass from one mating contact to the other, it must find a path through these points that make contact. If a large amount of current must be passed through a small conducting area, high, unwanted temperatures can be generated. Since the whole mating surface area does not conduct, for high levels of current flow, more surface is required. This is the reason why many switching supplies have multiple connector pins on a high current main output [1].

When connectors are mated, the contact surface (e.g. interface), on a microscale level, consists of high spots (asperities) that come into contact with each other, as shown in Figure 6.1. Since the asperities are small, they can deform plastically even at low applied loads. The asperities continue to deform until a contact area sufficient to support the applied force is created [2]. Thus, the actual contact area of the connector will depend upon the load applied (i.e., the normal force applied).

The asperities also affect the mechanical character of the interface, in particular, the friction and wear. This, in turn, affects the electrical characteristics. When there is a sliding motion between two surfaces, the asperities on the surfaces make contact and resist the motion, as shown in Figure 6.2 [3]. The shear force required to cause the interfaces to move is called the "frictional force." The frictional force influences the connector mating force and the mechanical stability of the connector interface [4]. The interfaces can be sheared off or deform, generating wear debris. The wear impacts the number of mating cycles a connector can perform without degradation.

Where the interface separates also depends on the asperity structure. The larger asperity has a greater probability of breaking within the bulk of the asperity since the interface has been work-hardened and may be stronger than the cohesive strength of the bulk material. This also depends on any oxides or intermetallics that might be formed by the contacting materials. When it breaks within the bulk (e.g. the wear process), a wear particle may be produced. The weaker asperity is more likely to break at or near the interface and generally will show less or no wear. These processes are described as adhesive wear and burnishing wear, respectively.

Electrical Connectors: Design, Manufacture, Test, and Selection, First Edition.
Edited by San Kyeong and Michael G. Pecht.

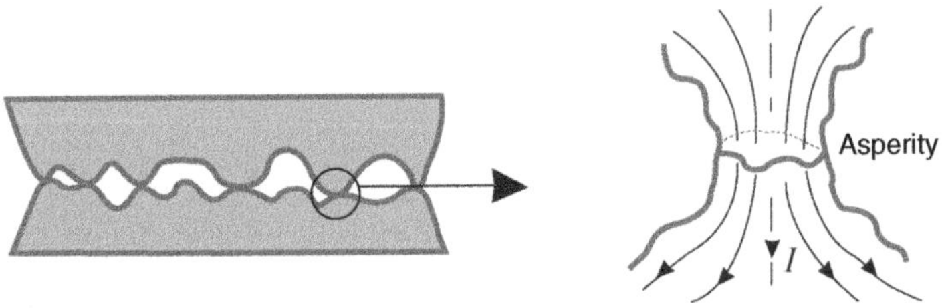

(a) Contact interface (b) Constricted current flow

Figure 6.1 Schematic illustration of (a) contact interface and (b) asperities and constricted current flow.

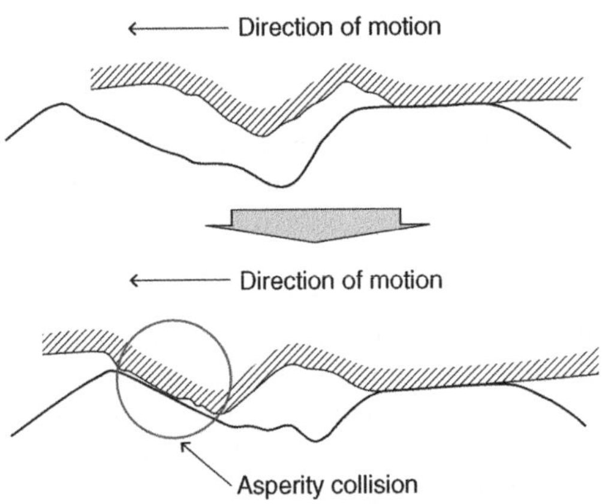

Figure 6.2 Asperity interaction [3].

6.1 Constriction Resistance

Due to their small size (the radii are measurable in micrometers), asperities deform plastically even at low applied loads. With increased loads, the asperities deform further, the contact area is enlarged, and more asperities come into contact. As noted previously, the number of asperities that come into contact depends on the surface roughness, material hardness, and magnitude of contact normal force.

Electrical current is restricted to flowing through the contacted asperities. This limited contact area results in a contact resistance called "constriction resistance." Figure 6.1 schematically shows the contact interface, asperities, and restricted current flow. Constriction resistance is dependent on the number, area, and distribution of contacted asperities by the following equation:

$$R_c = \frac{\rho}{na} + \frac{\rho}{D} \tag{6.1}$$

where ρ is the resistivity of the contact material (assuming the same materials), n is the number of asperities, a is the diameter of the average asperity, and D is the diameter of the area over which the contacts are distributed.

The area of asperities is determined by the applied load. The relationship between the applied load F_n and the expected contact area A_e can be expressed as

$$F_c = \varepsilon H A_e \tag{6.2}$$

where H is the contact hardness of metal, and ε is the pressure factor, which represents the deformation amount of the asperities. In practical systems, the pressure factor ε is usually distributed between 0.2 and 1 [5], whereas the hardness H is normally equal to three times the yield stress at which plastic deformation of the material occurs.

The constriction resistance can be expressed in terms of the contact normal force:

$$R_c = k\rho\left(\frac{H}{F_n}\right)^{1/2} \tag{6.3}$$

where k is a coefficient that includes the effects of surface roughness, contact geometry, and elastic/plastic deformation, which must be determined experimentally; H is the hardness of contact material; and F_n is the contact normal force.

In the previous equations, a pure metallic surface is assumed. However, in most applications, conditions of contact surfaces are not perfect; there may be surface films that occur initially or develop in the application environment. Furthermore, these surface films may be displaced or disrupted completely or partially or remain intact, depending on the applied contact force, applied bias, and film composition, structure, and thickness. Mating cycles and the application of a normal force may mechanically disrupt film (e.g. oxide) layers and expose metallic contacts. A voltage bias may also cause the electrical breakdown of surface films.

6.2 Contact Resistance

The overall contact resistance can be regarded as a combination of constriction resistance (due to asperity contact) and film resistance (due to oxide, corrosion, or other films accumulated on the contact surfaces). Mathematical models have been proposed to simulate the interface resistance due to constriction resistance and film resistance [5, 6]:

$$R_{\text{contact}} = R_{\text{constriction}} + R_{\text{film}} = k\rho\left(\frac{H}{F_n}\right)^{1/2} + \frac{\sigma_i H}{F_n} \tag{6.4}$$

where σ_f is the film resistivity. Although these equations are developed for metallic contacts, they can be applied to other types of contacts, such as conductive elastomer contacts, since the trend is generally true, and there are "fudge" factors that must be experimentally determined for each material set.

End users can increase the force on mechanical connectors with clamping action, while slide-on terminals are designed for fixed spring tension to provide engagement with mating pins. Since the user cannot increase spring tension, care must be taken to not exceed each pin's current carrying capability. Considerations have to be made if spring tension decreases over time due to heat and the number of mating cycles.

Another model that has been used to assess the contact loading conditions in terms of contact resistance involves the use of Hertz stress. Hertz stress takes into account contact load, contact geometry, and contact elastic modulus. As previously noted, due to the contact surface roughness, the effective contact area between two surfaces is much lower than the apparent area. For example, for a contact between two elastic spheres of radii R_1 and R_2, the effective contact area A is a circle of radius a, as shown in Figure 6.3.

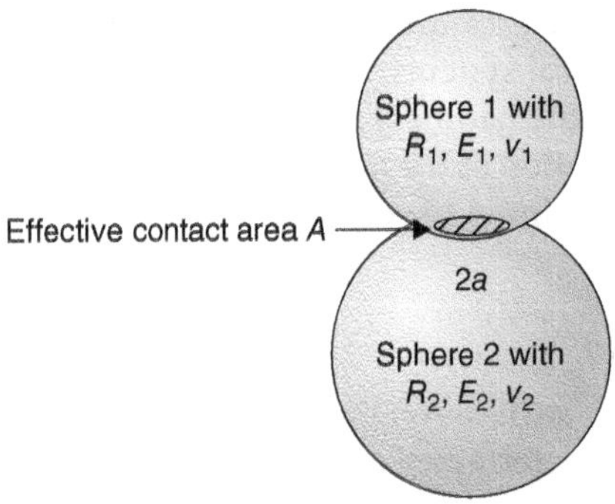

Figure 6.3 Contact between two spheres.

The radius a is related to the applied contact normal force F by the equation

$$a = \sqrt[3]{\frac{3F\left(\frac{1-v_1^2}{E_1} + \frac{1-v_2^2}{E_2}\right)}{4\left(\frac{1}{R_1} + \frac{1}{R_2}\right)}} \tag{6.5}$$

where E_1 and E_2 are the modulus of elasticity for spheres, R is the effective radius between two spheres, and v_1 and v_2 are Poisson's ratios, respectively. Hertz stress σ is calculated as $\sigma = F / A = F / \pi a^2$.

According to this equation, a high modulus of base metal and a small sphere diameter will yield a high Hertz stress, indicating potentially high contact wear. However, the Hertz stress model has serious drawbacks. In particular, a high Hertz stress may not necessarily mean a low contact resistance, since contact resistance is also a function of contact area. A sharp contact, even under a low force condition, may have a high Hertz stress but will yield a high contact resistance, because of its small contact area.

Contact force requirements are different for noble and non-noble plating systems. For non-noble electroplating, such as tin and solder, the contact force must exceed 100 g per contact to obtain a low and stable contact resistance. For noble electroplating, the required contact force is much lower. The usual specification for gold plating is around 30 ~ 50 g. For each material set, these normal force contact requirements should be determined experimentally.

A contact will deflect under an applied contact force; the extent of contact deflection depends on the contact design, applied force, and contact modulus. The working range of a contact defines the range of contact force or deflection in which a contact can work reliably in its lifetime applications. It determines the minimal force or deflection that is needed and the maximum force or deflection that a contact can sustain to achieve a stable contact interface.

Because of its small value, usually in the range of 10–100 mΩ, contact resistance cannot be measured accurately through the two-wire method. In a two-wire measurement, the lead resistance will cause a significant voltage drop, and the voltage (V_M) measured by the meter will not be the same as the voltage (V_R) directly across the contact interface. To eliminate the interference of the lead resistance, a four-wire (Kelvin) method is usually used. Figure 6.4 shows a schematic of the four-wire measurement.

In this measurement, a test current flows through a contact interface via one set of test leads, and the voltage across the contact interface is measured through another set of leads called "sense leads." The sense current (pA level) is much lower than the test

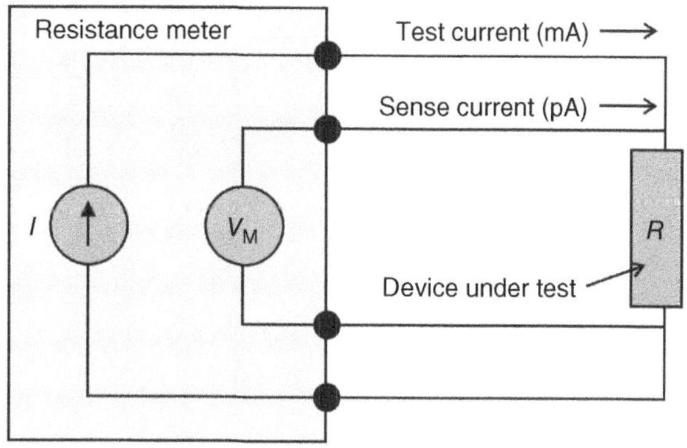

Figure 6.4 Schematic of four-wire measurement.

current (usually mA level); therefore, the voltage drop across the sense leads can be ignored, and the voltage measured by the meter is essentially the same as the voltage across the contact interface [7]. The contact resistance can be calculated as follows:

$$R = \frac{V_R}{I} \approx \frac{V_M}{I} \tag{6.6}$$

The contact resistance can be greatly affected by the interface conditions. A high current and voltage during the measurement may change the conditions, for example, by punctuating the oxide films. Accordingly, the testing result will be lower than the value if the interface remains intact, compromising the validity of testing results. Therefore, the contact resistance should be measured under "dry circuit" conditions, with an open voltage less than 20 mV and a short circuit current less than 100 mA. In dry circuit conditions, the physical properties of contact interface will generally not be affected.

There are some standards for measuring contact resistance. ASTM B539 provides techniques for measuring the resistance of static electrical connectors including wire terminations, friction connectors, soldered joints, and wrapped-wire connections [8]. EIA 364–06 and MIL-STD-1344 (Method 3004) are standard methods to determine the contact resistance of connectors by measuring the voltage drop across the contacts [9, 10]. EIA 364–23 focuses on measuring low-level contact resistance (mΩ level), considering the thermal electromotive force (EMF) that occurs at this level and causes a measurement error [11]. IEC 60512-2 contains general examination, electrical continuity/contact resistance tests, insulation tests, and voltage stress tests [12].

6.3 Other Factors Affecting Contact Resistance

Other factors that affect contact resistance include temperature, electrical current, and fretting. Generally, temperature rise increases the energy of the electrons. The electrons of high energy collide more frequently, increasing the electrical resistivity. However, the elevated temperature can decrease the contact resistance. Since the current flows through the asperities of the contact area, joule heating occurs at the localized area. As the current increases, the joule heating softens the asperities, increasing the contact area and reducing the contact resistance. As the current through the contact spot increases, joule heating softens the asperities. Figure 6.5 shows that a rise in the temperature of gold/gold contact leads to a decrease in contact resistance [13]. The electrical resistance

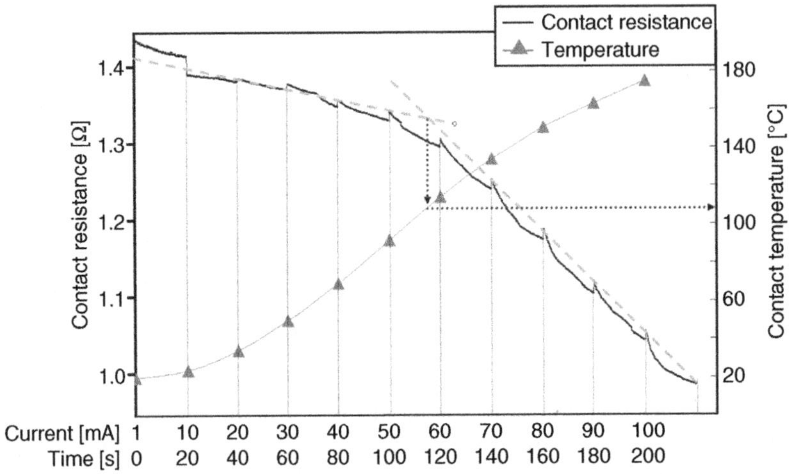

Figure 6.5 Time dependence of contact resistance by increasing the contact current for gold/gold contacts [13].

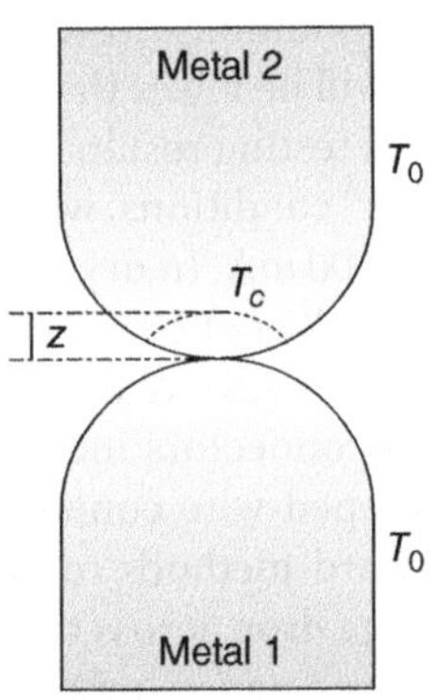

Figure 6.6 Temperature distribution in a constriction of a contact [13]. Graphically renewed.

of a gold contact is observed while the current level is increased by 10 mA for every 20 seconds at a constant contact force of 150 μN. At the first phase with a current level from 1 mA to 40 mA, there was a small decrease of contact resistance. The drop rate of contact resistance increased at 50 mA, indicating that thermal softening occurred and flattened the contact asperities at the contact temperature between 91 °C and 114 °C.

If two identical metals contact, the highest temperature can be observed at the contact spot. However, in the case of asymmetrical contact materials, the location of the highest temperature is not the contact spot. Broue et al. give a model, as shown in Figure 6.6 [13], to predict the temperature distribution at the contact constriction between two different metals, metal 1 and metal 2, which have different conductivities and mechanical hardness values. An assumption here is that the conductivity and hardness of metal 2 is greater than that of metal 1. The temperature distribution across the contact is dependent upon the nature of the materials in contact. The average distance of thermoelectric heat flow across the contact is dependent upon the connector.

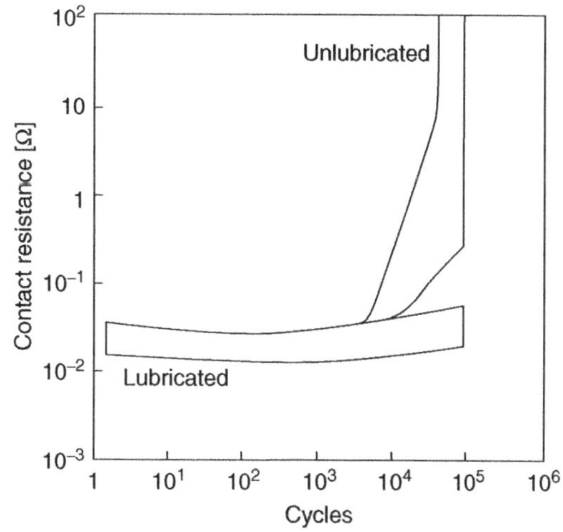

Figure 6.7 Contact resistance during fretting of gold-flashed palladium against DG-R156 [14].

The maximum temperature T_c will be within the less conductive material and at distance Δz from the contact. A mathematical relationship to predict the location of maximum temperature in a bimetallic contact is given as follows:

$$\frac{\Delta z}{a} = \frac{\rho_{\text{metal1}} + \rho_{\text{metal2}}}{2\rho_{\text{metal2}}} \left(\frac{\Delta V}{V_c}\right)\left(\frac{\pi}{2\sqrt{2}}\right) \tag{6.7}$$

where a is the effective radius of contact, ρ_{metal1} and ρ_{metal2} are the electrical resistivities of material 1 and material 2, respectively. The bulk temperature of both metal 1 and 2 is T_0. The maximum temperature is located within the less conductive material at a distance z from the physical interface. V_c is the voltage at contact, and ΔV is the voltage potential.

Fretting is the process where there is a degradation of material due to relative motion of two surfaces against each other. Contact resistance increases rapidly when fretting occurs, as shown in Figure 6.7 [14]. Contact resistance between gold-flashed surface on palladium electroplate and gold-diffused surface on DG-R156 (gold-diffused surface on Pd 60%/Ag 40% alloy) was measured during fretting motion. The unlubricated contact showed contact resistance drastically increased after 5000 cycles, whereas the lubricated contact stabilized its contact resistance below 15 mΩ after 100 000 cycles. This shows that contact resistance is dependent not just on the fretting but also whether the surface is lubricated.

6.4 Current Rating

Current rating, also called “current-carrying capacity,” specifies the maximum current that a conductor can carry safely. Due to current flow, joule heat is generated in a conductor, and the local temperature will be increased. The local temperature rise, compared with ambient temperature, depends on the heat balance between joule heat and heat dissipation to the neighboring regions. If a current flow is too large, excessive heat will be generated and accumulated, and the local temperature will rise so high that it may surpass the maximum operating temperature of the insulators that separate the connector contacts. The maximum operating temperature of the housing usually determines the maximum current flow through a contact.

Although current rating can be specified in terms of the transient current rating or overload current rating [4], the continuous current rating has generally been adopted by the industry. This current rating is based on the local temperature rise above the ambient, as induced by a current flow. It is commonly taken as the current that produces a 30 °C temperature rise, although other criteria can be used, such as a 10 °C or 15 °C T-rise. The criterion can be applied for both AC and DC current.

Current rating depends on contact size, contact pitch, contact type, and heat-sinking capability. A larger contact size is usually associated with a higher current rating. A small contact pitch of the contacts usually limits the applicable current rating. A high-conductivity contact can be adopted to compensate for a reduction in contact size and pitch to achieve the desired current rating. A contact with high conductivity not only generates less heat but also dissipates heat more effectively. The joule heat can be also be dissipated through conduction to the printed circuit board (PCB) traces or through airflow around the contacts. The shielding effect of the housing could reduce the heat dissipation, causing a great difference between free-air and in-housing current ratings. However, the current rating performance will be greatly improved if a heat sink is attached, as a heat sink greatly enlarges the area of heat dissipation.

There are some standards for measuring current rating. EIA 364-70 provides several methods to determine the current rating for electrical connectors, considering temperature rise during operation [15]. IEC 60512-5 addresses the current rating of connectors including conditions of elevated ambient temperature [16]. IEC 60512-10-4 describes a method to assess the performance of connectors with an electrical overload current [17].

6.5 Capacitance and Inductance

As a result of the presence of asperities in contact surfaces, a small capacitance at the metallic contacting interface exists, as shown in Figure 6.8 [18]. Cavities caused by asperities in the contact interface brings capacitance. There also can be parasitic inductances regarding different contact conditions including asymmetric contact distribution or presence of a loose contact [19].

Capacitance results from the interaction of the electric field around an active conductor with nearby conductors (mutual-capacitance) or with ground (self-capacitance). It defines the induced current flow generated by the change of charge due to changing voltage. In the case of two nearby conductors, the induced current flow due to voltage changing would be

$$i_1 = C_{11}\frac{dV_1}{dt} + C_{12}\left(\frac{dV_1}{dt} - \frac{dV_2}{dt}\right) \tag{6.8}$$

$$i_2 = C_{12}\left(\frac{dV_2}{dt} - \frac{dV_1}{dt}\right) + C_{22}\frac{dV_2}{dt} \tag{6.9}$$

where t is time, V_1 is the voltage in conductor 1, V_2 is the voltage in conductor 2, i_1 is the induced current flow in conductor 1, i_2 is the induced current flow in conductor

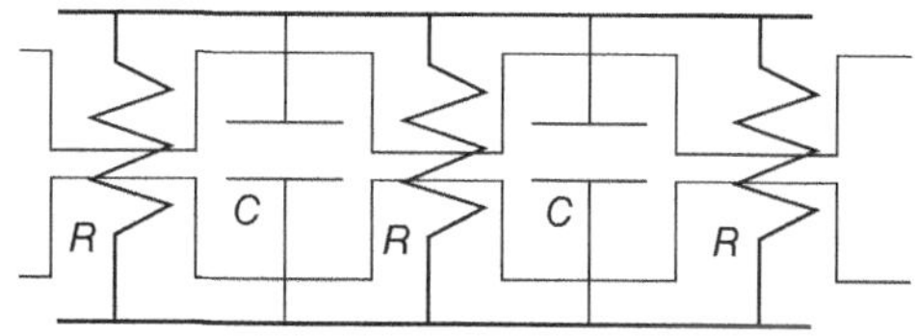

Figure 6.8 Schematic illustration for parasitic capacitance of an electrical contact [18].

2, C_{11} is the self-capacitance of conductor 1, C_{22} is the self-capacitance of conductor 2, and C_{12} is the mutual capacitance of conductor 1 and conductor 2.

Inductance results from the interaction of the magnetic field around an active conductor with nearby conductors (mutual-inductance) or with ground (self-inductance). Inductance determines the induced voltage generated by the change of magnetic flux due to a changing electrical current. Consider two adjacent conductors. The induced voltage due to current changing would be

$$V_1 = L_{11}\frac{di_1}{dt} + L_{12}\frac{di_2}{dt} \tag{6.10}$$

$$V_2 = L_{12}\frac{di_1}{dt} + L_{22}\frac{di_2}{dt} \tag{6.11}$$

where t is time, i_1 is the current in conductor 1, i_2 is the current in conductor 2, V_1 is the induced voltage in conductor 1, V_2 is the induced voltage in conductor 2, L_{11} is the self-inductance of conductor 1, L_{22} is the self- inductance of conductor 2, and L_{12} is the mutual inductance of conductor 1 and conductor 2. The magnitude of the capacitance and inductance depends on the dielectric medium between contacts and on contact and grounding geometry.

The following example illustrates how to calculate capacitance and inductance [20, 21]. Figure 6.9 shows two parallel conductors for the calculation of mutual capacitance and mutual inductance [22]. The mutual capacitance and inductance are calculated as follows:

$$C = \frac{\pi\varepsilon_r\varepsilon_0}{\ln\left[\frac{d}{2r}\left\{1+\sqrt{1-\left(\frac{2r}{d}\right)^2}\right\}\right]}l \tag{6.12}$$

$$L = \frac{\mu_r\mu_0 l}{2\pi}\left[\ln\left\{\frac{l}{d}+\sqrt{1+\left(\frac{l}{d}\right)^2}\right\}-\sqrt{1+\left(\frac{d}{l}\right)^2}+\frac{d}{l}\right] \tag{6.13}$$

where ε_r is the dielectric constant of the insulator, ε_0 is the dielectric constant of vacuum, μ_r is the permeability of the insulator, and μ_0 is the permeability of vacuum.

For example, for two pins of a connector with dimensions $l = 5.5$ mm, $d = 1.27$ mm, and $r = 0.38$ mm, the calculated mutual capacitance is as follows (for free air: $\varepsilon_r = 1$):

$$C = 0.14\,\text{pF}$$

The mutual inductance is as follows (for free air: $\mu_r = 1$):

$$L = 1.51\,\text{nH}$$

If the insulator is a thermoplastic, the previous value should be multiplied by the relative dielectric constant and relative permeability of the insulating thermoplastic to yield

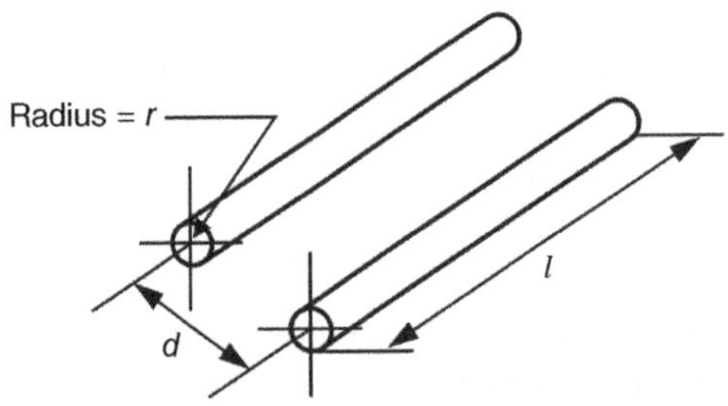

Figure 6.9 A model for calculation of mutual capacitance and mutual inductance: two long, circular conductors with insulator in between [22].

the true capacitance and inductance. This calculation presents an initial estimation of the order of capacitance and inductance. However, in practice, a simple calculation can never be reached; large pin count, complex contact configuration, number and configuration of grounding pins, and comparable value of contact size and contact pitch make direct calculation complicated. Therefore, these two parameters are usually determined experimentally.

Another concept is so-called "loop inductance." Considering the previous two pins forming a current flow loop, the current flow in one pin produces magnetic field lines around the other one, but in the opposite direction. The field lines around the entire loop can be calculated by counting the two pins together, as shown here:

$$L = L_{s1} + L_{s2} - 2L_m \tag{6.14}$$

where L_{s1} is the self-inductance of pin 1, L_{s2} is the self-inductance of pin 2, and L_m is the mutual inductance. If pin 1 and pin 2 are identical, then the loop inductance can be calculated as follows:

$$L = 2L_s - 2L_m \tag{6.15}$$

In general, the mutual inductance between two pins is only a small fraction of self-inductance of either one and drops off rapidly with the increase in contact pitch. Therefore, the loop inductance is largely determined by the self-inductance of contact pins.

Three phenomena are associated with capacitance and inductance: characteristic impedance, cross talk, and propagation delay. Characteristic impedance is given by

$$Z_0 = \sqrt{\frac{L_0}{C_0}} \tag{6.16}$$

where Z_0 is the characteristic impedance, L_0 is the inductance per unit length, and C_0 is the capacitance per unit length.

For a linear, homogeneous, isotropic dielectric propagation medium free of electric charge, the characteristic impedance is calculated as

$$Z_0 = \sqrt{\frac{\mu}{\varepsilon}} \tag{6.17}$$

where μ is the magnetic permeability of the insulating medium, and ε is the dielectric constant of the insulating medium. Characteristic impedance is a critical parameter to control signal reflection at the contact interface. The reflection coefficient is given by

$$\rho = \frac{Z - Z_0}{Z + Z_0} \tag{6.18}$$

where Z_0 is the characteristic impedance of a contact, and Z is the characteristic impedance of the PCB trace or the package pin. If $Z = Z_0$, the reflection coefficient is zero, and no signal reflection results. A uniform line terminated in its characteristic impedance will have no standing waves, no reflections from the end, and a constant ratio of voltage to current at a given frequency at every point on the line. If $Z < Z_0$, the reflection is negative, and the signal is reflected and inverted. If $Z > Z_0$, the reflection is positive; that is, the signal is reflected but not inverted. In both cases, the unmatched characteristic impedance produces signal reflection, causing signal distortion and attenuation.

"Cross talk" is a term for signal interference or coupled noise. It results from the coupling of the electromagnetic fields surrounding an active conductor with those of its adjacent conductors. Cross talk is another important source, besides unmatched characteristic impedance, for signal attenuation and distortion. The significance of cross talk depends on the mutual parasitic capacitance and inductance. In practice, cross talk can be minimized by increasing the contact pitch, reducing contact cross section and length, keeping circuits at right angles, and using balanced lines and grounding pins or planes.

Cross talk can be divided into capacitive cross talk and inductive cross talk. As literally implied, the former is mainly due to capacitive coupling between circuits, while the latter is primarily due to inductive coupling. The relative significance of capacitive and inductive cross talk depends on the impedance of the circuits [23].

Crosstalk can be indicated in terms of signal integrity and attenuation. It can be specified as the ratio between the amplitude of coupled noise over active-line signal amplitude or as the ratio of the output voltage of one channel (without signal input) over the output voltage of its nearby channel (with signal input) [23]:

$$K = 20 \log \left| \frac{E_{\mathrm{oa}}}{E_{\mathrm{ob}}} \right| \tag{6.19}$$

where K is the magnitude of crosstalk between circuits a and b in dB, E_{oa} is the output voltage of circuit a due to cross talk, and E_{ob} is the output voltage of circuit b with signal input.

Propagation delay is a measure of how long it takes for a wave to travel the length of a specific conductor. The propagation delay for a unit of length is expressed by

$$\tau_d = \frac{1}{v} = \frac{\sqrt{\varepsilon_r}}{C_0} = \sqrt{L_0 C_0} \tag{6.20}$$

where τ_{d} is the propagation delay for a unit length, v is the traveling velocity, C_0 is the velocity of light, ε_{r} is the relative dielectric constant of the insulator, L_0 is the inductance per unit length, and C_0 is the capacitance per unit length. The interaction of parasitic capacitance and inductance causes the propagation to increase by a factor of $(L_0 C_0)^{1/2}$.

The mutual capacitance and inductance can be measured between two adjacent pins, one of them grounded. The self-capacitance and inductance are usually measured with one pin to all surrounding grounded pins. The self-inductance can also be measured in free air.

There are some standards for measuring the capacitance and inductance of electrical connectors. EIA 364-30 provides a test method to measure capacitance (1 kHz ~ 1 MHz) for electrical connectors [24]. EIA 364-33 and EIA 364-69 describe test methods to measure high-level inductance (100 nH ~ 100 mH) and low-level inductance (10 nH ~ 100 nH) for electrical connectors, respectively [25, 26].

6.6 Bandpass and Bandwidth

A band refers to the frequency spectrum between two defined frequency limits. A bandpass is a fixed band of frequencies that a device can support. There are several definitions of bandwidth. Bandwidth can be defined as the frequency band within which a device performs with respect to some characteristics, such as signal attenuation and falls. Bandwidth can also refer to the maximum frequency a device can pass in which the responsivity is not reduced more than 3 dB from the maximum response.

The high operating frequency and large bandwidth of electronic circuitry today put more stringent requirements on connectors. At high operating frequencies, dielectric loss and skin effect become more pronounced. The dielectric loss results from the repeated atomic polarization imposed by the alternating electric field and is manifested primarily as dissipative heat. The skin effect describes a phenomenon in which the electronic conduction tends to occur only to a limited depth beneath the contact surface. The skin depth decreases with an increase in operating frequency. The skin effect is manifested by an increase in contact resistivity, as the conduction cross section is reduced. Therefore, dielectric loss and skin effect will cause signal attenuation. Furthermore, capacitance and inductance are functions of the operating frequency. Increasing the operating frequency inevitably causes an increase in capacitance and inductance and thus causes an increased cross talk. For high-frequency applications, to guarantee signal fidelity, the bandpass of a connector is usually several times larger than the operating frequency of electronic circuitry.

Bandwidth is expressed in terms of signal attenuation at a given frequency. Signal attenuation is a reduction in the amplitude of a signal. The degree of attenuation is often measured in terms of decibels (dB), the standard unit for expressing transmission gain or loss and relative power levels. A decibel equals 10 times the log of the ratio of the power out (P_o) to the power in (P_i) [27]:

$$\text{dB} = 10\log_{10}\left(\frac{P_o}{P_i}\right) \tag{6.21}$$

Bandpass is expressed as a range of applicable frequencies starting from DC to a maximum frequency. This maximum frequency is the frequency at which the signal attenuation is greater than −3 dB, per the second definition of bandwidth (as mentioned previously).

Bandpass and bandwidth are two critical parameters for maintaining the signal integrity. Short contact design, material selection, special grounding, and decoupling schemes are factors that expand the bandwidth and assure adequate signal fidelity.

There are some standards relating to measurement of contact electrical performances. IEC 60512-23-4 provides test methods to measure the reflections produced by a connector in a transmission line in the time domain. The connector under this test method is treated as a discontinuity in a transmission line with a controlled characteristic impedance [28]. The IEC 60512-25 series provides several methods to evaluate connector properties including cross-talk ratio, attenuation, propagation delay, and return loss [29–32].

References

1 Bock, E. and Whitley, J.H. (1974). *Fretting Corrosion in Electric Contacts*. Harrisburg, PA: AMP Inc.

2 Greenwood, J.A. and Williamson, J.B.P. (1966). Contact of nominally flat surfaces. *Proceedings of the Royal Society of London, Series A, Mathematical and Physical Sciences* 295 (1442): 300–319.

3 Moran, J., Sweetland, M., and Suh, N. P. (2004). Low friction and wear on non-lubricated connector contact surfaces. *Proceedings of the 50th IEEE Holm Conference on Electrical Contacts*, Seattle, WA (September 2004). IEEE.

4 Mroczkowski, R.S. (1998). *Electronic Connector Handbook*. New York: McGraw-Hill.

5 Holm, R. (1967). *Electrical Contacts: Theory and Application*, 4e. New York: Springer.

6 Bhagath, S. and Pecht, M. (Dec. 1993). Modeling the effects of mixed flowing gas (MFG) corrosion and stress relaxation on contact interface resistance. *Journal of Electronic Packaging* 115: 404–409.

7 Yeager, J. and Hrusch-Tupta, M.A. (1998). *Low Level Measurements*, 5e. OH: Keithley Instruments.

8 ASTM B539 (2018). *Test Methods for Measuring Resistance of Electrical Connections (Static Contacts)*. American Society for Testing and Materials.

9 EIA 364-06C (2006). *Contact resistance test procedure for electrical connectors*. Electronic Industries Alliance.

10 MIL-STD-1344 (Method 3004) (1971). *Test methods for electrical connectors - contact resistance*. US Dept. of Defense.

11 EIA 364-23C (2006). *Low level contact resistance test procedure for electrical connectors and sockets*. Electronic Industries Alliance.

12 IEC 60512-2 (2003). *Connectors for electronic equipment – tests and measurements: electrical continuity and contact resistance tests*. International Electrotechnical Commission.

13 Broue, A., Dhennin, J., Charvet, P.-L., Pons, P., Ben Jemaa, N., Heeb, P., Coccetti, F., and Plana, R. (2010). Multi-physical characterization of micro-contact materials for MEMS switches. *Proceedings of the 56th IEEE Holm Conference on Electrical Contacts*, Charleston, SC (October 2010). IEEE.

14 Antler, M. (1984). Effect of fretting on the contact resistance of palladium electroplate having a gold flash, cobalt-gold electroplate, and DG R156. *IEEE Transactions on Components, Hybrids, and Manufacturing Technology* CHMT-7 (4): 363–369.

15 EIA 364-70B (1997). *Temperature rise versus current test procedure for electrical connectors and sockets*. Electronic Industries Alliance.

16 IEC 60512-2 (2002). *Connectors for electronic equipment – tests and measurements: Current-carrying capacity tests*. International Electrotechnical Commission.

17 IEC 60512-10-4 (2003). *Connectors for electronic equipment – tests and measurements: Impact tests (free components), static load tests (fixed components), endurance tests and overload tests*. International Electrotechnical Commission.

18 Dervos, C. T., and Michaelides, J. M. (1997). The effect of contact capacitance on current-voltage characteristics of stationary metal contacts. In: Proceedings of the 43rd IEEE Holm Conference on Electrical Contacts; October 1997; Philadelphia, PA. IEEE.
19 Hayashi, Y., Wu, S., Fan, J., Mizuki, T., and Sone, H. (2010). Modeling connector contact condition using a contact failure model with equivalent inductance. *Proceedings of the IEEE International Symposium on Electromagnetic Compatibility*, Fort Lauderdale, FL (February 2010). IEEE.
20 Walker, C. (1990). *Capacitance, Inductance and Crosstalk Analysis*. Norwood, MA: Artech House.
21 Lau, J., Wong, C.P., Prince, J.L., and Nakayama, W. (1998). *Electronic Packaging Design, Materials, Process, and Reliability*. New York: McGraw-Hill.
22 Liu, W. and Pecht, M. (2004). Component socket properties. In: *IC Component Sockets* (eds. W. Liu and M. Pecht), 21–44. Wiley.
23 Carr, J. (1991). *Secrets of RF Circuit Design*. Blue Ridge Summit, PA: Tab Books.
24 EIA 364-30A (2002). *Capacitance test procedure for electrical connectors and sockets*. Electronic Industries Alliance.
25 EIA 364-33A (2002). *Inductance measurement test procedure for electrical connectors (100 nH ~ 100 mH)*. Electronic Industries Alliance.
26 EIA 364-69A (2002). *Low level induction measurement for electrical contacts of electrical connectors*. Electronic Industries Alliance.
27 Weik, M.H. (1996). *Communications Standard Dictionary*. New York: Chapman & Hall.
28 IEC 60512-23-4 (2001). *Connectors for electronic equipment – tests and measurements: transmission line reflections in the time domain*. International Electrotechnical Commission.
29 IEC 60512-25-1 (2001). *Connectors for electronic equipment – tests and measurements: crosstalk ratio*. International Electrotechnical Commission.
30 IEC 60512-25-2 (2002). *Connectors for electronic equipment – tests and measurements: attenuation (insertion loss)*. International Electrotechnical Commission.
31 IEC 60512-25-4 (2001). *Connectors for electronic equipment – tests and measurements: propagation delay*. International Electrotechnical Commission.
32 IEC 60512-25-5 (2004). *Connectors for electronic equipment – tests and measurements: return loss*.International Electrotechnical Commission.

7

The Back-End Connection

Chien-Ming Huang, San Kyeong and Michael G. Pecht

Center for Advanced Life Cycle Engineering (CALCE), University of Maryland, College Park, MD, USA

The interface between the connector and the connecting components (board or wire) is referred to as the "back-end" connection. The contact interface receives the most attention, but unless the connector has a reliable back-end connection with the connecting components, power or signal transfer across the connector can be compromised. The back-end connection area should be protected against the usage environment. Thus, the structure and assembly techniques for back-end connections should be considered not only by manufacturers that mass-produce electronic products equipped with connectors but also by end- users who use these products under various conditions.

The type of back-end connection depends on whether the connector is connected to a board or a wire. Soldering is the most widely used connection methodology [1]. An insulation displacement connection (IDC) or crimping connection can also be used for connector-to-wire configurations. Solderless press-fit technique is widely applied for connector-to-board configuration. Each of these methodologies has advantages and limitations. This chapter focuses on these four back-end connection types.

7.1 Soldered Connection

Soldering is a cost-effective methodology to attach connectors to a printed circuit board (PCB). Soldering technology has been well established for decades and is applicable from small sample preparation to mass production. There are two general soldering methodologies: pin-through-hole (PTH) and surface-mount technology (SMT). PTH requires a connector with pins on its back-end and a PCB that has electroplated holes. The connector pins are inserted into the holes and then soldered in place on the PCB. PTH soldering is mechanically durable, but the additional costs of drilling and electroplating holes should be considered [2]. Device miniaturization trends limit PTH soldering because the lead-hole connection occupies a large space. Figure 7.1 shows an example of connectors attached to a PCB with PTH soldering [3].

SMT soldering was developed in the 1960s and has been rapidly replacing PTH soldering in recent decades [4]. A connector is attached directly to a solder-coated PCB surface without pins and electroplated holes. A reflow soldering process follows. Compared to PTH, SMT can make a connection with more circuits and contacts in a smaller

Electrical Connectors: Design, Manufacture, Test, and Selection, First Edition.
Edited by San Kyeong and Michael G. Pecht.

(a) (b)

Figure 7.1 (a) Connectors embedded on a PCB using the PTH method and (b) a soldered pin [3].

(a) (b)

Figure 7.2 (a) Solder-pasted PCB and (b) connectors embedded on a PCB using the SMT method [3].

space, enabling the fabrication of more complex and compact products. Figure 7.2 shows connectors attached to a PCB using SMT soldering [3].

A typical SMT process follows: (i) the solder paste is applied to the PCB surface using a stencil, (ii) the excess paste is wiped off with a squeegee, (iii) a connector is positioned to the solder-pasted PCB surface, (iv) the assembled material is heated through a reflow oven, and (v) the soldered product is cooled. Figure 7.3 shows the reflow process.

Reflow ovens are commonly used for mass-production soldering. There are four phases in the typical soldering reflow profile [5, 6]. In the "pre-heat" phase, the temperature increases from room temperature (~25 °C) to about 150 °C. The heating rate is maintained at less than 2.5 °C/sec. The PCB and connector are heated slowly to minimize thermal shock, which can be caused by the subsequent reflow phase. Volatile substances evaporate during this phase. In the following "soak" phase, the heating rate is reduced to less than 1.0 °C/sec. Flux activates during this phase, and the temperatures of the PCB and connector become equivalent. Additionally, any oxides that still reside on the surface of the components are removed. In the "reflow" phase, the heating rate

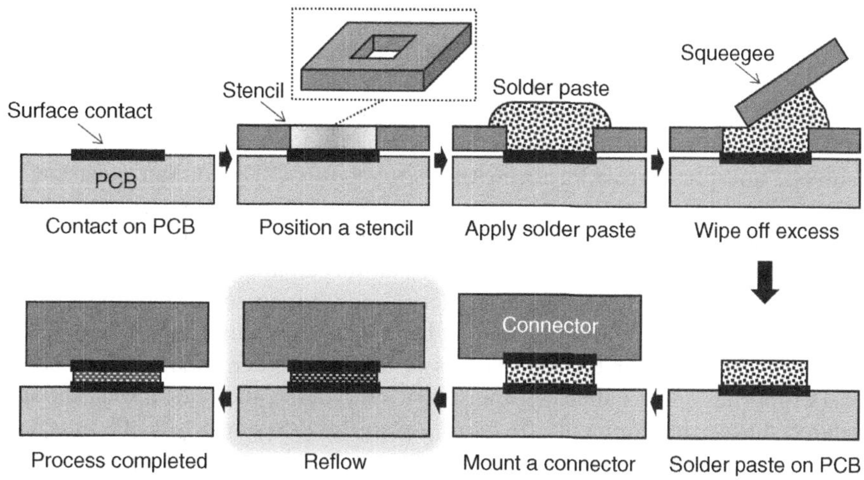

Figure 7.3 Typical SMT process.

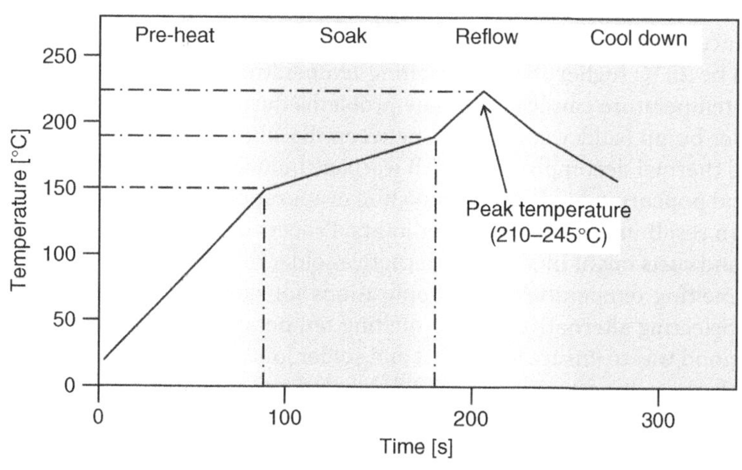

Figure 7.4 Example of a reflow profile. Graphically renewed [5, 6].

is about 1.5 °C/sec, the solder paste reaches its melting point, and the activated flux reduces surface tension at the metal surface. This results in metallurgical bonding with the PCB and the connector. The peak temperature is generally set to about 240 °C in the case of SAC305 solder but can be adjusted depending on the melting point of the solder to be used. It is recommended to set the peak temperature at least 20 °C above the melting temperature of the solder since the solder tends to have better wettability to metal when it is hotter. During the final "cool-down" phase, the PCB-connector assembly is cooled rapidly to room temperature to prevent any excess intermetallic forming. Figure 7.4 shows a reflow profile example [5, 6].

Tin-lead (SnPb) solders had been used to assemble electronics for more than 50 years owing to their soldering properties, manufacturability, reliability, and cost

[7]. Because of the health and environment concerns with lead, since the early 2000s, SnPb solders have been eliminated from consumer electronics starting in European Union (EU) markets and then China, Japan, and the United States [8, 9]. Instead of SnPb solders, tin-bismuth (SnBi) or tin-silver-copper (SnAgCu) alloy can be used. Sn-3.0wt%Ag-0.5wt%Cu (SAC305) is common among lead-free solders since SAC305 is near eutectic and possesses a lower melting temperature (217 °C) than most SnCu solders (227 °C) [10, 11].

SnCu-based solders and SAC solders with low silver content (<1%) are lower-cost alternatives to SAC305 [12]. The addition of trace nickel can improve the properties of SnCu solder for wave soldering, selective soldering, and hot air solder leveling. Other alternatives with different additives, including bismuth, germanium, cobalt, and manganese, are also available. For the alternatives to SAC305 solder, many factors, such as physical properties and reliability of the solder joint, costs, and regulations, should be considered during development. To transition to the lower-cost SAC305 alternatives, the development and qualification of the solder materials, processes, and products should be conducted gradually [13]. Process transfer and inventory purge should also be considered during the transition.

The melting temperature of the solder material can impact the soldering process and, thus, the performance and reliability of the electronics. The process temperature of the soldering could be 25 °C higher than the melting temperature of the solder materials. High process temperature can lead to many problems during the soldering process, including solder bump (solder connection between the silicon die and substrate) re-melting, material thermal decomposition, PCB warpage induced by the glass transition of the board, and popcorning induced by moisture evaporation. However, low process temperature can result in incomplete solder joints. Process or equipment changes might be required, and costs might increase if alternative solder materials are employed. Table 7.1 lists the melting temperatures and applications for common lead-free solder materials [12]. Selecting alternatives with a melting temperature similar to that of SAC305 could be a good way to ensure lower cost and solder joint reliability at the same time. Nevertheless, high-melting-temperature solder materials are still in demand due to specific high-temperature applications, including geothermal wells, deep oil and gas well logging, and electric aircraft [14–16].

Solder wettability is critical for the interconnection because of the compatibility between the solder materials and the finishes on the components and PCBs [15]. To obtain good wettability, chemical-contained fluxes are used in the solder pastes and wave soldering. Combining the fluxes with process and equipment optimization can ensure reliable interconnections are formed between components and PCBs. The mechanical strength of the alternative solders is also critical to withstand the loadings from the field applications. Possible failure mechanisms due to insufficient strength are interface delamination between the solder materials, silicon die, substrate, and PCBs, and bulk solder cracks [17]. Moreover, a trace amount of lead contamination on the solder materials should be considered because of the adverse effects. For example, mixing with lead and bismuth-contained solder materials can make the materials brittle and lower the mechanical strength. The reliability of SnAgBi and Sn58Bi solders can also be significantly reduced by mixing with lead [18]. For Sn58Bi solder, the solder joint could turn into powder under thermal cycling conditions if the solder was contaminated by the lead [18]. Lead contamination is a legitimate concern for electronics assembly because lead-based solders are still the main choices in reliability-critical applications [19, 20]

Table 7.1 Common lead-free solders [12].

Composition (wt%)	Melting Point (°C)	Applications	Concerns
SnAg (e.g. Sn3.5Ag, Sn2.3Ag, Sn1.8Ag)	~ 221	SMT, flip chip. Currently, most common binary lead-free solder for flip chip. Typically used in conjunction with electroplating.	Cu dissolution, excessive IMCs, voids
SnCu (e.g. Sn0.7Cu, Sn3Cu)	~ 227	PTH, flip chip.	Cu dissolution, wetting, excessive IMCs
SAC (e.g. SAC305, Sn3.8Ag0.7Cu, Sn3.9Ag0.6Cu)	~ 217	SMT, PTH, BGA, flip chip (limited), SOP.	Cu dissolution, excessive IMCs, voids
Sn80Au	280	Flip chip. Common for flux-free opto-electronic assembly on gold finishes, controlled standoff height.	High cost
Sn	232	Flip chip.	Tin whiskers

Cost is the major concern when selecting solder materials, including the materials (alloys and fluxes), patents, and the manufacturing process. Patent royalties also are a consideration when selecting the solder materials to substitute for SAC305 solder. Most of the lead-free solders applied in manufacturing are protected by North American and Japanese patents. For example, SN100C is patented by Nihon Superior. To manufacture the patented solder materials, manufacturers need to pay the patent fees or acquire the patents from their owners.

Solder materials can be acquired in many different forms, including sphere, bar, wire, and paste. Solder spheres are typically used in solder ball attachment through a reflow process. Solder bars and ingots are melted first to produce the desired constitution of the alloy and then molded to the desired form. Energy costs should be considered because of the energy consumption during manufacturing. For instance, a higher pot temperature is required for melting the low-Ag or no-Ag alloys. The cored solder wire requires formulating flux for the core, billet casting the alloy, extruding the cored wire, drawing the wire into a specific diameter, and spooling the wire on reels. The manufacturing costs of the solder wires can be higher than the material costs. Furthermore, solder pastes are more expensive to manufacture than other forms because of the complexity of processing, including powder forming, flux medium blending, paste blending, and packaging [21].

7.2 Press-Fit Connection

Press-fit connection, which was invented in the early 1970s [22], is a solderless connection methodology for mounting components onto the PCB. The back-end connection is made by assembling a connector terminal pin with an electroplated through-hole on

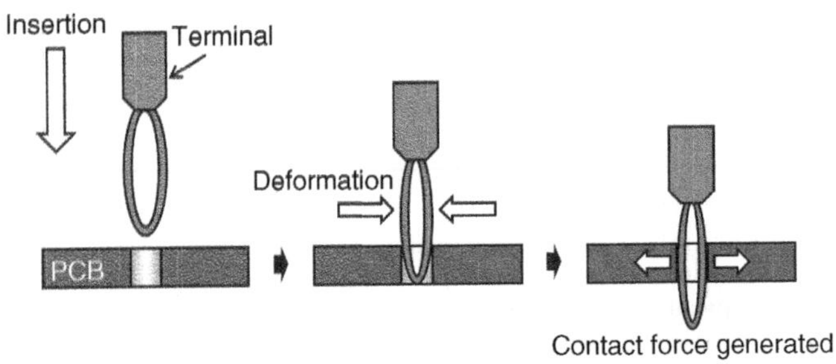

Figure 7.5 Schematic of press-fit connection [26]. Graphically renewed.

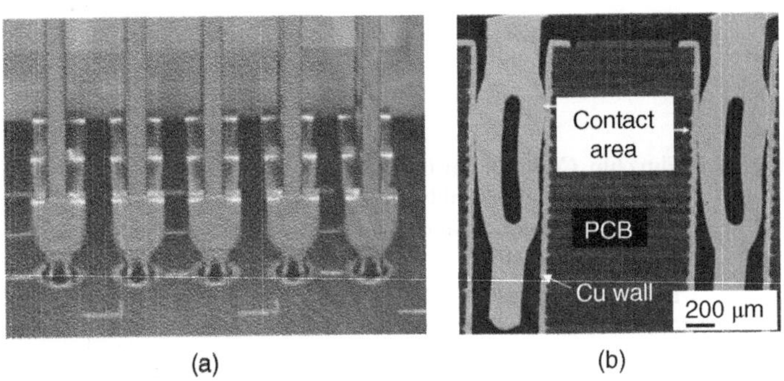

Figure 7.6 (a) Connection using press-fit [26] and (b) cross-section image of the contact area at the pin-to-PCB wall interface [28].

a PCB. When the electroplated terminal pin is inserted into the through-hole, the pin deforms and functions as a spring. This assembly is held by the frictional force between the inserted pin and the inner wall of the PCB through-hole. An electrical contact is also made through this assembly. Figure 7.5 shows an example of a press-fit connection[23].

Press-fit methodology has several advantages compared to conventional soldering. Assembly time and cost are far less than soldering methodology. It is free from the thermal shock caused by the reflow soldering process and from soldering-derived defects, including bridges, flux residuals, and solder balls. Since the press-fit connection does not bond the PCB and connector permanently, the connected PCBs can be separated again, enabling reusability and upgradeability of the end devices [24]. However, a press-fit connection requires PCBs that have electroplated through-holes, and its electrical contact area may be smaller than a soldered back-end connection. Also, the electroplated pin surface can be scraped off under a vibrational environment. Figure 7.6 shows a connector pin-PCB connection using press-fit and a cross-section image of the electrical contact between the connector pin [26] and the inner copper wall of a PCB through-hole [25]. Furthermore, similar to soldering, press-fit connections are vulnerable to tin whiskers [26]. See Chapter 8 for a detailed discussion of tin whiskers.

Press-fit pins are available as two structure types: solid press-fit and compliant press-fit. Compliant press-fit pins can be categorized as split-beam or deformable cross-section pins, as shown in Figure 7.7 [27]. Solid press-fit pins do not show elastic deformation during insertion, whereas compliant press-fit pins compress during

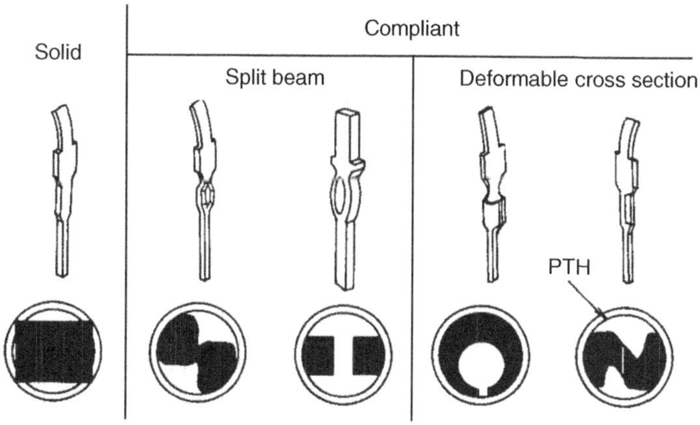

Figure 7.7 Types of press-fit pins [30].

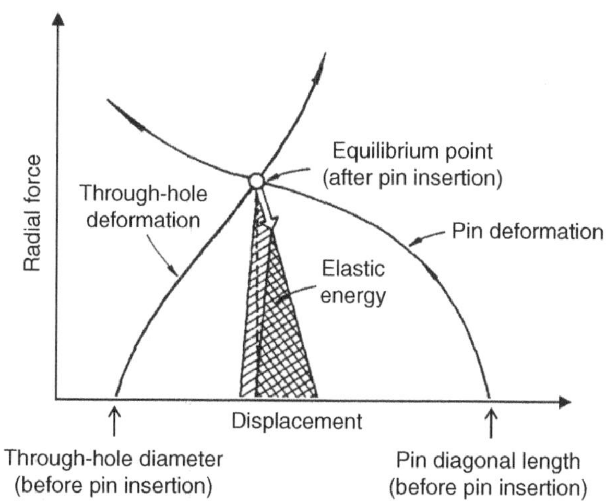

Figure 7.8 Model of compliant press-fit pin connection [30].

insertion. Solid press-fit pins can damage PCB through-holes, so compliant press-fit pins are generally preferred [28]. The additional benefit of using compliant press-fit pins is that the elastic strain energy stored in the pin retains the metal-to-metal contact even if stress relaxation occurs on the PCB through-hole. When a compliant press-fit pin is assembled with PCB through-holes, the relationship between the radial force and the deformation magnitude of both the pin and the through-hole can be plotted, as shown in Figure 7.8 [30].

7.3 Crimping Connection

Crimping is a connecting technique for wire-to-wire or wire-to-component (e.g. connector) connection. To connect a wire and a connector with the crimping method, the

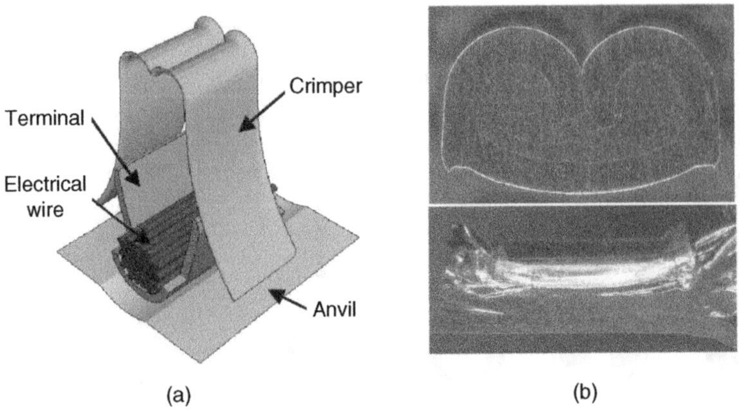

Figure 7.9 (a) Schematic diagram of the crimping method and (b) cross-section of the front and side after crimping [32].

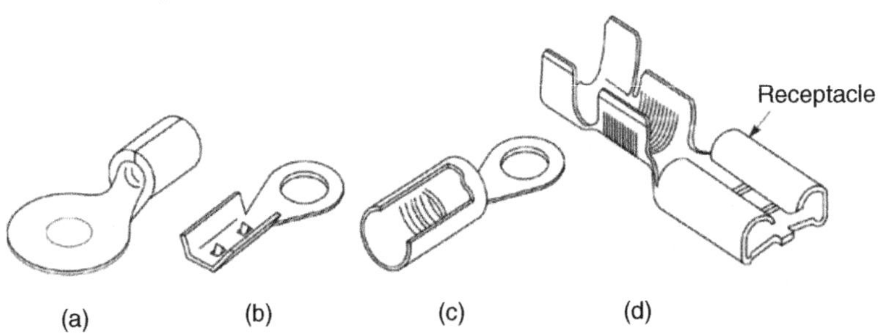

Figure 7.10 Crimp terminal configurations: (a) straight barrel, (b) open barrel with insulation-piercing teeth, (c) pre-insulated (nylon or polyvinyl chloride) terminal, and (d) quick disconnect terminal [1].

wire is first inserted into a barrel-shaped terminal. The terminal and wires are then compressed (crimped) together by a special tool that works as a crimper and an anvil, as shown in Figure 7.9a [29].

During the compression, the wire severely deforms inside the terminal barrel. After the crimping tool is removed, this assembly firmly maintains since the outer terminal provides radial compression, as shown in Figure 7.9b [32]. The radial compression also allows the inner wires to have a large contact area with the inner wall of the terminal, enabling low contact resistance. From both an electrical and a mechanical point of view, a crimping connection is good to use in harsh environments and has been widely used in automotive connectors [30]. In the case of a connector that carries a current of 10 A or more, the crimp connection is mostly applied [31]. Furthermore, the connecting process is quick and easy without requiring solder, heating, or flux. Figure 7.10 shows common crimp terminals [1].

After the crimping process is done and the crimp tool is released, there should be a large residual contact force between the deformed connector terminal and wires to maintain a reliable crimping connection. The residual contact force depends on the relative elastic springback of the terminal and the wires. Since the wires are constrained by

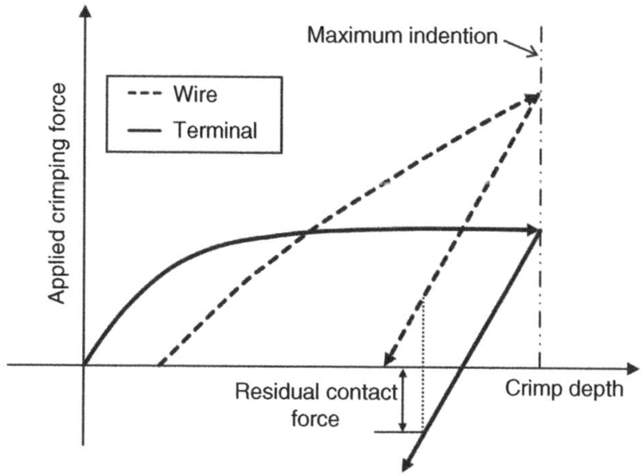

Figure 7.11 Residual radial contact force after crimping.

the terminal, they are under compressive stress, while the outer terminal is under tensile stress. As shown in Figure 7.11, mechanical stress on the terminal and wires increases during the crimping motion. When the crimping tool is released at the maximum indentation depth, the crimping stress on both the terminal and wires drops below the yield strength. The terminal and wires start to spring back according to their elastic modulus. This condition produces residual radial contact force between the crimped terminal and wires [32, 33].

When using the crimping tool, the degree of compression should be adjusted accordingly. Figure 7.12 shows the electrical and mechanical characteristics of the crimping connection [34]. A loose crimp connection (i.e. the crimp structure has a high crimp height) results in poor mechanical strength and low electrical conductivity. If the crimp connection is too tight, it may show a good electric conductivity but can damage the terminal or wires, reducing crimp tensile strength.

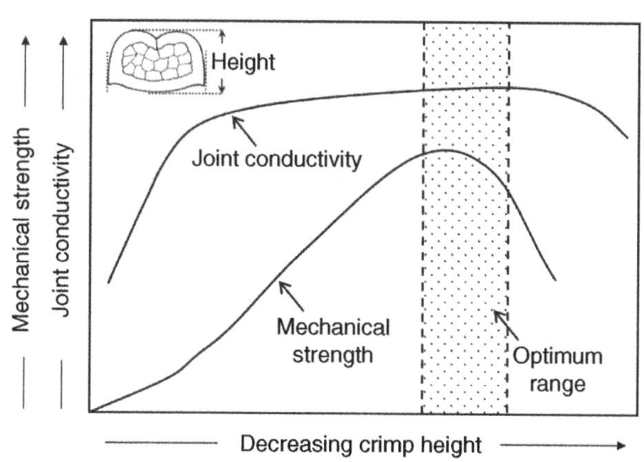

Figure 7.12 Crimp connection characteristics relative to the crimp height [34].

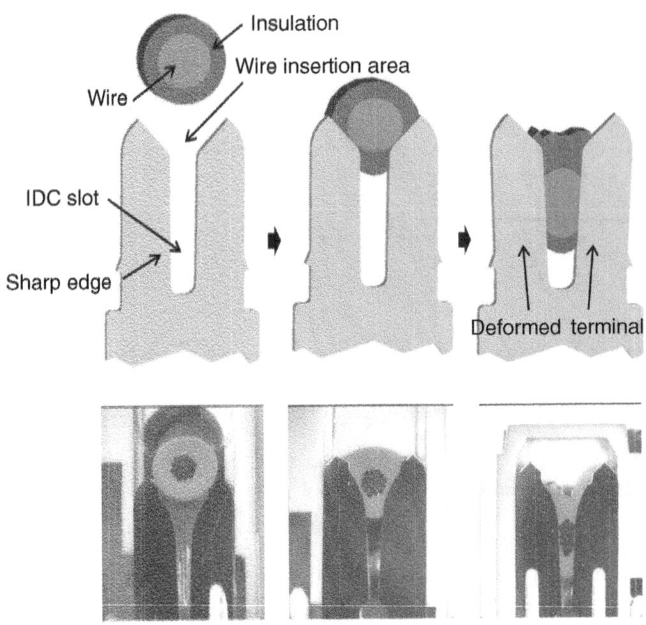

Figure 7.13 A typical IDC process [34].

IEC 60352–2, "Solderless connections – Part 2: Crimp connections – General requirements, test methods, and practical guidance," is an international standard for solderless crimp connections.

7.4 Insulation Displacement Connection

IDC was developed in the 1950s. In IDC, the conductive wires encapsulated within the insulation are pushed into a narrow conductive slot that has sharp edges in its inner wall. The edges of the slot displace the insulation and make electrical contact with the wires. During the assembly process, the terminal edges slightly deform and make contact pressure with the wires, forming a gas-tight connection [34].

Compared to the crimp connection, the IDC method does not require a special tool and can be performed without stripping the insulation of the wire end [35, 36]. The IDC method is widely used for power connectors that carry a current of less than 10 A or for signal connectors. IEC 60352–3, "Solderless connections – Part 3: Accessible insulation displacement connections – general requirements, test methods, and practical guidance," and IEC 60352-4, "Solderless connections – Part 4: Solderless non-accessible insulation displacement connections – General requirements, test methods, and practical guidance," are international standards for the solderless IDC connection.

References

1 Antler, M. (1994). Electrical connectors. In: *Kirk-Othmer Encyclopedia of Chemical Technology*, 4e, vol. v. 9, 38–60. New York: Wiley.

2 Vianco, P.T. and Feng, Y. (2001). Electronic packaging: solder mounting technologies. In: *Reference Module in Materials Science and Materials Engineering* (ed. S. Hashimi), 2705–2715.

3 Tsenev, V., Videkov, V., Stratev, A., and Georgiev, N. (2017). Improved soldering of components with different thermal masses by hybrid printing of solder pastes. *Proceedings of the 40th International Spring Seminar on Electronics Technology*, Sofia, Bulgaria (May 2017). IEEE.

4 Lau, J., Powers-Maloney, L.M., Baker, J.R. et al. (1990). Solder joint reliability of fine pitch surface mount technology assemblies. *IEEE Transactions on Components, Hybrids, and Manufacturing Technology* 13 (3): 534–544.

5 Rejewski, K. (1995). SMT process recommendations. Defect minimization methods for a no-clean SMT process. *Proceedings of the IEEE Technical Applications Conference and Workshops*. Portland, OR (October 1995). IEEE.

6 Pambudi, A. J. K., Hariadi, F. I., and Arsyad, M. I. (2017). Development of low-cost reflow oven for SMT assembly. *Proceedings of the International Symposium on Electronics and Smart Devices*, Yogyakarta, Indonesia (October 2017). IEEE.

7 Abtew, M. and Selvaduray, G. (2000). Lead-free solders in microelectronics. *Materials Science and Engineering: R: Reports* 27 (5–6): 95–141.

8 Ciocci, R. and Pecht, M. (2006). Impact of environmental regulations on green electronics manufacture. *Microelectronics International: An International Journal* 23 (2): 45–50.

9 Osterman, M. (2006). Being 'RoHS exempt' in a Pb-free world. *Proceedings of the Capital SMTA Chapter Pb-Free Tutorial Program*, Herndon, VA (May 2006). CALCE EPSC.

10 Liu, W. and Lee, N.C. (2007). The effects of additives to SnAgCu alloys on microstructure and drop impact reliability of solder joints. *JOM* 59: 26–31.

11 Biocca, P., and Rivas, C. (2008). Case study on the validation of SAC305 and SnCu-based solders in SMT, wave and hand-soldering at the contract assembler level. *Proceedings of the 12th International Symposium on Advanced Packaging Materials: Processes, Properties, and Interfaces*, San Jose, CA (January 2088). IEEE.

12 Cheng, S., Huang, C.M., and Pecht, M. (2017). A review of lead-free solders for electronics applications. *Microelectronics Reliability* 75: 77–95.

13 Hamilton, C., Kelly, M., and Snugovsky, P. (2007). Effects of SAC alloy copper dissolution rates on PTH processes. *Circuits Assembly* 18 (6): 34.

14 Hagler, P., Henson, P., and Johnson, R.W. (2010). Packaging technology for electronic applications in harsh high temperature environments. *IEEE Transactions on Industrial Electronics* 58 (7): 2673–2682.

15 Xu, X., Gurav, A.S., Lessner, P.M., and Randall, C.A. (2011). Robust BME Class-I MLCCs for harsh-environment applications. *IEEE Transactions on Industrial Electronics* 58 (7): 2636–2643.

16 Wondrak, W., Held, R., Niemann, E., and Schmid, U. (2001). SiC devices for advanced power and high-temperature applications. *IEEE Transactions on Industrial Electronics* 48 (2): 307–308.

17 Huang, M.L., Ning, Z.H.A.O., Shuang, L.I.U., and He, Y.Q. (2016). Drop failure modes of Sn – 3.0Ag – 0.5Cu solder joints in wafer level chip scale package. *Transactions of Nonferrous Metals Society of China* 26 (6): 1663–1669.

18 Woodrow, T. A. (2003). The effects of trace amounts of lead on the reliability of six lead-free solders. *Proceedings of the 3rd International Conference on Lead-Free Components and Assemblies*, San Jose, CA. IPC.

19 Menon, S., George, E., Osterman, M., and Pecht, M. (2015). High lead solder (over 85%) solder in the electronics industry: RoHS exemptions and alternatives. *Journal of Materials Science: Materials in Electronics* 26 (6): 4021–4030.

20 Zeng, G., McDonald, S., and Nogita, K. (2012). Development of high-temperature solders. *Microelectronics Reliability* 52 (7): 1306–1322.

21 IPC Solder Products Value Council (2007). Electrical/electronic enduser: Cost of lead-free solder materials. www.adhesivesmag.com/articles/86921-electrical-electronic-enduser-cost-of-lead-free-solder-materials (accessed Oct. 26, 2019).

22 Scaminaci, J. Jr., (1977). Solderless press-fit interconnections: a mechanical study of solid and compliant contacts. *IEEE Transactions on Manufacturing Technology* MFT-6 (2): 23–30.

23 Nomura, Y., Saitoh, Y., Furukawa, K., Minami, Y., Horiuchi, K., and Hattori, Y. (2006). Press-fit connector for automobile ECUs. *Proceedings of the 52nd IEEE Holm Conference on Electrical Contacts*, Montreal, QC (September 2006). IEEE.

24 Tohmyoh, H., Yamanobe, K., Saka, M. et al. (2008). Analysis of solderless press-fit interconnections during the assembly process. *Journal of Electronic Packaging* 130 (3): 031007 (6 pages).

25 Palaniappan, A., Li, L., and Lee, T. K. (2018). Impact of press-fit connector pin microstructure elastic response to PCB through-hole Cu wall interface long-term contact reliability. *Proceedings of the 20th IEEE Electronics Packaging Technology Conference*, Singapore, (December 2018). IEEE.

26 Liang, J., Dariavach, N., and Shangguan, D. (2008). Tin whisker nucleation and growth on Sn-Pb eutectic coating layer inside plated through holes with press-fit pins. *IEEE Transactions on Components and Packaging Technologies* 31 (1): 152–158.

27 Kanai, T., Ando, Y., and Inagaki, S. (1985). Design of a compliant press-fit pin connection. *IEEE Transaction on Components, Hybrids, and Manufacturing Technology* CHMT-8 (1): 40–45.

28 Mattson, J., Callies, T., and Kerchof, B. (July 2014). Press-fit technology. TE Connectivity.

29 Ogihara, S., Takata, K., Hattori, Y., and Yoshida, K. (2006). Mechanical analysis of the crimping connection. *Proceedings of the 52nd IEEE Holm Conference on Electrical Contacts*, Montreal, QC (September 2006). IEEE.

30 Prin, G. R., Courtin, T., and Boyer, L. (2002). A new method to investigate electrical conduction in crimp joints influence of the compaction ratio and electrical model. *Proceedings of the 48th IEEE Holm Conference on Electrical Contacts*, Orlando, FL (October 2002). IEEE.

31 Jorgens, S. (2004). Insulation displacement technology as technically equivalent and more cost-effective alternative compared to crimping technology. *Proceedings of the 50th IEEE Holm Conference on Electrical Contacts joint with the 22nd International Conference on Electrical Contacts*, Seattle, WA (September 2004). IEEE.

32 Rollings, R.W. (1975). Design of a reliable separable power connector with base metal contacts for telephone equipment applications. *IEEE Transactions on Parts, Hybrids, and Packaging* PHP-11 (1): 45–52.

33 Timsit, R. S. (2008). Formation of tubular crimp connections: elementary considerations. *Proceedings of the 54th IEEE Holm Conference on Electrical Contacts*, Orlando, FL (October 2008). IEEE.

34 Simon, E. P., Kallmayer, C., Schneider-Ramelow, M., and Lang, K. D. (2012). Development of a multi-terminal crimp package for smart textile integration. *Proceedings of the 4th Electronic System-Integration Technology Conference*, Amsterdam, Netherlands (September 2012). ESTC.

35 Prin, G. R., Courtin, T., and Boyer, L. (2002). A new method to investigate electrical conduction in crimp joints influence of the compaction ratio and electrical model. *Proceedings of the 48th IEEE Holm Conference on Electrical Contacts*, Orlando, FL (October 2002). IEEE.

36 Jorgens, S., and Taschke, H. (2005). Applications of the finite-element analysis for the calculation of an insulation displacement process. *Proceedings of the 51st IEEE Holm Conference on Electrical Contacts*, Chicago, IL (September 2005). IEEE.

8

Loads and Failure Mechanisms

San Kyeong, Lovlesh Kaushik and Michael G. Pecht

Center for Advanced Life Cycle Engineering (CALCE), University of Maryland, College Park, MD, USA

An electrical connector undergoes varying types of environmental and operating load conditions depending on the nature of the application. The automotive industry is one of the prime areas electrical connectors are used in. Other sectors include aeronautics, aerospace, telecommunications, and commerce.

The loads that can act on an electrical connector depend on a connector's "state of operation." These states are "in operation," "insertion," and "disconnection." For the "in operation" state, the connector is considered as a "mechanical assembly" because the separable parts of the connector make contact and operate as one assembly to provide continuity in the circuit. [1] Figure 8.1 shows a schematic of a typical connector. The three major parts to this assembly include the contact spring, the contact interface, and the connector housing. In this state of connector function, the loads that act on the connector are a combination of external and intrinsic forces.

When an electrical connector is switched on, the current flows through the contact spring, contact interface, and connector pins. Resistive joule heating at the contact interface can result in the following failure mechanisms: creep, welding of joints, silver migration, tin whisker formation, fretting, and corrosion. The current flow can cause silver migration and corrosion. Vibration can lead to fretting action of the contacts, tin whisker formation on the connector pins, and wear of the surface layer on the contact pins. Humidity can cause corrosion of the contact material, silver migration, and fretting corrosion of contact material. These topics are discussed in detail in this chapter.

On insertion of the contact spring to the connector housing, there can be wear at the surface of the connector and contact pin. The contact normal force can lead to surface layer cracking and bending of the connector pin due to misalignment in assembling or high insertion force. Because connectors that do not operate continuously will be inserted and disconnected multiple times, the contact spring may become fatigued. On disconnection of the contact spring from the connector housing, arcing can occur at the contact pin. Wiping on the contact surface can occur as a result of this motion and can lead to wear on the connector pin surface.

At the connector housing, "stress corrosion cracking" can occur as a result of residual stress generation from the external environment, which may contain corrosive, hot contaminants or gases at high pressure [2]. The details on this failure mechanism follow in

Electrical Connectors: Design, Manufacture, Test, and Selection, First Edition.
Edited by San Kyeong and Michael G. Pecht.

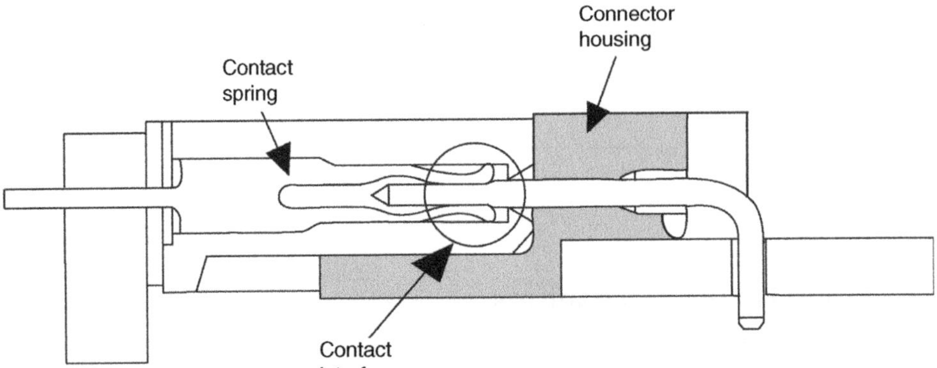

Figure 8.1 Schematic illustration of a typical connector [1].

Section 8.1. Differential heating, differential pressure, and/or vibrations at the connector housing can lead to fretting action or "micro-motion" of contacts at their common interface. If there is humidity or moisture present, the housing material can swell, which may lead to degradation of the electrical and mechanical performance.

8.1 Environmental Loads

External loads act on the connector housing. These loads are related to the environment surrounding the electrical connector. They can be high-temperature conditions due to hot environments or heat released from operating machines, humidity or contaminants in the air, or corrosive gases released from industries or vehicles. Vibration, including oscillation or displacement, can occur. External gases can also exert pressure on the connector. Air can seep in through the pores in the connector housing material [3]. Because the loads of temperature, humidity, contamination, corrosive gases, and pressure can act inside the connector, it is usually the contact interface of the two contact materials where the forces act. However, they may also act at other locations inside the connector.

8.1.1 Temperature

The temperature conditions under which a connector operates will affect its performance and reliability. The temperature of the connector is affected by the ambient environmental conditions as well as any nearby components that generate heat or provide thermal management of the product that the connector is used in (e.g. fans), and the heat generated by the connector due to current flow.

Lam et al. [3] studied the effect of temperature on the housing of connectors used at different locations in an automotive vehicle, as shown in Figure 8.2. The temperature load also varies with time. Different types of temperature profiles exist in real operation, such as thermal shock for sudden acceleration from the engine off and random temperature changes during acceleration/deacceleration of the vehicle.

Temperature can induce failures in an electrical connector. The presence of heat load will result in differential thermal expansion in materials across the electrical connector, and a subsequent cooling effect will lead to contraction of materials. The result

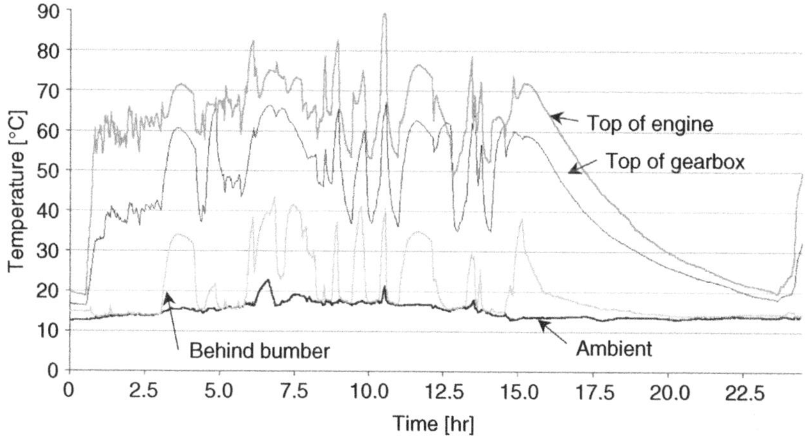

Figure 8.2 Temperature profiles for electrical connectors used at different locations [3].

is low-frequency micro-movements of contact surfaces that will cause fretting action. Thermo-mechanical stresses will be generated in the plating material from the heat load. These stresses can combine with the residual stresses from electrodeposition on the conductive connector body. The resulting effect is tin whisker formation on connector pins. The micro-movements or fretting can lead to wear action as a result of thermal cycling. Contact fretting can also occur as a result of rapid changes in the temperature of the housing material. High temperatures will control the corrosion mechanism by a diffusion process. Temperature is also responsible for creep failure in electric contacts under the load conditions.

8.1.2 Vibration Load

Vibration [4, 5] refers to displacement or oscillation of a body from a stable state of equilibrium, where the displacements can be periodic, non-uniform, or transient in nature. The elements that lead to vibration of a system include the mass or inertia element, which stores the kinetic energy, and the spring or elastic element, which stores the potential energy. As the mass of the oscillating body continues to oscillate about its mean position, the energy continually exchanges in between the kinetic and potential energy of the system.

Sources of vibration include contact normal force, periodic oscillation of the material in response to external forces of excitation such as engine vibration or transportation, and micro-motions due to differential thermal expansion of materials in the electrical connectors. The variables of vibration include frequency, amplitude, and time. It is the contact point of two mating surfaces inside the mechanical system that will move relative to each other as a response to vibration.

The effects of vibration are undesirable. Vibration can cause large deformation or stresses in the contact material. Accordingly, it can lead to fatigue, wear, or improper operation of the associated mechanical assembly. The same applies for the current application of electronic connectors. The critical location for failure of electronics is the contact between the socket and the connector pins. Vibration leads to rubbing action or back-and-forth motion at the contacts. This action leads to the removal of material from the plating layer on the surface of contacts, and the surface contact material itself wears out. Consequently, wear leads to an increase in contact resistance and to open circuits. This failure mechanism is called "fretting wear," and on formation of the corrosive layer, the mechanism is called "fretting corrosion." The failure mechanisms of contact fatigue, connector spring fatigue, and vibrations can sometimes lead to accidental disconnection and formation of an arc [6].

8.1.3 Humidity

When there is a high percentage of humidity in the atmosphere, water molecules can be adsorbed on contact surfaces. This adsorption could lead to galvanic corrosion [7], an electrochemical reaction that forms an oxidizing layer.

There are two forms of humidity [8]: absolute humidity and relative humidity. Absolute humidity is the water vapor mass in a given mass or volume of dry air. Relative humidity is the ratio of the partial pressure of water vapor to the equilibrium vapor pressure. The percentage of relative humidity is $\%RH = 100 \times (p/p_s)$, where p is the actual pressure of water vapor present in the atmosphere and p_s is the saturation pressure of water at the ambient temperature. In an open space, the %RH is found to be directly proportional to the total pressure.

Vapor pressure [9] is the exertion of pressure by the vapor on the gases in the surrounding atmosphere. The vapor pressure varies with temperature. The vapor pressure of water near its boiling temperature is higher than the normal condition of room temperature. At boiling, the vapor pressure of water is equal to the atmospheric pressure. At temperatures below the freezing point, the vapor pressure is lower. At lower atmospheric pressures, the vapor pressure rises. Consequently, water boils at a lower temperature at higher altitudes.

The following discussion explains the effect of temperature and total pressure on vapor pressure [9] in open and closed-space environments. In an open space, vapor is free to expand, so temperature does not affect the partial pressure. The partial pressure can vary only if the number of moles of gas molecules varies, namely, if vapor is being added or removed or if total pressure varies. In a closed container of fixed volume, the vapor occupies the entire volume of the container. With a change in temperature or a change in the amount of vapor, the partial pressure will vary accordingly.

In an open space, %RH decreases rapidly with an increase in temperature at a constant moisture level and constant pressure, whereas this change becomes slow in a closed-space environment. The %RH value is directly proportional to the pressure change.

The effect of humidity is deterioration of material properties over time. Humidity leads to corrosion in electrical connectors. Antler [10] provided results of the behavior of Ni specimens exposed to humidity in an enclosed environment for a period of more than one year. Each of the two Ni specimens was exposed to different humidity and exposure conditions. Specimen 1 was unsealed and stored outdoors from one summer to the next summer. The corroded Ni was produced as $12.5\ \mu g\ cm^{-2}$ from 1 mg of specimen 1 during the first summer, while almost none of the Ni corroded was gained during the following winter when the relative humidity was low. The corrosion rate increased again as the next summer started. Meanwhile, specimen 2 had been fully sealed and stored for the same period as specimen 1. Until one year had passed, a negligible amount of corroded Ni was observed.

The presence of humidity degrades the performance of an electrical connector. It can cause galvanic corrosion at contacts. Above an RH level of 50–60%, the rate of corrosion sharply increases. Corrosive gases consist of salts of either sulfides, sulfates, chlorides, nitrates, hydroxides, or ammonia [10]. The salts have a tendency to deposit on the surface of an electrical contact and can adsorb moisture present in the air. Most traditional types of corrosion can occur by local electrolytic cell formation on an electrical contact surface. In the presence of dust particles, crevice corrosion may occur at the contact spring. Corrosion behavior depends on the connector pin finish material. The corrosion rate of silver is greater than that for bronze.

8.1.4 Contamination

One of the main reasons for electric contact failure by the mechanism of wear is a high level of contamination from dust in the surrounding environment [11]. The surrounding environment of connector operation also contains pollutants, including sulfides, chlorides, nitrates, hydroxides, or ammonium salts. Connectors in automotive applications can pick up dust from highways, such as the chemicals for control of icy conditions in winter. Splash on roadways can contain ammonium compounds from chemical fertilizers and animal wastes. The industrial environment can contain contaminants of cement dust, oil films, ammonia, and nitric acid. Connectors for computers and telecommunications can be exposed to dust. Chemical plants may have high Cl_2 or HCl concentration, while oil refineries may have high H_2S and S_2 concentration. The military environment is another source of dust pick up. Molding compound (nylon) of the sealant material in connector housing is also a source of dust contaminants. The contaminations can be due to surface oxidation, dust deposition, or corrosion of contact material. Dust particles on the contact surface act as an insulator, and consequently, its deposition increases electrical contact resistance.

The characteristics of contaminants are an important variable in connector failure. The parameters of dust particles can be categorized as size, composition, and electric charge. Zhou [12] and Zhang [13] stated that the size of dust particles on a contact surface can range from 1 to 200 μm. The composition of contaminants on failed contacts can be dust particles, organic compounds, and worn contact materials with their oxide. X-ray powder diffraction (XRPD), electron probe micro-analysis (EPMA), and transmission electron microscopy (TEM) can be used to identify constituents in inorganic materials. Infrared spectrometry, gas chromatography (GC), and thermo-gravimetric analysis (TMA) can be used to identify constituents in organic materials. Dust materials with electric charge in a sequence from low to high could be solid particles with salt coverage, Al_2O_3, calcite, feldspar, quartz, mica, gypsum, TiO_2, or organic materials.

Most atmospheric corrosions follow electrochemical mechanisms in nature and involve the interaction of adsorbed moisture with various atmospheric gases at the metal surface. Dust contamination in the presence of humidity adsorption can lead to fretting corrosion on the contact surface [12, 13]. Stress corrosion cracking in copper alloys can occur in the presence of ammonia in an electrical connector. Contamination can also lead to silver electrochemical migration (ECM) in sockets. Dust contains materials of varying kinds. The materials can be rigid, such as quartz, feldspar, and

Figure 8.3 Wear tracks due to micro-motion [13].

mica, which may be embedded into the contact surface. They can also be soft, such as gypsum, calcite, and organic materials, which may be compressed and spread on the contact interface. Both types of dust materials can cause high or fluctuated contact resistance. Embedding of soft dust particles onto the contact surface causes an increase of resistance between contacts. The contact surfaces can be separated when there are hard dust particles between the connector spring and pin. The composition of dust contaminants consists of elements of Si, Na, Mg, Ca, and Al. Metallic elements of carbonates, nitrates, or oxides are also present.

As previously mentioned, the dust contamination can cause tarnish and abrasive wear [12, 13]. Wear is due to micro-motions between contact surfaces. Tarnishing and wear tracks on the contact surface can lead to an increase in contact resistance. Figure 8.3 shows the wear track of a failed connector.

To control contamination in electrical connectors, several methods can be adopted. Appropriate selection of materials can reduce tarnish rate. Alloy of pure contact materials can be a good example. Alloys of gold and silver with cadmium, indium, and palladium show a dramatic decrease of tarnish rate [10]. Modifying the design of electrical connectors also can inhibit contamination effectively [14]. For example, the electrical connector can be separated from the local flow conditions and concentration of contaminants by providing a shielding structure against the external environment. Contact wipe is another useful technique to avoid wear on a contact surface. Contact wipe refers to the relative motion between the mating contact surfaces during contact engagement or insertion. This motion disrupts the surface films and displaces contaminants and debris to ensure a metallic contact and consistent contact surface.

8.1.5 Differential Pressure

The atmospheric air exerts a differential pressure on the electrical connector housing and spring. The pressure can generate stresses inside the material. In technical terms, the definition of pressure refers to the force of action per unit area of surface. Air pressure varies with temperature. Gay-Lussac's law states that the pressure of a given mass of gas varies with the temperature of the gas when the volume is kept constant. For two gases at temperatures T_1 and T_2 and pressure P_1 and P_2, the interrelationship is given as $P_1 / T_1 = P_2 / T_2$. Pressure can build up within the connector housing as a result of sealing and waterproofing the electrical connector. With an increase in temperature, the pressure inside the electrical connector will increase. The final effect is fretting action at the contact surface.

Lam et al. [3] reported differential pressure inside and outside an electrical connector under several thermal shock cycles for electrical connectors with and without a seal. In the case of a sealed connector, with an increase in temperature inside the chamber from 0 to 90 °C, the internal pressure increased from 100.5 to 101.4 kPa. After it reached the maximum of 90 °C, the temperature decreased to 80 °C for the next couple of hours. The pressure decreased to 100.3 kPa and remained approximately at this level for the next couple of hours. With a further decrease in internal temperature to 5 °C, the internal pressure decreased to 98.7 kPa and then returned to its original level of 100.5 kPa as a result of constant internal temperature. The results for an electrical connector without any seals showed half increase and decrease of its pressure value compared to the sealed one under the same temperature manipulation. In both cases, the external pressure profile showed a minimal variation in its magnitude.

The measurements of pressure from experiments [3] prove that the theoretical predictions can be inaccurate since the pressure leakage of electrical connectors can happen in real operation. The result of change in internal pressure is fretting motion at the electrical contacts. An increase in internal pressure leads to the movement of contacts outwards. Similarly, the contacts would move inward when there is a decrease in internal pressure. Hence, a completely airtight connector is not desirable because there can be an acceleration in fretting motion.

8.2 Failure Mechanisms in Electrical Connectors

The reliability of an electrical connector is its ability to transmit electrical power and signals from the source to the electrical circuit under different loading conditions for a specific period of time without any damage to its mechanical structure. The time period for reliable operation of an electrical connector can vary with the nature of its application. To predict the reliability of electrical connectors, the underlying root cause of the failure mechanism must be investigated.

Electrical connector failures can be caused by mechanical, electrical, chemical, or a combination of these processes. A mechanical process refers to a degradation mechanism where a physical deterioration of the structural material occurs. Physical damage refers to the weakening of the mechanical properties at the surface or in the bulk material. This can be in the form of removal of the surface coating and subsequent damage to the surface finish layer and formation of dimples and pores, crack formation in the housing material, defects from manufacturing, generation of stresses, and spontaneous growth of structures from the contact pin surface.

An electrical process refers to a mechanism where electric current flow or voltage drop across the electrical contacts leads to a failure mechanism in an electrical connector. The flow of electric current across the contacts can generate Joule heating. The heat load may remain constant for the time period of action in the case of electrical connectors that

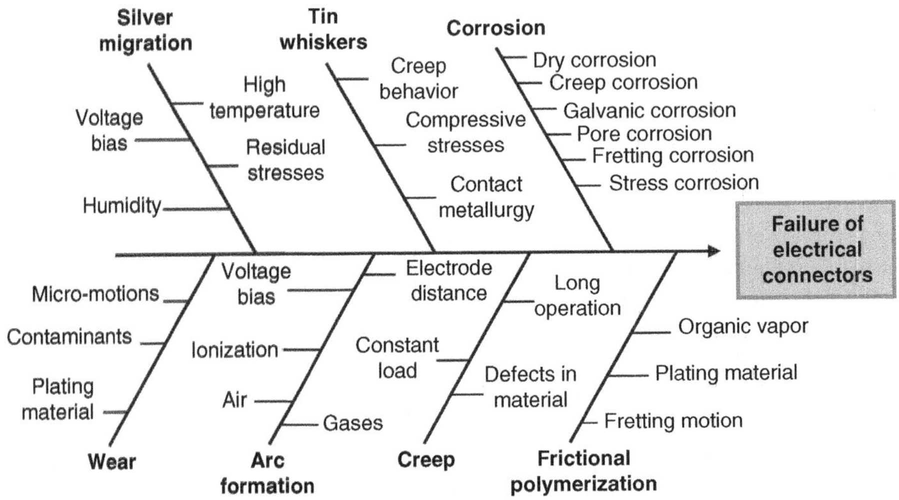

Figure 8.4 Fishbone diagram for failures in electrical connectors.

operate for long durations. This can result in contact joint welding, change in contact resistance, or change in surface properties of electrical contacts. Silver migration refers to the growth of dendrites on the surface of an insulator material in between two conductive lines as a result of the flow of electric current through the current lines and in the presence of humidity. Hot disconnection of an electrical connector forms an electric arc at the interface between the two contacts. "Hot disconnection" refers to a separation of the connector spring from its housing when the electrical connector is on.

A chemical process is an alteration in the chemical behavior of the material. It refers to a chemical reaction leading to the formation of a corrosive product layer. The chemical process usually follows a diffusion-controlled mechanism.

A cause-and effect-diagram or "fishbone" diagram schematically represents the causes of a failure mechanism, as shown in Figure 8.4. Failure in electrical connectors can occur by silver migration, tin whisker formation, corrosion, arcing, creep, and wear. The following sections discuss each failure mechanism.

8.2.1 Silver Migration

Silver migration refers to "dendrite formation" or film deposition on a dielectric surface. The process starts with silver ion dissolution, migration, and accumulation at the cathode from the anode. The failure mechanism can take place in the presence of humidity. The moisture adsorbs on the dielectric material and/or the conductive pathway (current flow paths) on the printed circuit board (PCB). The dielectric material refers to the layer of insulate in between the contact buttons or sockets. Thus, the presence of humidity and flow of electric current is necessary to trigger the failure mechanism of silver migration. This failure occurs more frequently in direct current (DC) than alternating current (AC). Temperature can also be a load that triggers silver migration.

Yang et al. [15] reported a silver migration failure in a land grid array (LGA) socket assembly. Figure 8.5 is an image of an electronics assembly. LGA is a surface mount packaging for integrated electronics. On the underside of the package, there are contacts

Figure 8.5 Land grid array socket assembly [15].

Figure 8.6 Electrochemical cell formation and growth of silver dendrites.

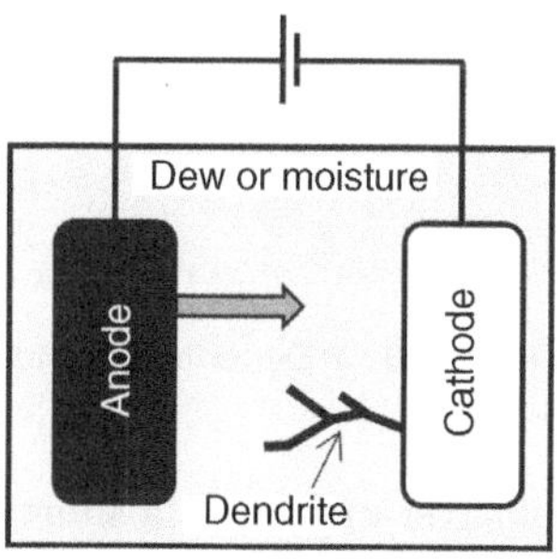

that mate with contacts on the PCB. The interconnection of an LGA with a PCB can be done by soldering material to the board or by using sockets. An LGA assembly can either have a ball grid array (BGA) or a pin grid array (PGA) kind of interconnection. The socket consists of an array of elastomer contact buttons, a polyimide insulator, and a thermoplastic frame. Silver particles are embedded in the elastomer matrix to make the contact button electrically conductive. The insulation polyimide provides the arrayed structure and holds the contact buttons.

Figure 8.6 schematically shows an explanation of dendrite formation. An electrochemical cell formation first takes place as a result of potential difference formation across the conducting surfaces (socket pins) and electrolyte formation from moisture present on the surface. The next step is the dissolution of metal ions from the anode surface in the electrolytic bath. The positive ions move toward the negative cathode, and there is a deposition on the surface of the cathode. The pile-up of metal ions on the cathode results in dendrite formation. Precipitation grows in a direction toward the anode. Figure 8.7 shows the chemical process specific to silver migration [15]. As can be seen from the chemical reaction, the electrolytic bath is a mixture of H_2O, which disassociates into OH^- and H^+ ions and positive ions of Ag^+. Ag^+ ions reduce to Ag on the surface of the cathode by electrons.

Two possible mechanisms of formation are ECM and dry migration, which occurs in the absence of humidity. The resulting effect of silver ECM on the degradation of electrical reliability is reduction in the surface insulation resistance (SIR). As a result

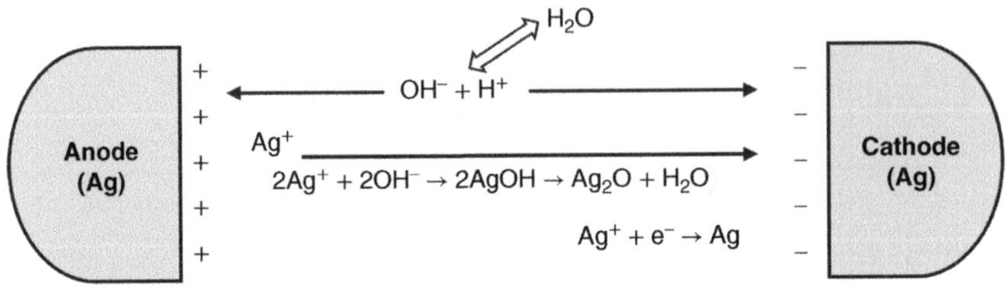

Figure 8.7 Electrochemical reactions on the insulating surface [15].

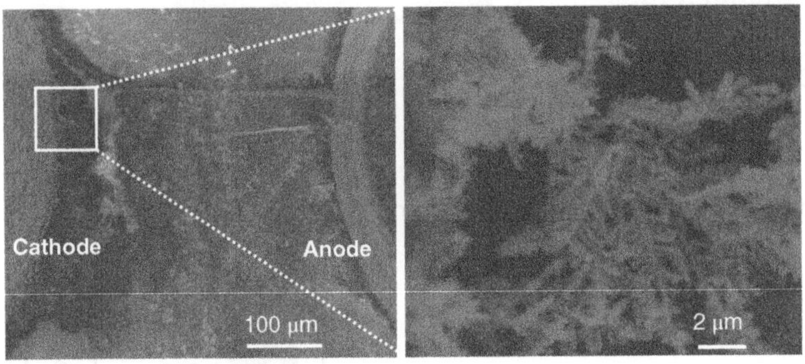

Figure 8.8 (a) Dendrite film and (b) dendrites in region a on a polyimide surface in between two IC sockets [15].

of silver migration, a short circuit can occur, and the connector may intermittently or permanently fail.

Dendrite formation can occur early in the life cycle of connectors placed in highly accelerated stress or loading conditions. Figure 8.8 shows a scanning electron microscope (SEM) image of actual silver film and dendrite formation on a polyimide surface in between two integrated circuit (IC) sockets [15]. The critical insulation resistance (IR) drop of the sample was observed after more than 30-day temperature-humidity-bias (THB) exposure (85 °C, 85%RH, 5 V). However, another sample from a highly accelerated stress test (HAST; 130 °C, 85%RH, 5 V) showed a critical IR drop within 10 minutes.

Defects on the surface of an electrical contact, undercoat, and substrate materials (Ni, Cu), and spacing in between electrodes are variables that cause ECM [16]. The surface roughness of the base metal and deposition of insulation substances on that surface can result in surface defects and pores during the electroplating process. Insulation materials that can deposit on the surface include oil, dust particles, residuals, and hydrogen bubbles released during the electroplating process. The surface defect density is dependent on the plating thickness of the Ag layer. Residual stress generation occurs in the plating material from the manufacturing processes of stamping and formation.

Yang and Christou [17] clarified the role of SIR degradation and dendritic growth with a time-to-failure (TTF) model. SIR degradation can occur prior to initiation of dendritic growth and thus was used as the basis of their model. To operate the silver ECM model, the ion accumulation rate must first be calculated. The ion accumulation time is necessary to reach a critical concentration that determines TTF. The analytical equation of the model develops a relationship in -between the surface resistivity between the anode and cathode as a function of concentration of water adsorption and ion concentration in the electrode region. The Brunauer–Emmett–Teller (BET) model provides an expression to calculate the amount of water adsorption. The TTF model is a function of temperature, RH, bias, metal material, insulating material, and surface geometry. THB tests are done to validate the model [18].

Ion migration time t_{IM} is given as $t_{IM} = L^2 / 2\mu V$, where L is the spacing between conductors, μ is the silver ion mobility, and V is the applied bias voltage. The quantity of silver discharge is given as $q = mnF / M = V\sigma_s t_{IM}$, where m is the mass of an element discharged at the anode, n is chemical valance, M is atomic weight, and F is Faraday's constant.

Surface conductivity is given as $\sigma_s = \alpha\upsilon\exp(-E_\sigma / RT)$, where υ is the concentration of water, E_σ is the activation energy or electrolyte conduction, α is constant, and R is the gas constant. Equation (8.1) gives an expression for surface conductivity in terms of RH,

$$\sigma_s = \frac{1}{R_s} = \frac{c\mathrm{RH}\exp(-E_\sigma/RT)}{\beta(1-\mathrm{RH})\{1+(c-1)\mathrm{RH}\}} \tag{8.1}$$

where $\beta = 1 / \upsilon_m\alpha$, υ_m is the concentration of adsorbed water as a complete monolayer on the insulation surface, $c = \exp[(Q_1 - Q_v) / RT]$, Q_1 is the heat of evaporation, Q_v is the heat of condensation, RH is relative humidity, and T is temperature.

Thus, a mathematical relationship for silver ECM TTF is given as:

$$\mathrm{TTF} = \frac{nFm\beta(1-\mathrm{RH})[1+(c-1)\mathrm{RH}]\exp(-E_\sigma/RT)}{MVc\mathrm{RH}} \tag{8.2}$$

As a summary, Figure 8.9 shows a fishbone diagram for silver migration. Silver migration can be triggered by environments including humidity and temperature, SIR degradation accompanied by silver ion, electrochemical cell formation, and defects.

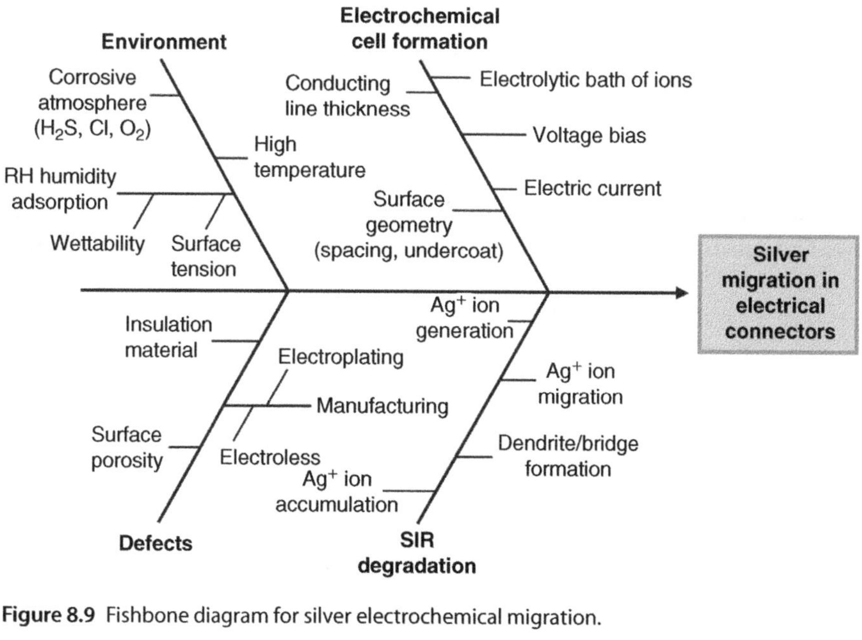

Figure 8.9 Fishbone diagram for silver electrochemical migration.

Various methods are available to reduce silver migration in connectors. As a material-involved technique, forming Ag/Ni/Cu layered conductors can reduce the rate of ECM. Also, secondary electroplating over the silver surface with less reactive metals can prevent silver ion migration. Modifying the design of contact point also works. For example, forming a thicker silver electroplating layer or widening line intervals might slow down ECM. Sealing connectors also will reduce the effects of moisture and ion contamination. When sealing is not available due to lack of space, coating hydrophobic material on the connector surface can work [16]. Another possible way to reduce silver migration may be to optimize load conditions to threshold values below the occurrence of silver migration.

8.2.2 Tin Whiskers

"Tin whisker" refers to the growth of elongated needle-like structures of pure tin from pure tin or tin alloy surfaces. It is a spontaneous growth under ambient conditions. The growth occurrence is a reliability concern for long-duration electronics. Whiskers tend to form an electric path. At high electric potentials, whiskers can melt or evaporate from the surface of the connector pin. As a result, intermittent failures may occur. In low voltages or high impedance circuits, a whisker will not melt and will result in short circuit or permanent failure.

An automotive electronics control unit (ECU) is a specific example related to the growth of tin whiskers in electronic connectors [19]. Figure 8.10 shows electrical connector pins on a PCB, and Figure 8.11 illustrates tin whisker growth on an ECU connector pin. The loading condition under which a whisker can grow is a temperature–humidity condition of 60 °C and 85% RH.

There is another example for tin whiskers generated at connection terminals of an acceleration pedal position sensor (APPS) board for automobiles [20]. Analysis of the terminals in the accelerator position sensors disclosed several tin whisker formations, as shown in Figure 8.12. Since tin whiskers are difficult to detect, analytic tools with high magnification including SEM and X-ray fluorescence (XRF) are required to reveal them.

The causes of tin whisker growth on connector pins include compressive stresses in the tin finish on electronics, weak local surface spots, low plating stiffness, stresses from electrodeposition, geometry constraints in the packaging assembly, coefficient of thermal expansion mismatch between tin plating and base metal, mechanical behavior of the underlayer and substrate materials, intermetallic compound formation in between the

Figure 8.10 Connector pins on board side with tin finish [19].

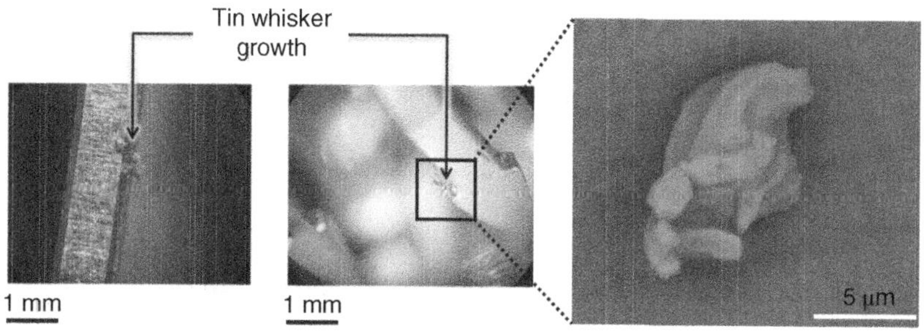

Figure 8.11 Tin whisker growth on a connector pin [19].

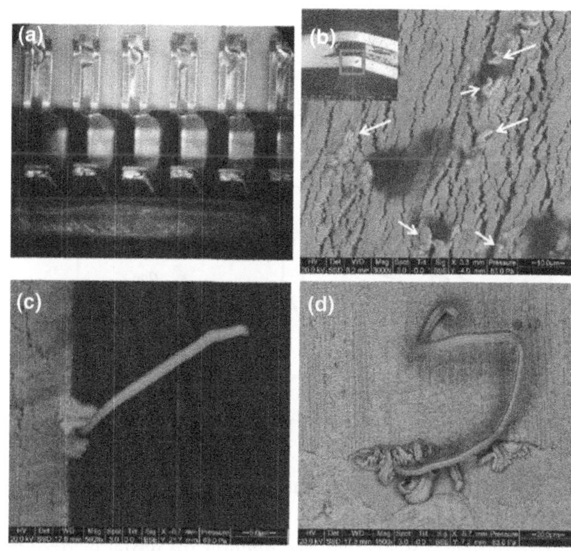

Figure 8.12 (a) Board connections of APPS and (b)–(d) tin whiskers on the connection terminals within [20].

plating material and substrate, high stiffness of substrate material, increase in external loads, and scratches on the electroplating surface from handling [19].

Smetana [21] proposed a possible underlying mechanism for tin whisker growth, as shown in Figure 8.13a. The driving force for tin whisker growth is compressive stress that is applied to a thin film. If there is oblique angle in a grain boundary, a stress to the grain boundary is less compared to other vertical grain boundaries. Hence, a stress gradient occurs and makes diffusion of Sn atoms to the base of the grain boundary, which has oblique angle. This causes grain boundary sliding, resulting tin whisker growth. This process will end when the compressive stress is relieved or when one of the grain boundaries is pinned. Figure 8.13b is an example of cross-section image of the oblique grain boundaries on a whisker grain.

Shibutani et al. [22] proposed a model for tin whisker initiation regarding contact load utilizing multiaxial creep theory. When a contact load is applied, the tin plating sandwiched between a rigid connector and a rigid substrate is pushed out due to creep. However, the sandwich structure of the contacting area and the film surrounding would restrict the tin plating deformation, which generates compressive multiaxial stresses, as

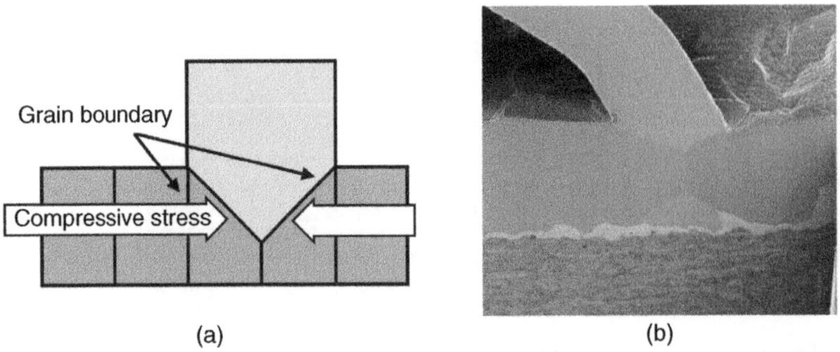

Figure 8.13 (a) Grain growth mechanism of tin whisker growth and (b) a typical tin whisker grain [21].

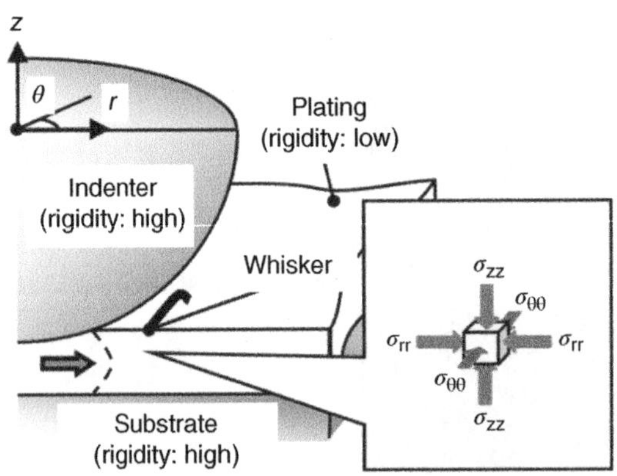

Figure 8.14 Schematic illustration of whisker initiation model by contact load [22].

shown in Figure 8.14. If the multiaxial stress exceeded a threshold, it would cause tin whisker initiation.

The external pressure can accelerate whisker growth. The growth rate was measured as 10 000 Å s^{-1} under 50 MPa of pressure [23]. This accelerated growth is a key reliability concern with lead-free connectors. The creep properties obtained from tin films can be used to screen electroplating finishes susceptible to pressure-induced tin whisker growth. The stress generated by electromigration is also able to induce tin whisker growth. Lin et al. [24] reported about tin whiskers affected by current density. When the current density was 3.6×10^5 A cm^{-2}, some whiskers formed in as little as one hour. Increasing the current density caused more whiskers to form and then to grow at a higher rate. When the current density was reduced to 4.5×10^4 A cm^{-2}, no whiskers were observed, even after hundreds of hours of current stressing.

Tin deposition methods are a variable in propensity to form tin whiskers, as shown in Table 8.1 [25]. For example, tin films deposited by electron beam evaporation or DC-sputtering show immediate tin whisker formation, while tin films formed by resistive evaporation or bright electroplating method have highly resistant property against tin whisker formation. Matte electroplating or electroless plating-utilized tin films show a tin whisker in propensity of somewhere between the two groups.

Figure 8.15 is a fishbone diagram for tin whisker formation in electrical connectors. Residual stresses in plating material, its creep behavior, and loadings on the

Table 8.1 Propensity to form tin whisker by various deposition methods [25].

Tin Deposition Method	Thickness (μm)	Age at First Observed Whisker
Matte electroplating (thin)	5.7	9 weeks
Matte electroplating (thick)	54.95	N/A (no whiskers observed)
Bright electroplating (thin)	19.34	N/A (no whiskers observed)
Bright electroplating (thick)	54.85	N/A (no whiskers observed)
Electroless plating	10.16	9 weeks
Electron beam evaporation	2.7	0 weeks
Resistive evaporation	3.8	N/A (no whiskers observed)
DC sputtering	4.27	0 weeks

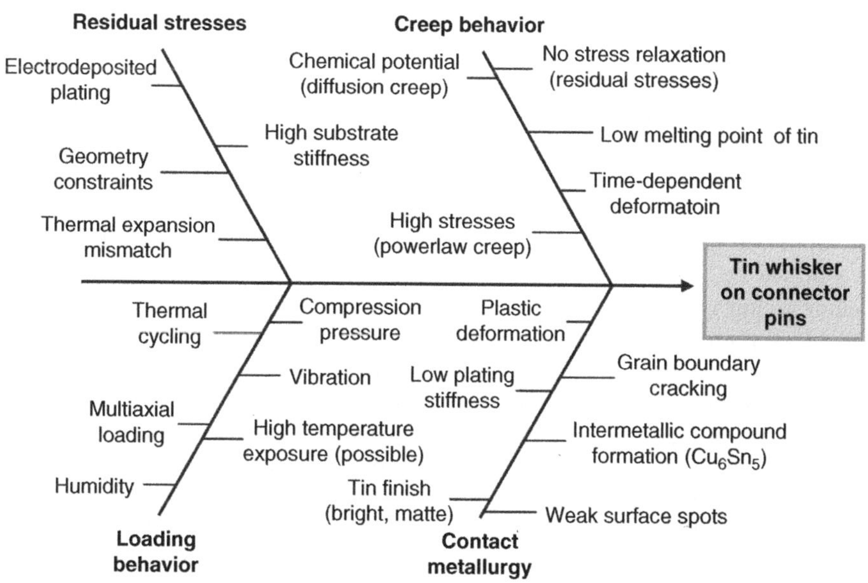

Figure 8.15 Fishbone diagram for tin whiskers.

electro-mechanical assembly and metallurgy of plating material are some causes for this uncontrollable structural growth from the surface of a contact pin.

The growth incubation period of tin whiskers ranges from a few seconds to years [26]. The whiskers can grow up to 18 mm in length and have a diameter range between 0.006 and 10 μm. The tin whisker growth rate varies according to the purity of the tin. The rate is around 0.01–0.1 Å/s in pure tin film under ambient conditions, whereas it is 2–6 Å s^{-1} in tin-manganese films. The whiskers are conductive and can cause short circuits in electrical components by creating bridges in the gap between two electrical conductors. Thus, tin whiskers are the cause of a number of costly failures in lead-free, tin-finished metal surfaces. Lead added as an alloying element reduces the length and number of tin whiskers. This has led to regulations banning lead-free tin in critical environments.

Table 8.2 Whisker growth length of various types of solder dip [30].

Sample Lead Finish	Average Whisker Length (μm)	Longest Whisker Observed (μm)	Density (mm^{-2})
Sn	26 ± 9	57	254 ± 149
Sn-Bi dipped	15 ± 5	37	1041 ± 365
SAC dipped	13 ± 6	50	524 ± 249
Sn-Pb dipped	7 ± 3	15	1009 ± 704

The failure modes caused by tin whiskers can be grouped into several categories as follows: permanent short circuits in low-current applications; transient short circuits in which the current is high enough to cause the whisker to fuse open; metal vapor arcing in vacuum; and contamination due to vibration, which frees loose whiskers that lead to interference with optical surfaces or with bridge-exposed electrical conductors. A study of tin whisker risk assessment using Monte Carlo simulation showed that the risk of an electrical short is present from a tin whisker spanning from a conductor to an adjacent conductor [27]. The maximum current that a whisker can carry before fusing open is up to 10 mA for whiskers with a 1 μm diameter, up to 30 mA for whiskers with a 2.5 μm diameter, and up to 75 mA for whiskers with a 4 μm diameter.

Metal vapor arc can be formed by an electrically bridged tin whisker in a vacuum at a biased DC. In summary, tin whiskers present risks for electrical shorting, metal vapor arcing, and mechanical obstruction failure. There are industry-standard tests for whisker growth that have been adopted by manufacturers. In particular, GEIA-STD-0005-2 is a guide for determining tin whisker risk tolerance; it also covers approaches for addressing whisker failure risk [28].

A discussion on mitigation and control of tin whisker growth follows. Application of Ni underlayer can reduce compressive stresses built up in the plating layer due to formation of Cu_6Sn_5 intermetallic compound. It is recommended to replace pure tin finish with alternatives of NiPd, NiAu, NiPdAu, and Ag-based finishes [19]. For electronic parts including hybrids, relays, and commercial off-the-shelf parts for NASA spacecraft, the tin plating surface should be covered with conformal coating such as Uralene 5750 [29]. The application of a material layer delays the growth of tin whiskers. The growth of tin whiskers on a brass surface without a coating of conformal material will not show this behavior. Copper flash (protective sheath) over brass material before plating of tin material can reduce nucleating site density and growth of tin whiskers in comparison to brass specimens that have only bright tin plating over their surface. Whisker formation density is higher for specimens at room temperature than those at approximately 50 °C.

Solder dipping contact pins can also be an effective strategy to mitigate tin whisker formation. Mathew et al. [30] provided the results of their work on solder dipping processing. As per their research, the descending order of longest whisker growth is Sn > Sn-Ag-Cu dipped > Sn-Bi dipped > Sn-Pb dipped for temperature cycles of 1000 and a temperature range between −55 and 85 °C. This testing was done for tin-based lead-free finished surfaces for alloy 42 lead frames. Table 8.2 shows the results for different types of solder dip.

Substrate stress influences whisker formation. Substrate stress can be treated by pretreatment methods such as descaling or cathodic cleaning. It was also shown that substrate pretreatment could affect tin crystal orientations and could make the tin layer more resistant against whiskers [31]. Any alloy of tin could be plated with either tensile or compressive stress depending on the pretreatment process.

No whiskers occur for very pure electroplated bright tin deposited onto copper substrates at ambient conditions [32]. Controlled additions of copper to the electrolyte were easily co-deposited into the tin film, and whiskers were observed for all samples with copper plating concentrations greater than 30 ppm

The annealing process can retard this whisker growth. It is seen that the annealing process is effective in reducing the maximum whisker length, especially in bright tin plating [33].

8.2.3 Corrosion Failure

Corrosion is undesirable deterioration of material properties of a metal, an alloy, or a nonmetal on interaction with its environment [34]. The various kinds of corrosion relevant to failure of an electrical connector are as follows. Dry corrosion is oxide product formation by diffusion of gas onto a metal surface in the absence of moisture. Creep corrosion is diffusion of metal across a noble-metal contact pair. Fretting corrosion refers to the sliding motion at the contact interface in conjunction with corrosion. Galvanic corrosion is corrosion of metal due to dissimilar metals used as a contact pair. If stress corrosion occurs, materials can crack on formation of corrosion product due to stress concentrations and oxidizing environment. Pore corrosion is diffusion of a material from the internal pores onto a metal surface in the presence of an oxidizing atmosphere. Corrosion migration, a subset of galvanic corrosion, is the transfer of a corrosion product from one surface to another due to chemical potential between oxide and metal. [35] Figure 8.16 shows a fishbone diagram for the causes of different types of corrosion. The following subsections discuss each type of corrosion in detail.

8.2.3.1 Dry Corrosion

Dry corrosion occurs in the presence of oxidizing gas such as oxygen or sulfur. Dry corrosion of copper forms a porous, electrically insulating film whose thickness is not self-limited. Three types of phenomena pertaining to dry corrosion include oxidation corrosion, liquid metal corrosion, and corrosion by other gases [36]. Oxidation corrosion refers to a chemical reaction, where the metal surface is in direct contact with oxygen (CO_2 and SO_2) without any moisture present in the atmosphere. In general, oxidation corrosion occurs at room temperature. When the surface is encountered by oxygen molecules, a chemical reaction takes place resulting in formation of an oxide layer. The pores on the metal surface provide an active region for corrosion formation. As long as the oxygen molecules can diffuse to the metal surface through the pores, corrosion product formation will take place. After a certain period of time, the oxidation reaction stops because the oxidation layer formed inhibits this reaction. Liquid metal corrosion refers to a phenomenon wherein the corrosion reaction takes place when liquid metal reacts with another solid metal surface to form a corrosion product [37]. In the third type of dry corrosion, called "corrosion by other gases," a gas that does not has

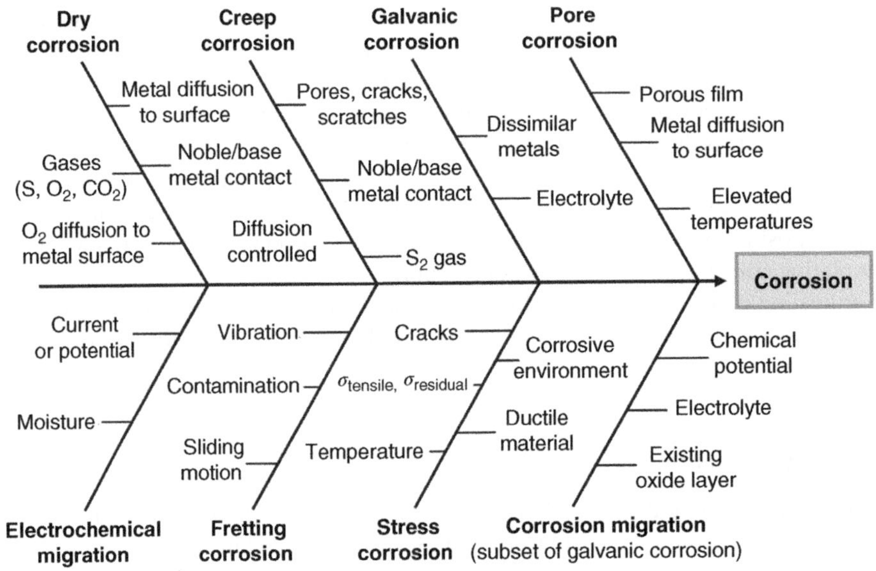

Figure 8.16 Fishbone diagram for corrosion in electrical connectors.

oxygen atoms reacts with a metal. For example, when a chloride gas contacts a silver surface, a film of AgCl can form.

8.2.3.2 Galvanic Corrosion

Galvanic corrosion occurs when a metal is electrically coupled to a more noble metal in the same electrolyte. Galvanic corrosion results in serious degradation in a bimetallic system [38]. When different metals are contained in the solutions of ionized salts, galvanic corrosion will occur. The requirements for galvanic corrosion are (i) materials possessing dissimilar surface potentials, (ii) common electrolyte, and (iii) an electrical path.

The drive force to flow electrons is the potential difference between the two metals. The direction of electron flow depends on which metal is more electrochemically active, as shown in Figure 8.17a. The more active metal becomes anodic, and the noble metal becomes cathodic. Hence, galvanic corrosion occurs at the surface of the anodic metal. If aluminum is coupled to copper in the same electrolyte, aluminum, which becomes the anodic component, dissolves and is deposited at the copper cathode as a complex hydrated aluminum oxide. Simultaneously, an evolution of hydrogen occurs at the copper cathode. This process will continue until all the aluminum has been consumed. In most of cases, however, the corrosion at the surface might be limited by the pile-up of corrosion products such as aluminum oxide.

Aluminum-to-copper connections can result in corrosion-mediated failures in two ways [39]: electrical failure by reduction of the contact area or mechanical failure by severe connector corrosion. Both results can occur simultaneously. The severity of galvanic corrosion is determined by humidity. Galvanic corrosion can be prevented by several measures such as a metal plating of intermediate galvanic potential and transition washers. Lubricating the surface can block the current flow, inhibiting contact degradation by galvanic corrosion, as shown in Figure 8.17b.

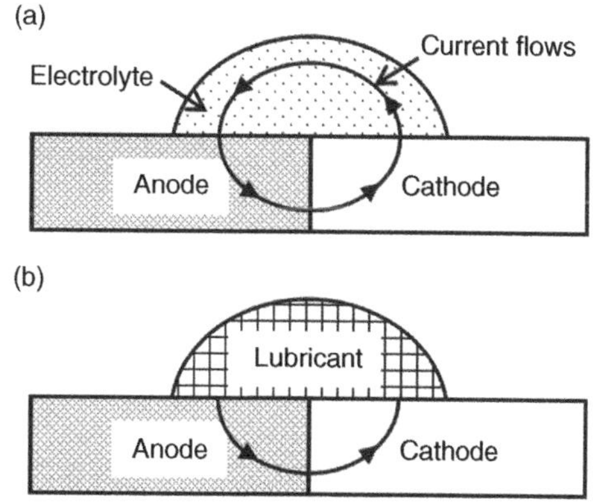

Figure 8.17 Schematic of (a) galvanic corrosion in anode-to-cathode joint and (b) protection via lubricant.

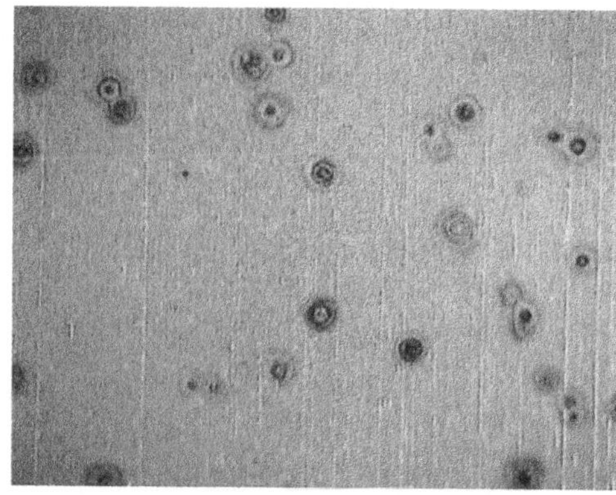

Figure 8.18 Example of pore corrosion in a gold-plated contact [41].

8.2.3.3 Pore Corrosion

Pore corrosion can be found in the structure of a reactive base metal covered by a thin layer of noble metal. The noble metal is applied to the surface to protect the base metal from corrosion [40]. However, when the base metal is exposed through pores of the surface metal, it can react chemically or electrochemically with atmospheric corrosives. This detrimental phenomenon is called "pore corrosion." The products generated by pore corrosion can reduce stability, accelerate wear on the surface, increase contact resistance, and ultimately lead to the total failure of a connection. Figure 8.18 shows an SEM image of pore corrosion on a contact surface [41].

Figure 8.19 shows a schematic of the mechanism of pore corrosion [42]. If the surface noble metal has pores under a corrosive environment, an electrolytic path can form between the noble metal and the base metal. This causes galvanic corrosion, and the base metal migrates to the gold surface.

8.2.3.4 Creep Corrosion

Creep corrosion can occur when a reactive base metal adjoins a noble metal plating. The creep corrosion spreads over the surface of the adjacent noble metals as a liquid form and then becomes immobilized and forms insoluble sulfates [43, 44]. Sulfur is the major cause element to result creep corrosion. This corrosion process is usually associated with copper sulfide and silver sulfide corrosion films. Since the resulting corrosion products are fairly resistive, failures by creep corrosion are not detected early

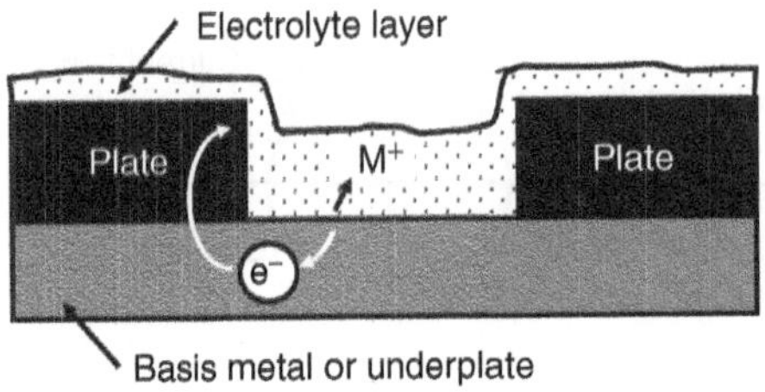

Figure 8.19 Mechanism of pore corrosion [42]. Graphically renewed.

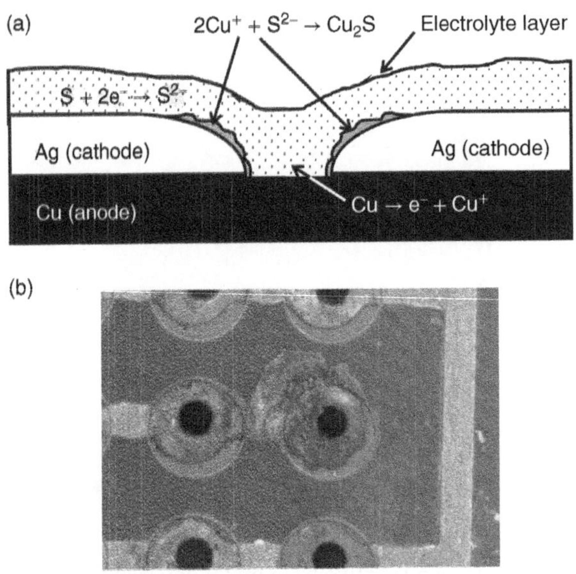

Figure 8.20 (a) Mechanism of creep corrosion [46] and (b) example of creep corrosion [47].

[45]. Figure 8.20a shows the mechanism of creep corrosion [46]. Initially, a metal sulfide forms on the surface and then precipitates. Since sulfur compounds dissolve in water and create a weak sulfuric acid, the metal oxide of the surface could diminish and expose the underlying metal. A thin layer forms on the contact surface due to sulfide driving the exposed metal to migrate over the surfaces of the noble metal and electrically insulating them. Figure 8.20b shows an example of creep corrosion [47]. Copper creep corrosion occurred on the surface of a PCB sample with a silver finish after soldering with organic acid flux.

Gold is the most susceptible to creep corrosion [43]. The creep corrosion rate is higher when a reactive substrate metal is located close to the noble metal/alloy plating. The substrate metal corrosion products creep over the noble metal surface.

Other types of base metal alloy also affect creep corrosion [48]. The rate of creep depends on the propensity of the base metals to form stable oxide films, the stability of these films against a reactive chloride, and the reactivity of the base metals to form Cu_2S. Nickel corrosion products did not show significant creep rates. Hence, thick nickel barriers can be effective for mitigating the migration of copper corrosion products onto gold.

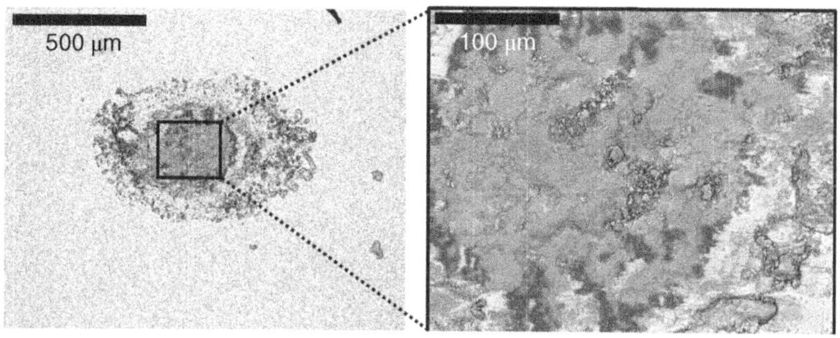

Figure 8.21 SEM image of fretting corrosion at a gold-plated copper alloy surface [50].

Figure 8.22 Mechanism of fretting corrosion in electrical connectors [51].

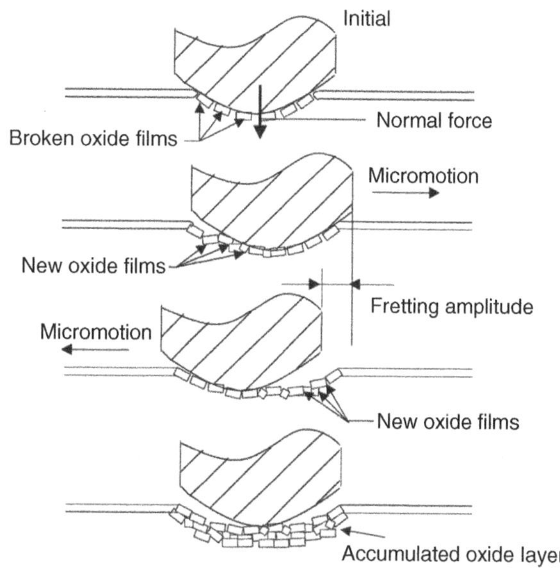

8.2.3.5 Fretting Corrosion

The main reasons for fretting corrosion are thermal cycling and micro-motions caused by micro-vibrations in a system [49]. Environmental conditions, vibration frequency, normal force, the presence of lubricants, and contact materials are all factors that affect fretting corrosion. Figure 8.21 shows the typical fretting corrosion [50].

Figure 8.22 shows the mechanism of fretting corrosion [51]. A relative movement between the two contact surfaces initiates the mechanism. Vibration is a type of loading condition and leads to removal of material deposited on the contact over a period of time. Under the presence of atmospheric oxygen, oxidation will be observed on the contact surface. These deposited oxide particles or oxide film formation will lead to an increase in contact resistance at the interface. This affects the operational performance of the electrical connector. The corrosion product formed at the surface contact can be observed as small reddish-brown spots, which corresponds to rust formation. High-amplitude vibration environments inducing fretting corrosion in high-speed electrical connectors include automotive, aerospace, medical, manufacturing, instrumentation, space exploration, storage and networking, and energy industries.

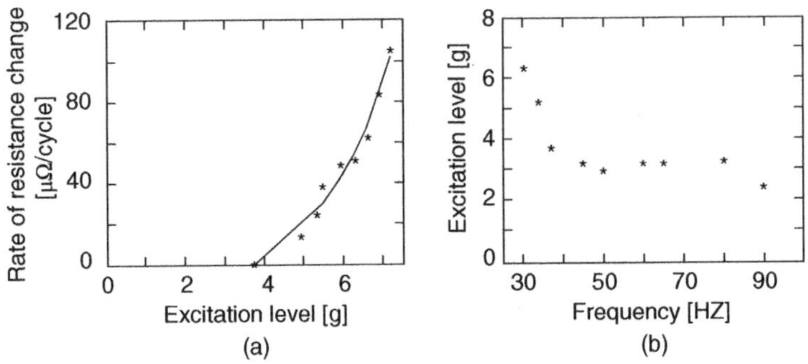

Figure 8.23 (a) Resistance change with excitation level under fixed frequency and (b) threshold excitation g-levels according to the frequency [53].

Fretting corrosion mainly occurs in metal connectors. The surface layer of metal can be oxidized, and the resulting metal oxides will accumulate, covering the contact area [52]. The oxidized film formed due to fretting corrosion is insulative, and the reduction in conductive area constricts the flow of current. The increase in contact resistance also increases the temperature across the contact surface. The temperature rise accelerates the increase in the oxidization process; that is, more insulative film is formed, which leads to more increase in contact resistance. This constant loop eventually renders the electrical connectors useless.

The dependence of fretting corrosion rate on vibrational factors such as excitation level and frequency has been studied using samples of 25-pin connectors [53], as shown in Figure 8.23. The resistance change was measured with the excitation g-level and applied frequency. In terms of the excitation levels, Figure 8.23a shows a threshold below where resistance is maintained for the duration of the test under a fixed frequency condition, 36 Hz. When the excitation increases above the threshold level, the resistance increases. As the frequency increases, the threshold excitation g-level decreases, as shown in Figure 8.23b. There is a plateau between about 40–80 Hz of frequency, showing that the damping behavior that occurs in the resonant region may affect the mechanical system.

8.2.4 Arc Formation

An electric arc forms by interaction of electrons with gas molecules present between two contact surfaces [54]. Power voltage, distance between contacts, atmospheric pressure, separation speed of the male part from the female part, current intensity, load (inductive or resistive), and the nature of contact materials are parameters that control formation of arc [55].

When an electron is introduced into gas molecules located between open contacts that possess a voltage U and are a distance d apart, it gains energy of $E\lambda_e = U\lambda_e / d$, where λ_e is the average distance an electron travels when it collides with a gas molecule. If the traveling electron has enough energy to remove a weakly bound electron from the outer shell of the colliding gas molecule, ionization can occur. Ionization directly results in arc formation. The first Townsend ionization coefficient gives the number of ionizing collisions per meter. The mathematical expression of the coefficient is α. Collisions are

Table 8.3 Values of ionization potential.

Gas	Air	Ar	CO_2	H	Pd	Cd	O	Cu	Ag	Al
Ionization potential (V)	14	15.7	14.4	13.5	8.3	9.0	13.5	7.7	7.5	5.9

inelastic in nature.

$$\alpha = \frac{f(E\lambda_e)}{\lambda_e} = Ap\exp\left(\frac{-AV_i}{E/p}\right) \tag{8.3}$$

where A is a constant depending on the gas type, p is gas pressure, V_i is the ionization potential of gas, and E is the electric field [56, 57]. Table 8.3 gives several values of ionization potentials.

The number of electrons arriving at the anode from the cathode can be given by $n_1 = n_0\exp(\alpha d)$, and the corresponding equation for current is $I = I_0\exp(\alpha d)$; the number of electrons arriving at the anode is n_1, and the number of electrons generated from the cathode is n_0, respectively. I and I_0 correspond to n_1 and n_0, respectively. This collision generates ions and electrons. Ions bombard on the surface of the cathode and generate extra electrons. This gives rise to the secondary ionization process.

Electric current during the ionization process is described as follows:

$$I = \frac{I_0\exp(\alpha d)}{1-\gamma\exp(\alpha d)} \tag{8.4}$$

where γ is the second Townsend coefficient, which refers to the average number of electrons emitted from a surface by an incident positive ion. As the electric field in the gap increases, there is an increase in current, which results in a sudden transition from dark discharge to a discharge that sustains for a period of time. This breakdown is referred to as "Townsend breakdown" [54] and results in formation of a spark that can lead to formation of an electric arc. It can be assumed that the contact gap breaks down when $\gamma\exp(\alpha d) \to 1$, which means $\ln(1/\gamma) = \alpha d$. Since the energy that the electron gains at breakdown would become $E = V_B/d$, where V_B is the breakdown voltage at which arc formation occurs, Eq. (8.3) could be rearranged as:

$$V_B = \frac{-AV_i pd}{\ln\left[\frac{Apd}{\ln(1/\gamma)}\right]} \tag{8.5}$$

The breakdown voltage is given as the product of pressure of gas p and distance d between contacts. According to the ideal gas law, the gas pressure p can be described as Nk_BT, where k_B is the Boltzmann constant. Thus, the equation, which is known as Paschen's law, can be derived as follows:

$$V_B = \frac{-AV_iNk_BTd}{\ln\left[\frac{ANk_BTd}{\ln(1/\gamma)}\right]} \tag{8.6}$$

Paschen's law, however, is difficult to apply for small gaps under about 10 μm in air. It also loses its accuracy for a vacuum environment. In the case of a short gap or vacuum circumstance, field emission from the cathode surface should be considered to evaluate

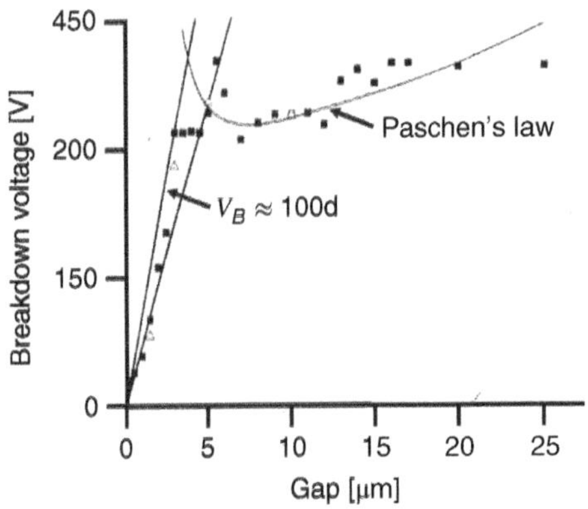

Figure 8.24 Breakdown voltage according to the contact distance between Fe cathode and Ag anode [59].

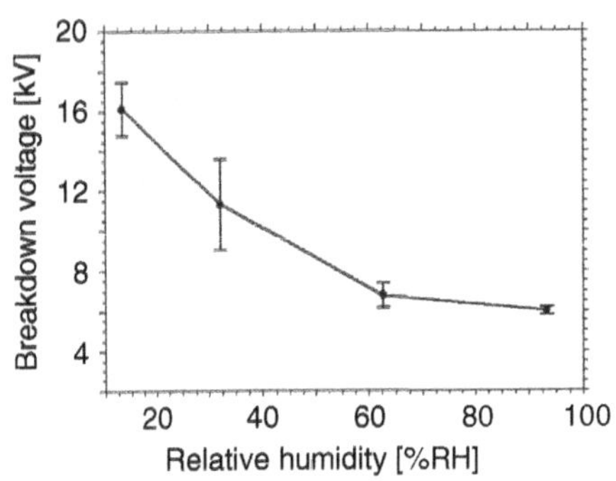

Figure 8.25 Breakdown voltage versus relative humidity [61].

the breakdown voltage [58]. For contact gaps of less than about 4 μm, when the electric field on the cathode surface is high enough to produce a current density with a certain critical point, a high-density plasma can be formed. This plasma is released as a metal vapor. When the plasma arrives at the surface of the anode, there is an electrical breakdown, which is similar to the process between open electrodes in vacuum. Thus, as shown in Figure 8.24 [59], the breakdown voltage in the short gap of air shows a linear function proportional to the gap distance d and is given by [54]:

$$V_B = \frac{E_c}{\beta} \tag{8.7}$$

where E_c is critical breakdown field regarding various metals and β is a factor that depends on the structure variation of the cathode surface.

There are various factors influencing the flashover of connectors. A distortion of the electric field by the contacting condition of the electrodes, for example, could induce the flashover. Conducting particles, including fine metal powders that are present around the connectors, could decrease the flashover voltage [60]. Breakdown is also more likely to occur under humid conditions. Figure 8.25 shows that a considerable decrease of breakdown voltage between electrodes is observed under the high RH. The high humidity could generate a thin water layer on the surface, which could be easily ionized [61].

Contamination on the surface of a connector enhances the possibility of flashover. In 1958, Obenaus proposed a mathematical model for the arc process on polluted electrode, as shown in Figure 8.26 [62, 63]. According to this model, the critical voltage for

Figure 8.26 A flashover model on polluted electrode, proposed by Obenaus [62].

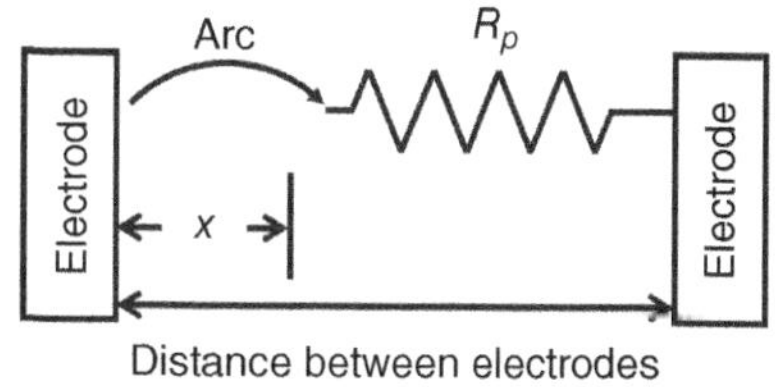

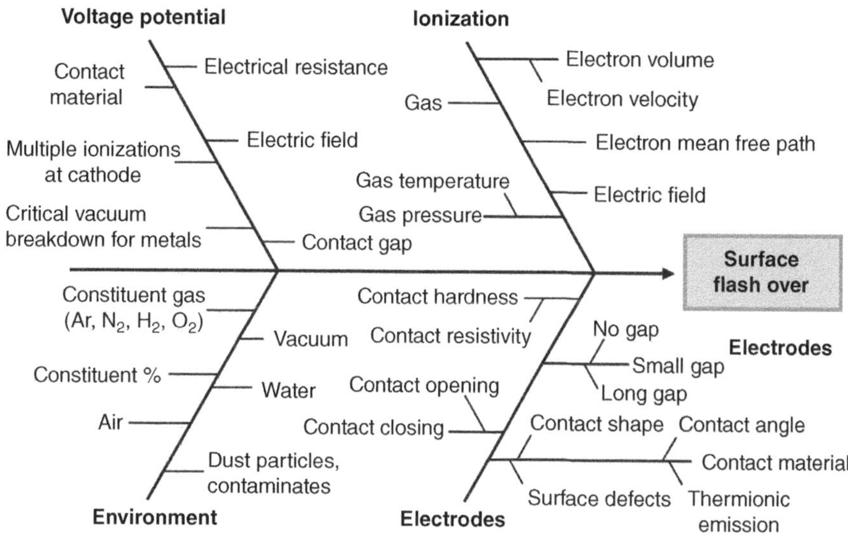

Figure 8.27 Arc formation in electrical connectors.

the flashover can be deduced as follows:

$$V = AxI^{-n} + R_p I \tag{8.8}$$

where V is the peak value of the applied voltage, x is the length of the arc, I is the peak value of the arc current, R_p is the resistance of the remaining polluted layer, and A and n are the arc characteristic constants.

The fishbone diagram in Figure 8.27 shows various factors that induce electric arc in connectors. Surface flashover is an intricately intertwined phenomenon caused by various mechanisms and conditions.

Contact erosion can occur when the temperature of the connectors under the arc reaches the boiling point of the contact material. Erosion in DC circuits can cause a net transfer between the opposing contact materials. This net transfer can form a pip and a crater, as shown in Figure 8.28. During actual use, these pips and craters normally cause uncontrolled relay failures [64]. As the contact resistance increases to a critical point, the contact may melt and result in welding.

Since the temperature of the high current due to the arc is high, it sometimes changes not only the contact material but also the surrounding gases. When the arc occurs at the contact surface, the air nearby the area could generate several products such as oxide, nitrides, and carbonates, which are not favorable for the electrical operation. If

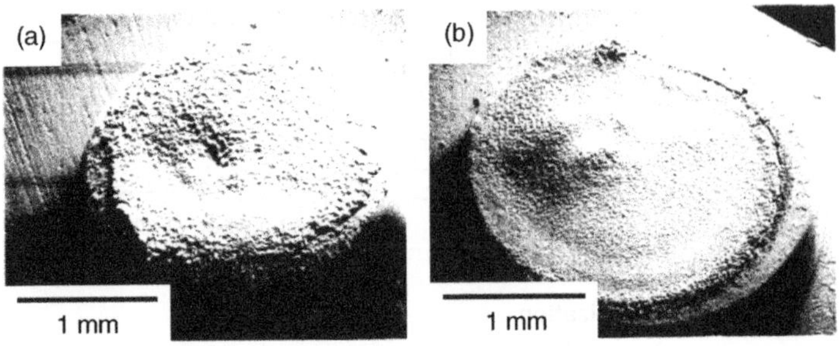

Figure 8.28 (a) Crater and (b) pip on a contact of electrodes [64].

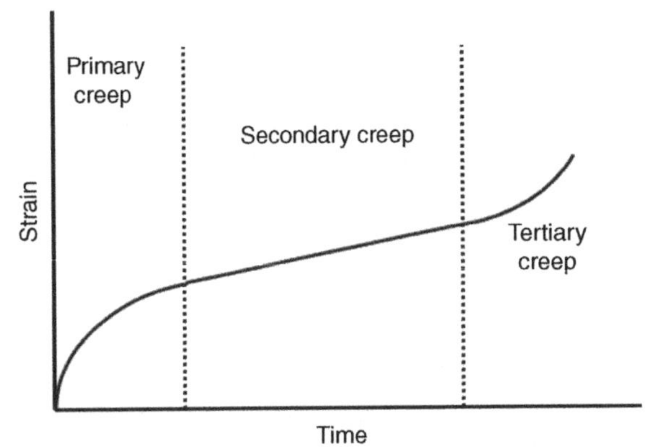

Figure 8.29 Creep curve with constant load [68].

the connector is surrounded by polluted gas or dust, worse products including sulfides or chlorides can form [65].

8.2.5 Creep Failure

Creep is defined as a time-dependent deformation of materials and is mainly affected by temperature with constant stress [66]. Time and stress affect the creep rate differently, depending on the temperature. At room temperature, the creep rate of most materials mainly depends on external stress and is highly independent of time. On the other hand, as the temperature is elevated, the influence of time increases [67].

Generally, creep phenomenon has three stages, as shown in Figure 8.29 [68]. In the first stage, primary creep, the creep rate decreases with time because the creep resistance of the material increases as the deformation occurs. In the next stage, secondary creep, the creep curve is observed linearly due to the constant creep rate. In this regime, a balance is achieved between strain hardening and recovery of the material. This stage is also called "steady-state creep." The duration of this stage could be varied by the applied stress σ. The creep rate increases in the final stage, tertiary creep, reaching failure. Among the three creep stages, the minimum creep rate $\dot{\varepsilon}_{ss}$ is observed at the steady-state creep stage and described as $\dot{\varepsilon}_{ss} = K_1\sigma^n$, where K_1 and n are material constants [69].

Creep models are established on the observations of microstructural mechanisms. The creep mechanisms are mainly described as two types depending on the temperature

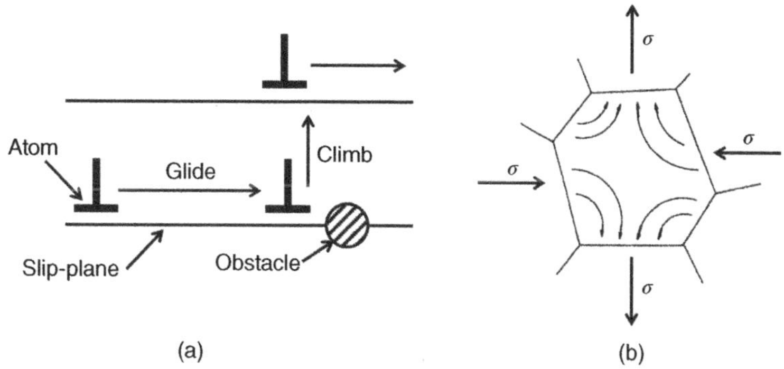

Figure 8.30 Schematic diagrams of (a) dislocation creep and (b) diffusional creep [71].

level: dislocation creep and diffusion creep [70]. Both mechanisms are governed by the diffusion of atoms. Figure 8.30a shows a schematic diagram for dislocation creep. When an atom glides along its slip-plane and meets an obstacle such as precipitate, a diffusion driven by mechanical force makes the atom climb the obstacle and glide along the upward slip-plane. Since diffusion is highly dependent on temperature, this dislocation of the atom usually occurs when the temperature exceeds 0.3 T_M, where T_M is the melting point of the material. The diffusion coefficient is given as [71]:

$$D = D_0 \exp\left(\frac{-Q}{RT}\right) \tag{8.9}$$

where D_0 is the maximal diffusion coefficient at infinite temperature, and Q is the activation energy.

Hence, the empirical relationship between the steady-state creep rate and temperature can be described as:

$$\dot{\varepsilon}_{ss} = A\sigma^n \exp\left(\frac{-Q_c}{RT}\right) \tag{8.10}$$

where A is an empirical constant and Q_c is the activation energy for creep. As shown in Eq. (8.9), there is a power law relationship between $\dot{\varepsilon}_{ss}$ and σ. Under high stresses, n has a value between 3 and 8 [69].

As temperature is elevated, the second mechanism, diffusion creep, also contributes to deformation. In here, the diffusion of atoms occurs from one grain to another. At high temperature, the atom diffuses through the bulk crystal itself, whereas diffusion along the grain boundary is more preferable at low temperature, as shown in Figure 8.30b. Both diffusions could induce grain elongation and relieve the applied stress. As the stress is reduced, the relationship between the creep rate and stress changes from power law to linear, $n = 1$. The relationship can be described as follows:

$$\dot{\varepsilon}_{ss} = BD_0 \frac{\sigma \exp(-Q/RT)}{d^2} \tag{8.11}$$

where B is an empirical constant and d is the grain size.

The dominant mechanism of deformation differs for each combination of temperature and stresses. Figure 8.31 shows a typical deformation mechanism map with various

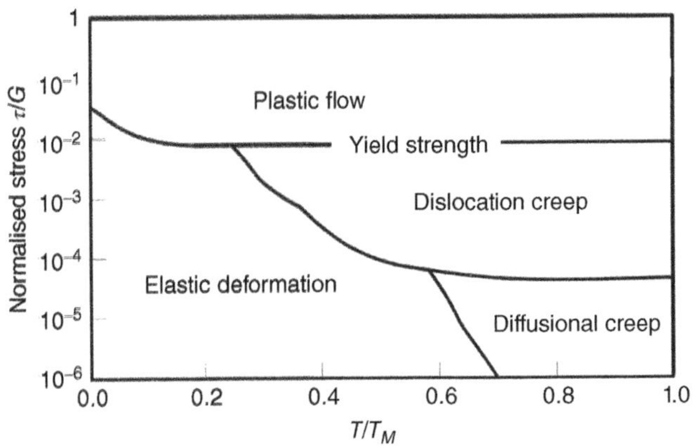

Figure 8.31 Creep deformation mechanisms with various temperature and stresses [72].

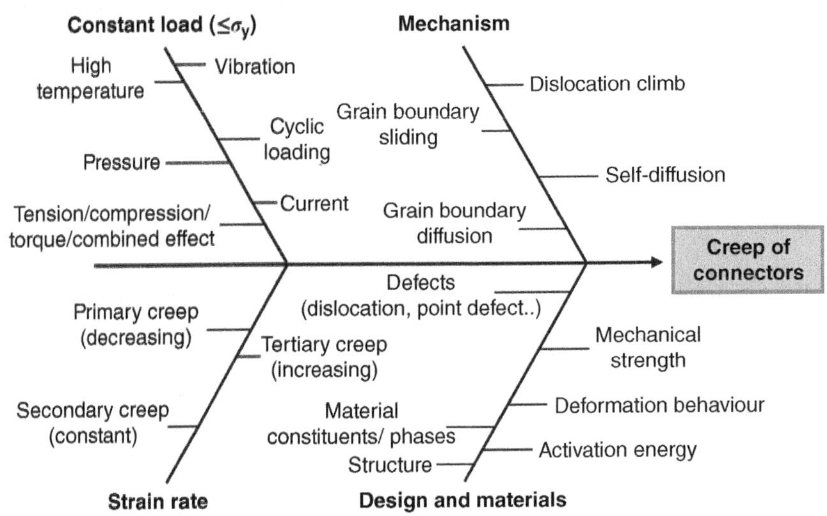

Figure 8.32 Fishbone diagram of creep failure in electrical connectors.

stress and temperature [71, 72]. Most ceramics and metals follow this diagram with minor adjustments.

Figure 8.32 shows a fishbone diagram of creep failure. Temperature and external force are major factors for the creep deformation. The creep rates also vary by the creep stage.

Creep rupture is major critical failure that originates from creep deformation. Creep rupture is a result of coalescence of creep cavities that are generated along grain boundaries, as shown in Figure 8.33a [73]. The creep cavities are formed at grain boundaries where the self-diffusion rate is not high enough to release the stress. These voids, which can be produced as shown in Figure 8.33b, weaken the material and lead to creep rupture [71].

During creep, the deformation of metal occurs slowly but continuously, and damage accumulates. At a certain point when the damage is piled up enough, the creep-induced strain of substrates would decrease the contact pressure of the connector, resulting in

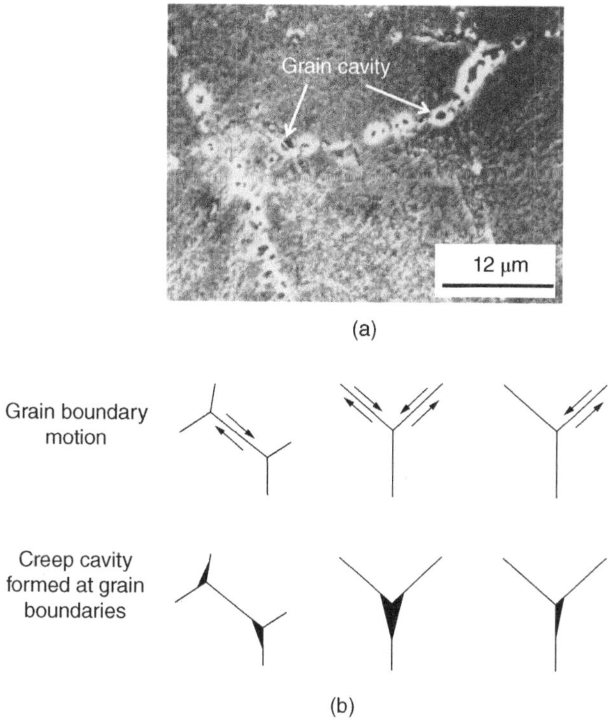

Figure 8.33 (a) Creep cavities along grain boundary [73] and (b) schematic of creep cavity nucleation [71].

increase of contact resistance [74]. The increased resistance induces heat generation in the contact area of connectors and accelerates the creep rate. The contact point can also be oxidized due to the generated heat. These states will result additional reduction of contact pressure, making a vicious circle toward the critical total failure of connector.

It is important to choose proper materials that are resistant to deformation [69, 72]. Choosing materials with high T_M is considered as a basic strategy because creep mechanisms are dependent on temperature. To impede dislocation movement, foreign atoms of high T_M can be added to the matrix atoms. The resulting alloy usually has strong bonds, so the dislocation mobility can be limited. Alloying to produce precipitates also is a useful technique to restrain grain boundary sliding. Since the creep rate is inversely proportional to the grain size, as shown in Eq. (8.10), materials with large grain size would be more resistant to the diffusion creep.

8.2.6 Wear

Wear is defined as "the progressive loss of substance from the operating surface of a body occurring as a result of relative motion at the surface" [75] and determined as the volume loss of solid surfaces in moving contacts. Because of wear, connectors can be out of service even if their working condition is well controlled without exceeding stress limit [76]. A contact surface protected by a lubricant should be free from wear problems, but many practical loads including repetitive start/stop or dusty environments could

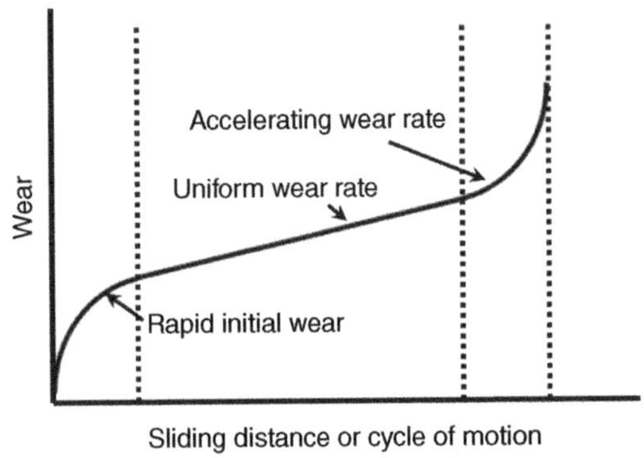

Figure 8.34 Typical curves of wear process [77]. Graphically renewed.

damage the lubricant film and result in wear. The wear process can be divided into three stages, as shown in Figure 8.34 [77]. In the first stage of wear, rapid wear dominates. During the next stage, materials turn into debris via various mechanisms, showing a uniform wear rate. As the wear progresses, the amount of debris subsequently increases between the contact. When the magnitude of wear reaches a certain limit point, the wear process enters the third stage in which the rapid wear rate results in critical failure of the system. In this stage, the generated debris can break away completely or pile up on the contact area [76].

The relationship between friction and wear is quite close, although the two phenomena need to be distinguished. They both contribute to each other. Wear mechanisms and wear rates depend on the shear loading from friction and temperature rise caused by friction heat, whereas friction is deprived of its energy by the wear process. It is hard to define a universal correlation between the coefficient of friction and wear, but the wear rates are obviously affected by the coefficient of friction in most wear mechanism cases [78].

Since wear is a result of combinations of physical and chemical mechanisms, it is difficult to present a simple model to describe it. Moreover, wear behaviors could differ by various factors such as roughness of surface, hardness of materials, and thickness of oxide film [78]. According to Burwell's work [79], the principal mechanisms of wear could be categorized as four major types: adhesive, abrasive, corrosive, and fatigue. In addition to these types, there is fretting wear [80], which is a combination of adhesive and abrasive wear, and mostly involves the corrosion mentioned in Section 8.2.3. Each type of wear mechanism is discussed as follows.

Any relative motion due to sliding or fretting, presence of contamination, or plating design can result in surface wear. Figure 8.35 shows the different mechanisms of wear in electrical contacts. Wear is a process that evolves with time. The wear of a friction pair is imaged graphically as a function of time (or friction path) at constant load and sliding speed conditions.

8.2.6.1 Adhesive Wear

Wear due to localized adhesion at the contact of materials is called "adhesive wear." When the adhesive strength at the contact is larger than the cohesive strength of the bulk material, atoms located on the surface of the material separate from the bulk material and adhesive wear occurs, as shown in Figure 8.36. Figure 8.37 [81] shows an example of adhesive wear debris, which is made at the Au–Rh metal contacts. The separated debris

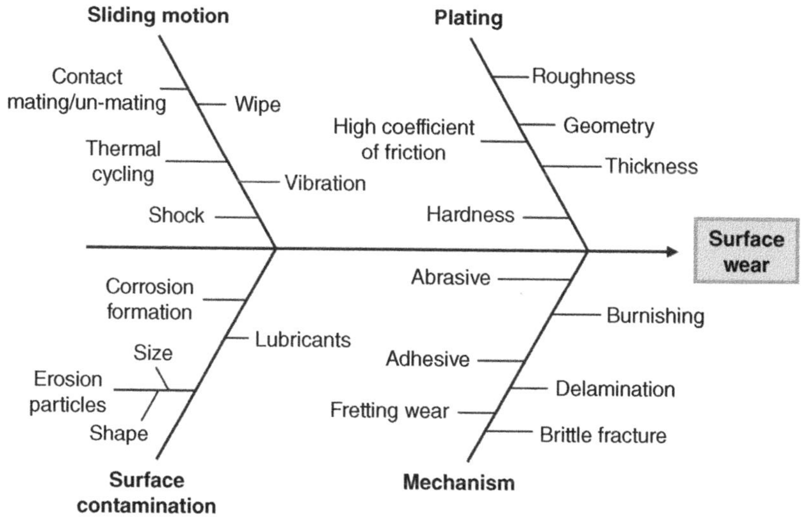

Figure 8.35 Failure mechanism of wear in electrical connectors.

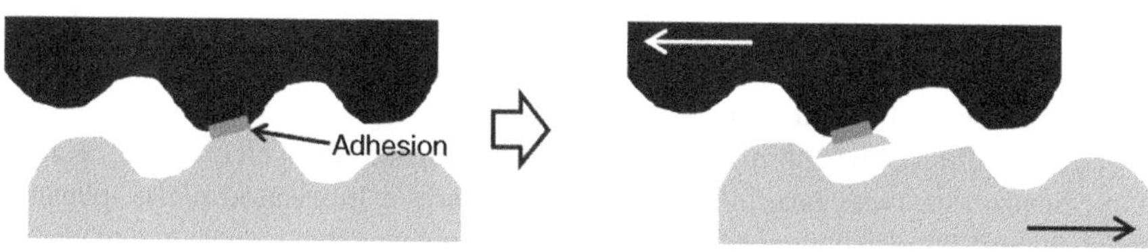

Figure 8.36 Schematic illustration of adhesive wear.

Figure 8.37 Adhesive wear debris [81].

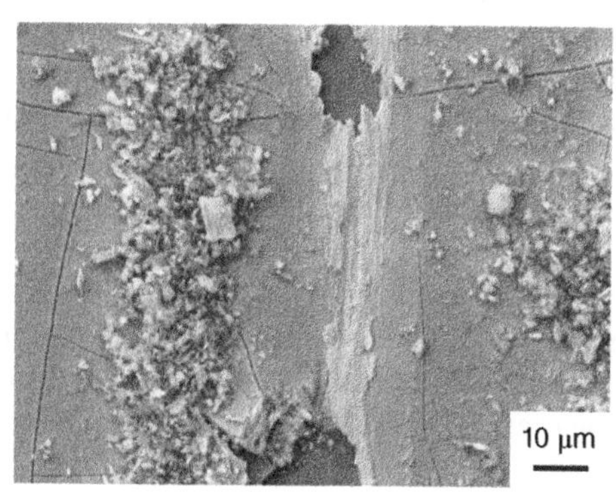

is usually from the softer materials. The adhesive wear rate highly depends on the surface energy [82], and the surface energy is varied by the oxide layers or lubricant film on the contact surface. Hence, adhesive wear is sensitive to the surface condition.

8.2.6.2 Abrasive Wear

Abrasive wear occurs between hard materials and soft materials. In this case, hard materials should be at least 1.3 times harder than the soft materials [78]. When a rubbing motion is introduced, wear debris of the soft material is generated and displaced from

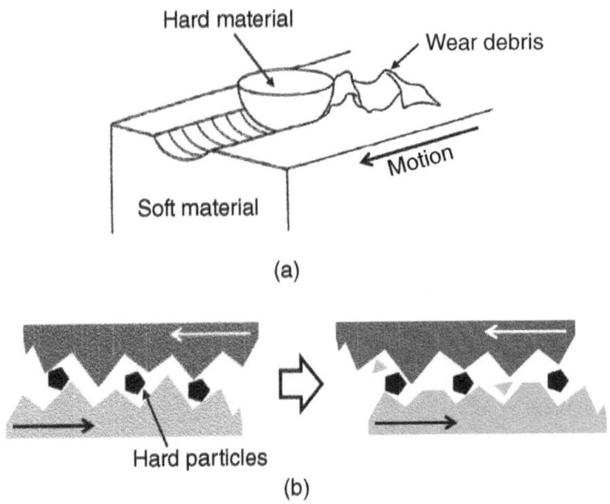

Figure 8.38 Schematic illustrations of (a) two-body abrasion [83] and (b) three-body abrasion.

the surface by hard material. Abrasive wear can be divided into two types: two-body abrasion and three-body abrasion. Two-body abrasion occurs when asperities of a hard material scrape the surface of a soft material [83]. Three-body abrasion is caused by hard particles that are trapped between the surfaces of materials. The particles make abrasive wear while they roam in the contact interface [84].

Antler examined a nickel-based modeling study for the influence of electroplating and lubricant application on wear rate [85]. Through the study, a lubricated nickel-plating reduced friction and adhesive wear of the contact surfaces, yet the occurrence of abrasive wear was still observed proportionally to the magnitude of roughness. When both surfaces of the contact were electroplated with hard gold or gold-flashed palladium, it was found that there was negligible wear under a lubricated condition and the surface roughness had less effect. On the contact surfaces without lubrication, adhesive wear was observed dominantly whether the contact surfaces were electroplated or not.

8.2.6.3 Fatigue Wear

The major cause of fatigue wear is deformation accumulated by repetitive frictions between two surfaces. Under conditions of cyclic sliding or rolling motions, the material surface is shifted in the sliding direction because of the deformation, while the area beneath the surface is left undeformed. Hence, the strain induced by the sliding shear exerts stress on the original grain structure of the materials [86]. When the shear stress exceeds a certain fatigue limit, the propagation of fatigue crack occurs as shown in Figure 8.39, and wear fragments are formed [87]. Fatigue wear is barely observed in the early stages, because it requires a large amount of cycles to accumulate damages. However, when it occurs, the size of the pit produced by removal of wear debris is relatively large compared to the other scars induced by several wear mechanisms [79].

8.2.6.4 Corrosive Wear

Wear often includes not only mechanical damage but also chemical damage. For example, surfaces can be exposed to a corrosive environment while rubbing motion is ongoing, which is why it's called "corrosive wear." The reaction products by corrosion

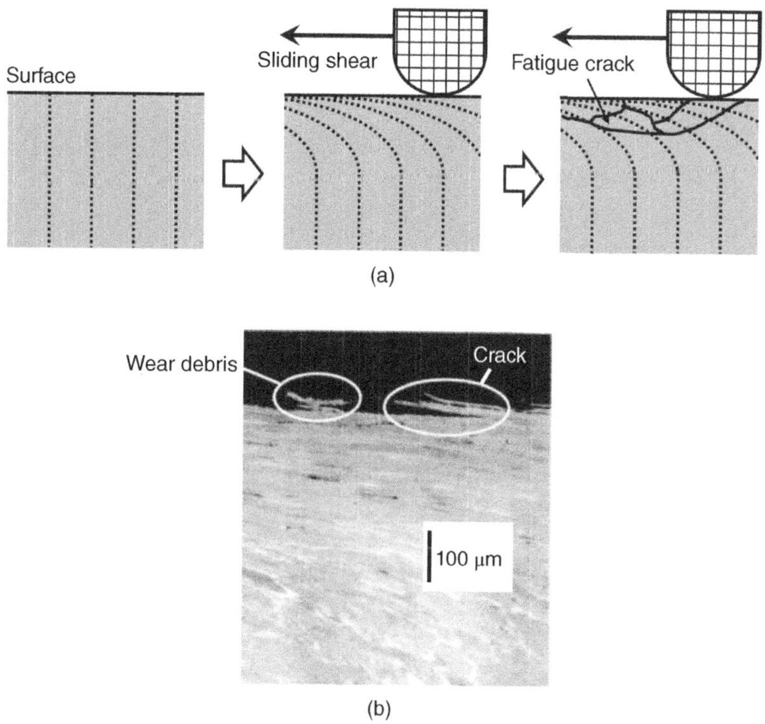

Figure 8.39 (a) Schematic of fatigue wear and (b) cross-section of worn steel [87].

are formed on one or both surfaces and then subsequently removed by the rubbing motion. The corrosion debris may be regarded as hard particles between the surfaces, so corrosive wear can be regarded as three-body abrasion. As the corrosion debris is removed, the next layer below the surface would be damaged by corrosive wear, and the cycle of corrosion debris formation and removal continues [88].

8.2.6.5 Fretting Wear

When connectors are subject to vibrational motion, fretting wear can occur between the interfaces of the contact pair. If the surrounding environment is corrosive and hot, fretting wear can accompany fretting corrosion. This synergetic event can result in a substantial increase in resistance. Similar to the other wear mechanisms, as more fretting cycles repeat, more degradation occurs. In the early stage of vibration cycles, large-sized wear debris generated from one or both surfaces of a contact can be trapped between the contact surface as a three-body mode and then is oxidized. As the vibration cycles increase, the debris fragments into smaller pieces, and their oxidation is accelerated. When the number of the vibration cycles reaches thousands, the fine and fully oxidized debris causes electrical failure of connectors [89]. As an example, the resistance of Au1.2 (probe)/BNi (coupon) contact pair was measured under various loading conditions and fretting cycles [90]. Figure 8.40 shows the results.

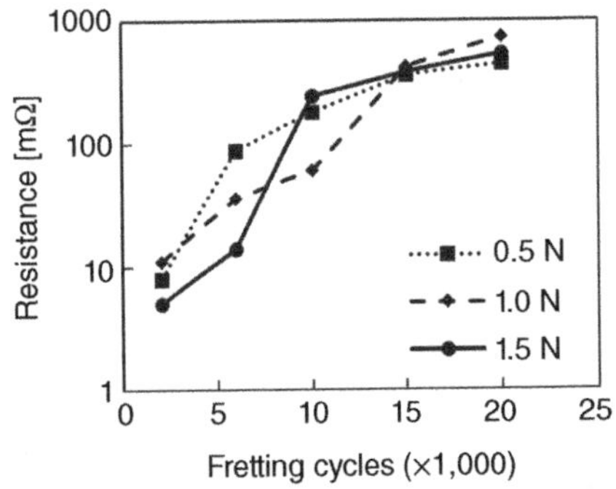

Figure 8.40 Contact resistance of contact pair versus fretting cycles [90].

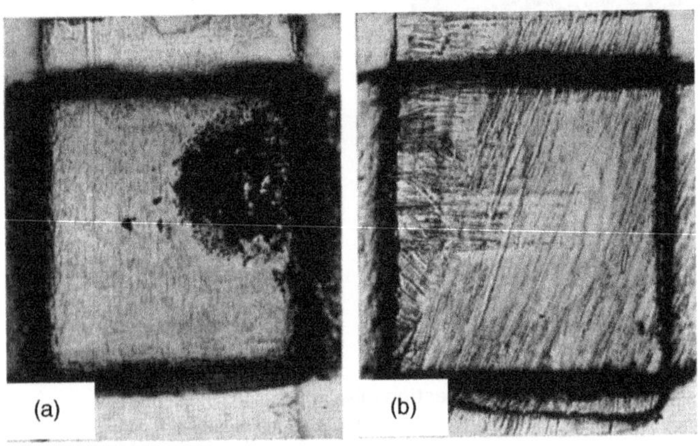

Figure 8.41 Vinylite replica of (a) palladium contact with organic deposit and (b) silver contact that shows absence of organic deposit [91].

8.2.7 Frictional Polymerization

When contact surfaces made of catalytic metals are surrounded by organic vapor under operating conditions, a deposition of organic polymer can be observed on the surfaces. This phenomenon is called "frictional polymerization." Frictional polymerization can cover the surface of a contact and cause high-contact resistance. In the first report of frictional polymerization in 1958 [91], insulating polymer compounds were produced when metals such as palladium, platinum, ruthenium, rhodium, gold, and their alloys experienced rubbing motion under an environment with low concentrations of benzene or hexane vapor. On the other hand, similar organic deposit was not observed in the silver contact, as shown in Figure 8.41.

Although the mechanisms of frictional polymerization had not been fully understood, Shinichi et al. [92] found that the oxidation process should be involved in the frictional polymerization-derived contact failure. Palladium-plated contact surfaces surrounded by benzene-saturated dried air or ambient air were exposed to a fretting load condition. A small amount of polymer formed on both surfaces. There was an increase of electric resistance in the case of the ambient air condition, whereas resistivity of the contact surfaces under the dried air stayed low and stable, as shown in Figure 8.42.

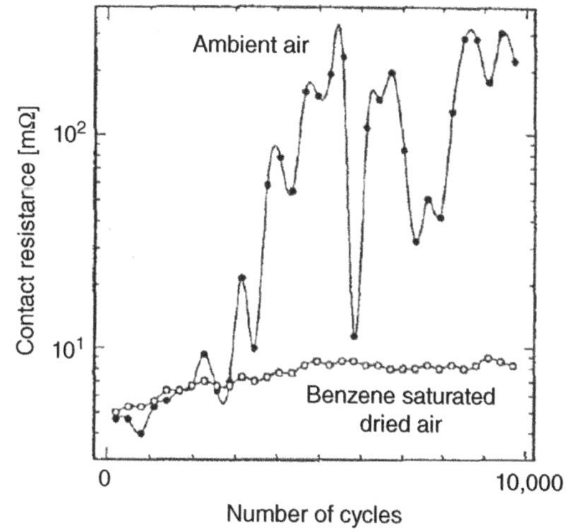

Figure 8.42 Contact resistance of palladium surfaces in benzene-saturated dried air and in ambient air versus fretting cycles [92].

The formation of frictional polymerization is found to be effectively inhibited under the presence of organometallic compounds such as tetraethyl lead. Organic vapor added with iodine-derivatives also generates a lower amount of frictional polymer. Since tetraethyl lead and iodine-derivatives are known as free radical chain terminators, it is assumed that the major cause of frictional polymerization is free radical polymerization [93].

Application of fluidic lubricants is an effective solution to minimize the increase of electrical resistance, which is derived from frictional polymerization [94]. On the contact made of palladium, the fluidic lubricants interrupted the adsorption of polymers and kept them dispersed, so the electrical resistance of the contact stayed low and stable. On the other hand, applying a solid lubricant such as wax is not effective to impede frictional polymerization because the fretting motion can easily push out the solid lubricants from the contact spot.

8.3 Case Study by NASA: Electrical Connectors for Spacecraft

Selection and applications of electrical connectors for a build-up of spacecraft should be carefully conducted to avoid catastrophic failures. The electrical connectors are exposed to harsh environments including extremely high temperature and strong vibration, which are caused by the spaceship lift-off sequence. Connector failures are mostly detected during the final inspection phase prior to the launch of a spacecraft. Delaying a spacecraft launch due to anomaly detection results in serious budget waste. If a space shuttle lift-off is called off after the fuel filling is completed, NASA spends approximately one million US dollars for the wasted liquid hydrogen and oxygen fuels and for paying the employees [95]. Several shuttle lift-offs have been canceled on the launch day because the main engine cut-off (ECO) sensor detected anomaly. For example, on December 9, 2007, NASA postponed the Shuttle Atlantis launch because a reading error was produced by the ECO sensor [96]. In a subsequent inspection, it was revealed that an electrical open occurred at the liquid hydrogen external tank (ET)

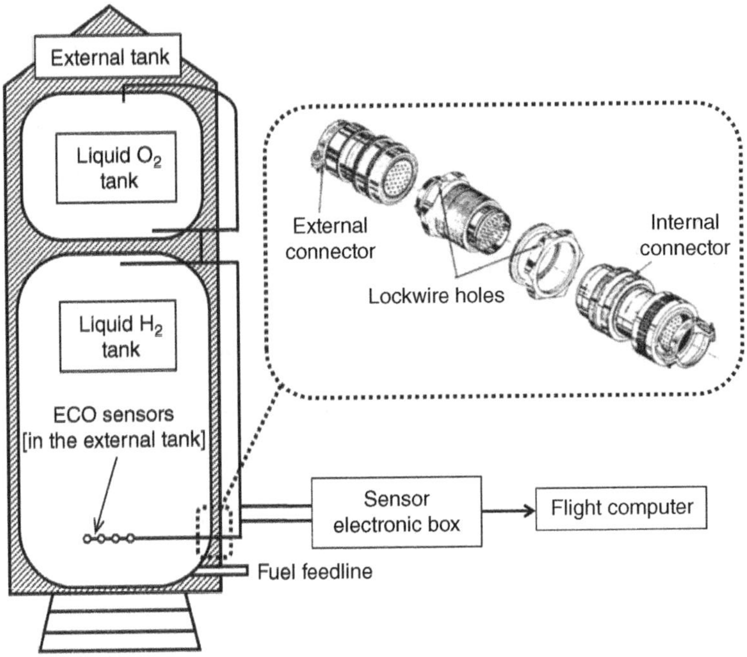

Figure 8.43 Overall schematic of the ECO system and ET liquid hydrogen cryogenic feed-through and mating connectors [98]. Graphically renewed.

feed-through connector [97]. Case studies for the ET feed-through connector failure had been conducted by NASA engineers [98, 99].

Figure 8.43 shows the overall ECO system for the space shuttle. This system is built to trigger a shutdown of the main engine if the fuel in the ET runs low. Four ECO sensors for liquid hydrogen are installed in the bottom of the ET. When the fuel level becomes low, the ECO sensors are set to indicate "dry" via increasing their electric resistance. The feed-through connector penetrates the tank wall and mates the internal connector with the external connector. The signals from these sensors pass through the ET feed-through connector and reach the sensor electronic box. The sensor electronic box supplies a constant electric current to the sensors through the ET feed-through connector, reads the signals, and gives an output signal to the flight computer. The flight computer collects and polls the signals to decide whether to start the ECO sequence. This process prevents mistakes such as an undetected low-fuel situation.

As mentioned previously, the shuttle launch that was scheduled for December 6, 2007, was canceled owing to errors occurred from circuits. Two of the ECO circuits failed to signal "wet" indication. Three days later, the next launch trial was also canceled because of a similar failure observed from one ECO circuit. To determine the root cause of this problem, a test was conducted to reproduce the failure behaviors of ECO circuits. An ECO circuit showed a similar phenomenon as the former failures, indicating that there is an open circuit. An open circuit was observed at the pin-to-socket connections in the ET feed-through connector. The electrical contact of the ET feed-through connector restored about 30 minutes after the liquid hydrogen was drained out. It was found that there were combinations of several sources for the failure [98]. A thin layer of frozen

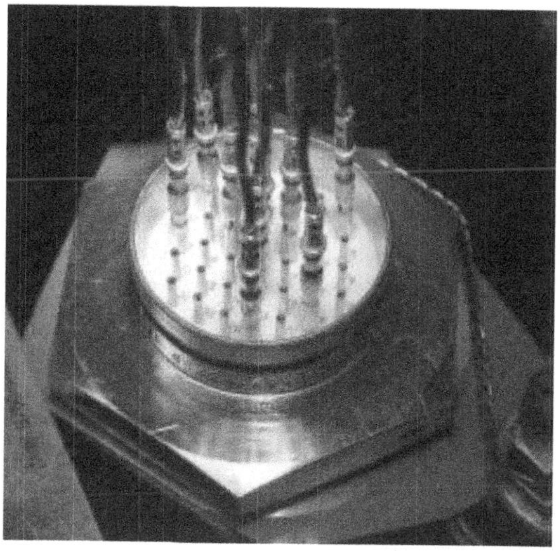

Figure 8.44 Soldered pins and sockets [99].

air and moisture was generated around the connector under the cryogenic temperature. This layer could make an open circuit. Silicone oil produced by silicone grommets of the external connector or the feed-through connector was another source of nonconductive contamination. Krytox®, a lubricant composed of fluorocarbon ether polymers, could contaminate the surface of the pin and socket during its application as grease on the connector interfaces. Since all three substances can produce nonconductive contamination, it could be explained that the ECO failures had occurred only during the fueling sequence and then disappeared after the drain sequence [99].

After the cause of failure was analyzed, troubleshooting proceeded. First, the pins and the sockets were soldered together to form a firm connection between the external connector and the ET feed-through connector, as shown in Figure 8.44. Second, the external connectors that utilized silicone grommets were fully replaced with ones that utilized Teflon grommets. The redesigned connectors passed the cryogenic cycle and cryogenic vibration test and were installed in the ECO circuit of Shuttle Atlantis. The shuttle launched successfully on February 7, 2008.

References

1 Mroczkowski, R. S. (2004). A perspective on connector reliability. *Proceedings of the 50th IEEE Holm Conference on Electrical Contacts, joint with the 22nd International Conference on Electrical Contacts Electrical Contacts*, Seattle, WA. IEEE.

2 Mueller, T. and Burns, N. (2009). Analysis and subsequent testing of cracked brass connector housings. *Journal of Failure Analysis and Prevention* 9 (5): 466–469.

3 Lam, L., Maul, C., and McBride, J. (2004). Temperature, humidity and pressure measurement on automotive connectors. *Proceedings of the 50th IEEE Holm Conference on Electrical Contacts*, joint with the *22nd International Conference on Electrical Contacts*, Seattle, WA (September 2004). IEEE.

4 Yang, B. (2002). Fundamentals of vibration. In: *Encyclopedia of Vibration* (eds. S. Braun, D. Ewins and S.S. Rao), 1290–1299. Academic Press.
5 Braun, S., Ewins, D., and Rao, S.S. (2002). *Encyclopedia of Vibration*. Academic Press.
6 Abdi, R.E., Elmanfalouti, A., and Benjemaan, N. (2006). Study of electric arc parameters during the extraction phase for automotive connectors. *Journal of Automobile Engineering* 220: 1581–1587.
7 Fontana, M.G. (1986). *Corrosion Engineering*, 41. Singapore: McGraw Hill.
8 Rübner, K., Balköse, K., and Robens, E. (2008). Methods of humidity determination. Part I: hygrometry. *Journal of Thermal Analysis and Calorimetry* 94 (3): 669–673.
9 Speight, J.G. (2019). *Natural Gas: A Basic Handbook*, 325–358. Gulf Professional Publishing.
10 Antler, M. and Abbot, W.H. (1975). The effect of environment on electrical connects: a discussion at the 1973 holm seminar. *IEEE Transactions on Parts, Hybrids, and Packaging* PHP-11 (1).
11 Zhang, J.G. and Wen, X.M. (1986). The effect of dust contamination on electric contacts. *IEEE Transactions on Components, Hybrids, and Manufacturing Technologies* CHMT-9 (1).
12 Zhou, Y. (2015). Failure mechanisms of gold-plated contacts in mobile phones under dust contamination jointing with micromotion condition. *IEEE Transactions on Components, Packaging and Manufacturing Technologies* 5 (3): 337–344.
13 Zhang, J. G. (2007). Effect of dust contamination on electrical contact failure. *Proceedings of the 53rd IEEE Holm Conference on Electrical Contacts*, Pittsburgh, PA (September 2007). IEEE.
14 Liu, W. and Pecht, M. (eds.) (2004). Component Socket Properties. In: *IC Component Sockets*, 29–31. Wiley.
15 Yang, S., Wu, J., and Pecht, M. (2005). Electrochemical migration of land grid array sockets under highly accelerated stress conditions. *Proceedings of the 51st IEEE Holm Conference on Electrical Contacts*, Chicago, IL (October 2005). IEEE.
16 Zhou, Y. and Huo, Y. (2016). The comparison of electrochemical migration mechanism between electroless silver plating and silver electroplating. *Journal of Materials Science: Materials in Electronics* 27 (1): 931–941.
17 Yang, S. and Christou, A. (Mar. 2007). Failure model for silver electrochemical migration. *IEEE Transactions on Device and Materials Reliability* 7 (1): 188–196.
18 Brunauer, S., Emmett, P.H., and Teller, E. (1938). Adsorption of gases in multimolecular layers. *Journal of the American Chemical Society* 60 (2): 309–319.
19 George, E. and Pecht, M. (2014). Tin whisker analysis of an automotive engine control unit. *Microelectronics Reliability* 54 (1): 214–219.
20 Sood, B., Osterman, M., and Pecht, M. (2011). Tin whisker analysis of Toyota's electronic throttle controls. *Circuit World* 37 (3): 4–9.
21 Smetana, J. (2007). Theory of tin whisker growth: "the end game,". *IEEE Transactions on Electronics Packaging Manufacturing* 30 (1): 11–22.
22 Shibutani, T., Yu, Q., Yamashita, T., and Shiratori, M. (2006). Stress-induced tin whisker initiation under contact loading. *IEEE Transactions on Electronics Packaging Manufacturing* 29 (4): 259–264.
23 Fisher, R., Darken, L., and Carroll, K. (1954). Accelerated growth of tin whiskers. *Acta Metallurgica* 2 (3): 368–373.

24 Lin, Y.W., Lai, Y.-S., Lin, Y.L. et al. (2008). Tin whisker growth induced by high electron current density. *Journal of Electronic Materials* 37 (1): 17–22.

25 Piñol, L.A., Melngailis, J., Charles, H.K. Jr., et al. (2011). Influence of tin deposition methods on tin whisker formation. *IEEE Transactions on Components, Packaging and Manufacturing Technology* 1 (12): 2028–2032.

26 Courey, K.J., Asfour, S.S., Bayliss, J.A. et al. (2008). Tin whisker electrical short circuit characteristics – part I. *IEEE Transactions on Electronics Packaging Manufacturing* 31 (1): 32–40.

27 Alagarsamy, K., Kohani, M., Fortier, A., and Pecht, M. (2018). Risk of tin whiskers in medical devices. *Journal of Biomedical Engineering and Research* 2 (101).

28 SAE-GEIA-STD-0005-2 (2012). *Standard for Mitigating the Effects of Tin Whiskers in Aerospace and High Performance Electronic Systems.* Society of Automobile Engineers.

29 Kadesch, J. S., and Leidecker, H. (2000).Effects of conformal coat on tin whisker growth. *Proceedings of the IMAPS Nordic Annual Conference,* Helsinger, Denmark (September 2000). IMAPS.

30 Mathew, S., Wang, W., Osterman, M., and Pecht, M. (2011). Assessment of solder-dipping as a tin whisker mitigation strategy. *IEEE Transactions on Components, Packaging and Manufacturing Technology* 1 (6): 957–963.

31 Suganuma, K. (2003). *Lead-Free Soldering in Electronics: Science, Technology, and Environmental Impact,* 157–158. CRC Press.

32 Moon, K.-W., Williams, M. E., Johnson, C. E., Stafford, G. R., Handwerker, C. A., and Boettinger, W. J. (2001). The formation of whiskers on electroplated tin containing copper. *Proceedings of the 4th Pacific Rim Conference on Advanced Materials and Processing,* Honolulu, HI, Japan Institute of Metals.

33 Fukuda, Y., Osterman, M., and Pecht, M. (2006). The effect of annealing on tin whisker growth. *IEEE Transactions on Electronics Packaging Manufacturing* 29 (4): 252–258.

34 Shreir, L.L. (ed.) (1976). Principles of corrosion and oxidation. In: *Corrosion,* 2e, vol. 1, 3–22. Boston: Butterworth.

35 Liu, W. and Pecht, M. (2004). *IC Component Sockets,* 140. Hoboken, NJ: Wiley.

36 Slade, P.G. (1999). *Electrical Contacts: Principles and Applications,* 117–118. CRC Press.

37 Manly, W.D. (1956). Fundamentals of liquid metal corrosion. *Corrosion* 12 (7): 46–52.

38 Slade, P.G. (1999). *Electrical Contacts: Principles and Applications,* 176–177. CRC Press.

39 Braunovic, M., Myshikin, N.K., and Konchits, V.V. (2017). *Electrical Contacts: Fundamentals, Applications and Technology,* 212–214. CRC Press.

40 Sun, A. C., Moffat, H. K., Enos, D. G., and Glauner, C. S. (2005). Pore corrosion model for gold-plated copper contacts. *Proceedings of the 51st IEEE Holm Conference on Electrical Contacts,* Chicago, IL (September 2005). IEEE.

41 Ming, S., Pecht, M., and Natishan, M.A.E. (1999). A comparative assessment of gold plating thickness required for stationary electrical contacts. *Microelectronics Journal* 30 (3): 217–222.

42 Krumbein, S.J. and Antler, M. (1968). Corrosion inhibition and wear protection of gold plated connector contacts. *IEEE Transactions on Parts, Materials and Packaging* PMP-4 (1): 3–11.

43 Abbott, W. (1969). Mechanism of tarnishing of precious-metal contact alloys. *IEEE Transactions on Parts, Materials, and Packaging* 5 (4): 156–160.

44 Williams, D.W.M. (1988). The effect of test environment on the creep of base metal surface films over precious metal inlays. *IEEE Transactions on Components, Hybrids, and Manufacturing Technology* 11 (1): 36–42.

45 Savolainen, P., and Schueller, R. (2012). Creep corrosion of electronic assemblies in harsh environments. *Proceedings of the IMAPS Nordic Annual Conference*, Helsingor, Denmark (September 2012). IMAPS.

46 Wang, W., Choubey, A., Azarian, M.H., and Pecht, M. (2009). An assessment of immersion silver surface finish for lead-free electronics. *Journal of Electronic Materials* 38 (6): 815–827.

47 O'Malley, G. and Fu, H. (2012). Investigation of factors that influence creep corrosion – iNEI project report. *Proceedings of the 4th Electronic System-Integration Technology Conference*, Amsterdam, Netherlands (September 2012). IEEE.

48 Braunovic, M., Myshikin, N.K., and Konchits, V.V. (2017). *Electrical Contacts: Fundamentals, Applications and Technology*, 323–324. CRC Press.

49 Tristani, L., Zindine, E.M., Boyer, L., and Klimek, G. (2001). Mechanical modeling of fretting cycles in electrical contacts. *Wear* 249 (1–2): 12–19.

50 Malucci, R. D. (2003). Fretting corrosion degradation, threshold behavior and contact instability – Electrical contacts. *Proceedings of the 49th Holm Conference on Electrical Contacts*, Washington, DC (September 2003). IEEE.

51 Ganesan, S. and Pecht, M. (2006). *Lead-Free Electronics*, 648–654. New York: Wiley.

52 Park, Y.W., Narayananb, T.S.N.S., and Lee, K.Y. (2008). Fretting corrosion of tin-plated contacts. *Tribology International* 41: 616–628.

53 Flowers, G.T., Xie, F., Bozack, M.J., and Malucci, R.D. (2004). Vibration thresholds for fretting corrosion in electrical connectors. *IEEE Transactions on Components and Packaging Technologies* 27 (1): 65–71.

54 Slade, P.G. (1999). *Electrical Contacts: Principles and Applications*, 554–565. CRC Press.

55 Abdi, R.E., Elmanfalouti, A., and Benjemaa, N. (2006). Study of electric arc parameters during the extraction phase for automotive connectors. *Journal of Automobile Engineering* 220: 1581–1587.

56 Townsend, J.S. (1910). *The Theory of Ionization of Gases by Collision*, 19–22. Constable and Company.

57 Davydov, Y.I. (2006). On the first Townsend coefficient at high electric field. *IEEE Transactions on Nuclear Science* 53 (5): 2931–2935.

58 Hourdakis, E., Simonds, B.J., and Zimmerman, N.M. (2006). Submicron gap capacitor for measurement of breakdown voltage in air. *Review of Scientific Instruments* 77 (3): 034702.

59 Slade, P.G. and Taylor, E.D. (2002). Electrical breakdown in atmospheric air between closely spaced (0.2 m–40 m) electrical contacts. *IEEE Transactions on Components and Packaging Technologies* 25 (3): 390–396.

60 Nitta, T., Shibuya, Y., Fujiwara, Y. et al. (1978). Factors controlling surface flashover in SF_6 gas insulated systems. *IEEE Transactions on Power Apparatus and Systems* PAS-97 (3): 959–968.

61 Krile, J.T., Neuber, A.A., Dickens, J.C., and Krompholz, H.G. (2004). DC flashover of a dielectric surface in atmospheric conditions. *IEEE Transactions on Plasma Science* 32 (5): 1828–1834.

62 Sundararajan, R. and Gorur, R.S. (1993). Dynamic arc modeling of pollution flashover of insulators under dc voltage. *IEEE Transactions on Electrical Insulation* 28 (2): 209–218.

63 Obenaus, F. (1958). Contamination flashover and creepage path length. *Dtsch. Elektrotechnik* 12: 135–136.

64 Leung, C. H., and Lee, A. (1990). Contact erosion in automotive DC relays. *Proceedings of the 36th IEEE Conference on Electrical Contacts, joint with the 15th International Conference on Electrical Contacts*, Montreal, Canada (August 1990). IEEE.

65 Slade, P.G. (1999). *Electrical Contacts: Principles and Applications*, 659–657. CRC Press.

66 Poulain, C., Boyer, L., Sainso, P., Maitournam, M. H., Houze, F., Leclercq, M., Guery, J. P., and Charpentier, J. P. (1995). Experimental and theoretical study of creep effects in electrical contacts. *Proceedings of the 41st IEEE Holm Conference on Electrical Contacts*, Montreal, Canada (October 1995). IEEE.

67 Jones, D.R.H. and Ashby, M.F. (2012). *Engineering Materials I: An Introduction to Properties, Applications, and Design*, 311–312. Elsevier.

68 Limaye, P., Vandevelde, B., Vandepitte, D., and Verlinden, B. (2009). An integrated ceep, crack growth and thermo-mechanical fatigue model for WLCSP assemblies soldered with SAC 405. *Proceedings of the 59th Electronic Components and Technology Conference*, San Diego, CA (May 2009). IEEE.

69 Igwemezie, V.C., Ugwuegbu, C.C., and Anaele, J.U. (2015). Review of physical metallurgy of creep steel for the design of modern steam power plants – fundamental theories and parametric models. *Journal of Engineering Science and Technology Review* 8 (5): 84–94.

70 Weertman, J. (1955). Theory of steady-state creep based on dislocation climb. *Journal of Applied Physics* 26 (10): 1213–1217.

71 Li, J. and Dasgupta, A. (1993). Failure-mechanism models for creep and creep rupture. *IEEE Transactions on Reliability* 42 (3): 339–353.

72 Kovačević, I. F., Drofenik, U., and Kolar, J. W. (2010). New physical model for lifetime estimation of power modules. *Proceedings of the 2010 International Power Electronics Conference – ECCE ASIA*, Sapporo, Japan (June 2010). IEEE.

73 Zhang, J., Zhang, G., Guo, J., Luo, C., and Zhang, Y. (2009) On ther characteristics of creep damage and rupture of dissimilar heat-resistant steel welded joint. *Proceedings of the 2009 Asia-Pacific Power and Energy Engineering Conference*, Wuhan, China (March 2009). IEEE.

74 Slade, P.G. (1999). *Electrical Contacts: Principles and Applications*, 305–306. CRC Press.

75 OECD Research Group on Wear of Engineering Materials (1969). *Glossary of Terms and Definitions in the Field of Friction, Wear and Lubrication; Tribology*. OECD.

76 Zmitrowicz, A. (2005). Wear debris: a review of properties and constitutive models. *Journal of Theoretical and Applied Mechanics* 43 (1): 3–35.

77 Yuan, H., Song, J., and Schinow, V. (2018). Simulation methodology for prediction of the wear on silver-coated electrical contacts with a sphere/flat configuration. *IEEE Transactions on Components, Packaging and Manufacturing Technology* 8 (3): 364–374.

78 Jiménez, A.E. and Bermúdez, M.D. (2011). *Tribology for Engineers: A Practical Guide*, 33–63. Woodhead Publishing.

79 Burwell, J.T. (1957). Survey of possible wear mechanisms. *Wear* 1: 119–141.

80 Waterhouse, R.B. (1984). Fretting wear. *Wear* 100: 107–118.

81 Holzapfe, C., Heinbuch, P., and Holl, S. (2010). Sliding electrical contacts: wear and electrical performance of noble metal contacts. *Proceedings of the 56th IEEE Holm Conference on Electrical Contacts*, Charleston, SC (October 2010). IEEE.

82 Rabinowicz, E. (1961). Influence of surface energy on friction and wear phenomena. *Journal of Applied Physics* 32 (8): 1440–1444.

83 Antler, M. (1996). "Sliding wear and friction of electroplated and clad connector contact materials: Effect of surface roughness. *Proceedings of the 42nd IEEE Holm Conference on Electrical Contacts*, Chicago, IL (September 1996). IEEE.

84 Song, J., Schinow, V., and Yuan, H. (2017). "Third bodies in electrical contacts – wear and electrical performance. *Proceedings of the 63rd IEEE Holm Conference on Electrical Contacts*, Denver, CO (October 2017), IEEE.

85 Antler, M. (1996). Sliding wear and friction of electroplated and clad connector contact materials: effect of surface roughness. *Proceedings of the 42nd IEEE Holm Conference on Electrical Contacts and the 18th International Conference on Electrical Contacts*, Chicago, IL (September 1996). IEEE.

86 Rigney, D.A. and Hirth, J.P. (1979). Plastic deformation and sliding friction of metals. *Wear* 53: 345–370.

87 Fletcher, D.I., Franklin, F.J., and Kapoor, A. (2003). Image analysis to reveal crack development using a computer simulation of wear and rolling contact fatigue. *Fatigue & Fracture of Engineering Materials & Structures* 26 (10): 957–967.

88 Liskiewicz, T., Neville, A., and Achanta, S. 2006. Impact of corrosion on fretting damage of electrical contacts. *Proceedings of the 52nd IEEE Holm Conference on Electrical Contacts*, Montreal, Canada (September 2006). IEEE.

89 Sauger, E., Fouvry, S., Ponsonnet, L. et al. (2000). Tribologically transformed structure in fretting. *Wear* 245: 39–52.

90 Lin, X. Y., Xu, L. J., Shao, Y. C., Luo, G. P., and Zhang, H. X. (2011). Research on fretting resistance and fretting wear property of Ni-Au contact pair. *Proceedings of the 57th IEEE Holm Conference on Electrical Contacts*, Minneapolis, MN (September 2011), IEEE.

91 Hermance, H.W. and Egan, T.F. (1958). Organic deposits on precious metal contacts. *The Bell System Technical Journal* 37 (3): 739–776.

92 Shinchi, A., Imada, Y., Honda, F., and Nakajima, K. (1995). A tribochemical investigation of the reaction products on Pd plated contacts. *Proceedings of the 41st IEEE Holm Conference on Electrical Contacts*, Montreal, Canada (October 1995). IEEE.

93 Chaikin, S.W. (1967). On frictional polymer. *Wear* 10 (1): 49–60.

94 Antler, M. (1983). Effect of lubricants on frictional polymerization of palladium electrical contacts. *ASLE Transactions* 26 (3): 376–380.

95 Santini, J. L. 2009, "Space shuttle delays cost NASA $ 4.5 million," in *Sunday Morning Herald*, Jul. 14, 2009.

96 Curie, M., and Thomas, C. (2007). NASA postpones Shuttle Atlantis Launch. *NASA Newsroom* (Dec. 9).

97 Bergin, C. (2008). STS-122: Engineers install ET-125's modified connector. *NASA spacefilght.com*(Jan. 13).

98 Iannello, C., Davis, M., Kichak, R., and Slenski, G. (2009). Spacecraft electrical connector selection and application process. *Proceedings of the 2009 Aerospace Electrical Interconnect System Conference*, Portland, OR (October 2009). SAE.

99 Martinez, H. E., and Welzyn, K. (2011). Lessons learned from the space shuttle engine cutoff system (ECO) anomalies. *Proceedings of the 47th AIAA/ASME/SAE/ASEE Joint Propulsion Conference and Exhibit*, San Diego, CA (July 2011). AIAA.

9

Fretting in Connectors

Deepak Bondre and Michael G. Pecht

Center for Advanced Life Cycle Engineering (CALCE), University of Maryland, College Park, MD, USA

Connectors are an essential part of electronic systems. Compact and efficient design demands of electronic systems require connectors to be more efficient and reliable than ever. As a result of extensive research on materials and other factors concerned with efficiency, connectors are now designed for the highest efficiency and reliability. Despite all efforts, however, electronic system failure is not rare. System failures are often attributed to "no fault found," where connector failure is the culprit [1].

A connector's main functions are separability and electrical conductibility [1]. Electrical conductibility is a connector's ability to serve as a low-resistance path for current flow. Contact resistance is different than the electric resistance of materials. As explained in Chapter 6, the asperities formed on the contact surfaces offer constriction resistance, and insulating films such as oxidation and polymers between contacting surfaces offer film resistance. The combined effect of constriction resistance and film resistance is called "contact resistance." Connector failure is mostly associated with reduced conductibility, which is indicated by a gradual or intermittent increase in contact resistance of connectors. Degradation of a connector's conductibility is a contact failure.

Swingler [1] divided contact failure into four broad types: chemical, mechanical, physical, and a combination of these failures. Chemical failure is caused mainly by the chemical reactions around connectors. Oxidation is one of the most common chemical failure mechanisms. Depending on the environment and contact material, frictional polymer formation can also take place. An oxide layer, as well as a polymer layer, can prevent metal-to-metal contact, thus reducing the effective contact area. A reduced contact area is considered as a reason for connector failure. Mechanical failure involves metal wear. Wear can be caused by intentional movements between contacts (during separation and mating) or unintentional movements (due to external vibrations or thermal cycles). Under small motions, the latter category is usually referred to as "fretting." Fretting of connectors, the primary objective of this chapter, will be discussed in detail in the upcoming sections. Physical failure includes volume self-diffusion [2]. Asperity peaks are flattened as a result of a type of capillary action. Flattening of the surface causes a reduction in metal contact and results in failure. The fourth failure type, a combination of the first three types, has been postulated by many researchers for connector failure over time. It is a synergic effect of previous types of failures. Most researchers explain contact failure types as a combination of chemical, mechanical, and/or physical

Electrical Connectors: Design, Manufacture, Test, and Selection, First Edition.
Edited by San Kyeong and Michael G. Pecht.

Table 9.1 Fretting sources [1].

Source	Description	Fretting Cycle Frequency
Vibration	External vibration from motors or knocks	>1 kHz
Alternating magnetic field	High current carrying alternating current (AC) bus bars result in alternating magnetic fields	10–100 Hz
Differential thermal expansion	Temperature changes in the environment or device itself cause the material to expand/contract	<100 mHz

failures. Chemical factors such as oxidation layer formation combined with mechanical wear can cause abrasive material formation. This type of failure is explained in detail in Section 9.1.

Fretting is micromotions between contacting surfaces. Many researchers have used the terms "fretting," "fretting wear," and "fretting corrosion" interchangeably, creating ambiguity. To avoid similar ambiguity, this chapter follows the standard definitions by ASTM Standard G40-17, excerpted as follows:

> **Fretting, n**—in tribology, small amplitude oscillatory motion, usually tangential, between two solid surfaces in contact.
> DISCUSSION—Here the term fretting refers only to the nature of the motion without reference to the wear, corrosion, or other damage that may ensue. The term fretting is often used to denote fretting corrosion and other forms of fretting wear. Usage in this sense is discouraged due to the ambiguity that may arise.
> **Fretting corrosion, n**—a form of fretting wear in which corrosion plays a significant role.
> **Fretting wear, n**—wear arising as a result of fretting [3].

We also use the term "fretting failure" to include fretting corrosion and fretting wear or any other type of failure occurring due to fretting motion. Practically, there is no minimum amplitude limit for fretting because failure can occur even at amplitudes of 10^{-4} μm[4]. Micromotions are mostly the result of external vibrations and/or displacements occurring in metal due to temperature fluctuations. Table 9.1 gives examples of sources and typical frequencies of micromotions. The continuous motion between two metal surfaces results in wear. Wear is basically a degradation of the metal surface, which results in instability of desired parameters such as contact resistance and corrosion resistance. Fretting can cause signal data value distortion. Fretting is a well-known failure mechanism in mechanical systems in automobiles (as well as biomedical implants) for affecting components such as bearings, the gear/shaft interface, and bolted connections. However, automobile electronic systems also predominantly suffer from fretting in connectors. Fretting due to vibration is especially observed in the proximity of a vibration source such as an engine in a car, airplane, satellite, or military vehicle. These sources create vibrations that often cause small, unnoticeable micromotions. Fretting failure of multiple connectors can cumulatively cause the whole electronic system to fail. Therefore, it is important to study aspects of fretting failure in detail.

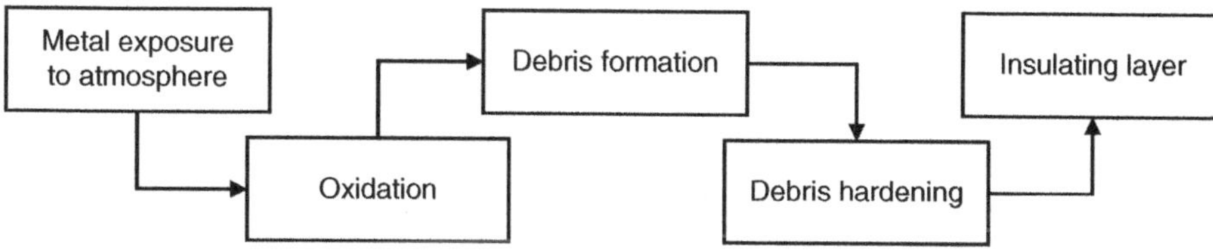

Figure 9.1 Summarized fretting failure process.

According to the factors that affect fretting, there are two types of failures: fretting fatigue and fretting corrosion. As the name suggests, fretting fatigue is a failure wherein the formation of a crack takes place, and eventually, the crack propagates to form fatigue failure. Fretting corrosion is a type of failure where chemical reactions play a major role, particularly oxidation. However, the fretting failure process does not always conform to a single type of failure; rather, it is a combination of both types. This observation can be supported by the following theory. In the first step, micromotions in contacting surfaces degrade the oxide layer, which is protecting the metal surface, causing the exposure of metal to the surrounding environment. Exposed metals are susceptible to oxidation and get re-oxidized to form oxidation products. The formation of oxidation products can be attributed to fretting corrosion. An oxidation product acts as an abrasive material between contacts and causes surface degradation. The second step can be attributed to fretting fatigue. Section 9.1 discusses various fretting mechanisms in detail.

9.1 Mechanisms of Fretting Failure

Fretting is a complex phenomenon that is difficult to describe as a single mechanism. As we will see later in the chapter, fretting is affected by multiple factors. These factors can be the type of material used in the connector, the fretting frequency, the amount of contact force, or the humidity in the environment. At any given point, more than one factor can be active. As different factors might lead to distinct mechanisms (e.g. humidity often leads to corrosion, and higher contact load often leads to wear), most of the time, fretting failure is the result of multiple mechanisms.

Ahmad [4] termed the mechanism as fretting corrosion and divided fretting corrosion into three steps. In the first step, the author explained the breakdown of the protective oxide layer. This oxide layer is usually present on nonworking contact surfaces because of its prolonged exposure to the atmosphere. The oxide layer acts as a protective layer and prevents further exposure of the metal to the atmosphere. When contacts meet, this layer can prevent metal-to-metal contact. This protective oxide layer is removed by motion between surfaces during fretting. After the removal of the oxide layer due to motion, the second step is oxidation and generation of debris. Debris is generated due to worn-out particles of metal. The author acknowledged two opposite perspectives on the sequence of the oxidation process and the surface metal removal process. The first perspective is that metal oxidizes before the surface metal removal, and the second perspective is that the sequence is reversed. In both cases, the material removed during the process is called "debris." The actual composition of debris depends on the connector materials. The loose particles in the debris cause the rate of wear to increase by damaging the surface to a greater extent. The third step is crack initiation. Debris in-between the moving contacts causes stresses in the asperities to rise and crack initiation to occur, which is called "fretting fatigue." This crack damages the surface of materials in contact.

Tomlinson [6] refers to "fretting corrosion," which includes the phenomena of fretting wear, fretting fatigue, and fretting corrosion. The differences between these three categories are mostly subject to the nature of the specific problem or objective of the study than the actual fundamental physical difference between them [5]. Uhlig [6] explained that fretting corrosion is the effect of mechanical and chemical factors. Rubbing of the asperities in the contacts exposes the track of pure metal. The exposed part easily adsorbs other gases or gets oxidized. As slip between the contact metals continues, the surface asperities wipe off the previous oxidation layer and expose the pure metal. This increases the possibility that the contact materials will be oxidized. The more mechanical part of this mechanism is illustrated by the penetration effect of asperities in one metal surface, causing wear due to welding or shearing action. Uhlig [6] used this model to study the effects of load, frequency, and amplitude on fretting failure.

Waterhouse and Taylor [7] used the delamination theory of wear proposed by Suh [8] to explain the fretting of smooth surfaces. The initial step in this theory is subsurface crack initiation. Crack initiation takes place at voids below the dislocation-free surface layer. These subsurface cracks propagate parallel to the surface, creating detached flake-like sheets of metal. The initial thickness of these particles varies from 0.1 to 20 μm, depending on the dislocation-free layer of the material. Due to grounding between moving surfaces, the metal flakes break down further into smaller sizes. Suh [8] explained this theory as high-cycle fatigue, which essentially means that a large number of cycles are required to initiate a subsystem crack in small-amplitude fretting. High-cycle fatigue explains the phenomenon of low wear rate in lower amplitudes.

Amid the presence of multiple theories for fretting failure mechanisms, Braunovic et al. [5] summarized the fretting failure process in the following steps (Figure 9.2):

1. Exposure of the metal layer to the atmosphere by removal of the protective oxide film due to mechanical fretting action
2. Oxidation of exposed metal part due to reactions with oxygen and other gases
3. Removal of metal due to adhesion wear, delamination, or micro weld shear
4. Hardening of debris due to oxidation of debris particles, which will continue to damage the surface with higher wear rates
5. Formation of a debris layer or thick insulating oxide layer, which affects the performance of electric contacts

As conductivity is the most important function of a connector, the degree of connector failure is assessed by the connector's decreased ability to conduct, in other words, its increase in contact resistance. Due to fretting mechanisms, contact resistance typically rises as fretting cycles continue. The threshold contact resistance value defines the maximum acceptable contact resistance value above which a connector is considered as failed. The threshold contact resistance value may vary according to the application of the connector. If variation of contact resistance is plotted against fretting cycles, the x-axis value corresponding to threshold contact resistance gives the life of the connector (in number of cycles) before it fails. In a simplified version, contact resistance can be assumed to gradually increase until it reaches the threshold value, but in real-life appli-

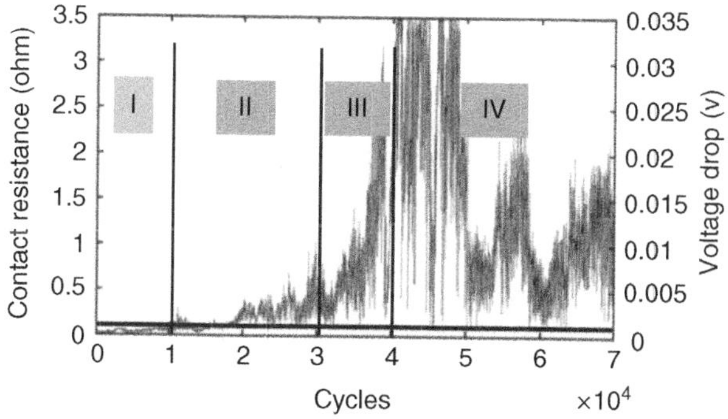

Figure 9.2 Actual variation of contact resistance over fretting cycles in gold-plated copper connectors [9].

cations, the variation of contact resistance is complex and fluctuating. To understand this complexity, an example study of contact resistance variation of gold-plated copper contacts is discussed.

The rise in contact resistance is not a straightforward trend. Ren et al. [9] studied the rather complex variation of contact resistance in gold-plated copper contacts. Figure 9.2 plots the variation of contact resistance. The figure also shows the voltage drop across the connector as contact resistance increases. The author divided the rise in contact resistance into four stages. In the first stage, contact resistance remained almost stable below 10 000 cycles. In the second stage, between 10 000 to 30 000 cycles, a slight increase in contact resistance was observed. After 30 000 cycles, in the third stage, a dramatic increase in contact resistance was observed until 40 000 cycles. After a sudden increase in contact resistance, a huge fluctuation region was observed after 40 000 cycles. Values as low as 100 mΩ and as high as 3500 mΩ were observed in this region. A similar relationship between fretting cycles and contact resistance was reported by other researchers as well [10–13].

This variation in contact resistance can be attributed to metal wear, metal transfer, and oxidation. The schematic representation in Figure 9.3 illustrates the physical processes in the four stages. In the first stage, gold is in contact with gold, which ensures the low and stable contact resistance. As the fretting motion continues, the gold coating wears out eventually, exposing the underlying copper metal. The worn-out gold accumulates in the form of debris. Because copper is easily oxidized in air, oxidation layers form and reduce the conductivity by increasing contact resistance, which is evident in the second stage. In the third stage, as the fretting motion continues, contact resistance increases sharply as a result of contact sliding over the accumulated debris. More debris and corrosion products form on the surface as more copper is worn out and exposed. The fluctuation of contact resistance in the fourth stage after 40 000 cycles is anomalous and can be attributed to nonuniform layers of debris. As shown in Figure 9.3, the surface is contaminated with nonuniform layers of gold debris, copper debris, and oxidation products. Contact resistance is high when the debris occurs between sliding connectors, whereas it is lower when connectors slide in an area where less debris is present, which allows more metal-to-metal contact and subsequently less contact resistance.

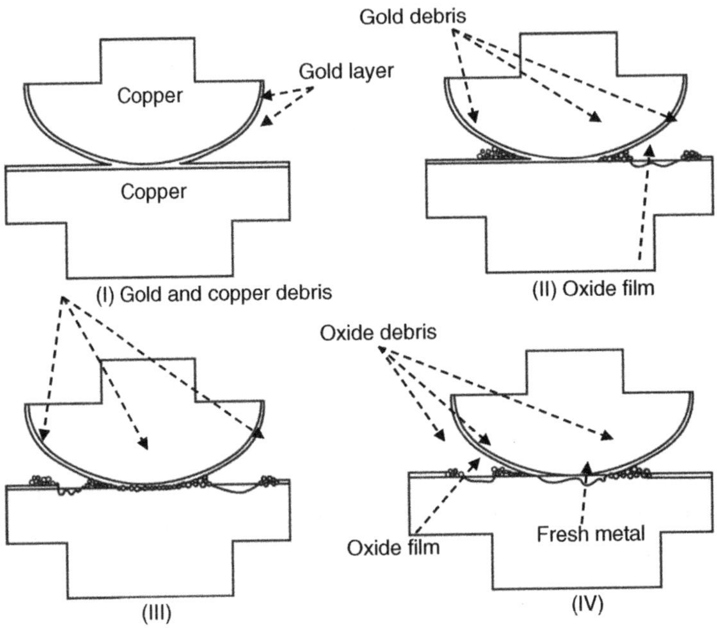

Figure 9.3 Schematic representation of the degradation process [9].

As stated earlier, the fretting mechanism is controlled by many factors. For example, copper connectors in an automobile may degrade at a different rate than similar copper connectors in computers owing to the different vibration conditions. The rate of degradation of a copper connector and a gold-plated copper connector in the same automobile may also differ owing to the different types of material. Similarly, several examples can be given with different factors. Hence, to reduce fretting, it is necessary to study the effects of factors in detail. Sections 9.1.1–9.1.3 discuss a few major types of factors that affect fretting. The large class of factors can be divided into material factors, operating condition factors, and environmental factors.

9.1.1 Material Factors That Affect Fretting

The factors that are dependent on the contact material and its properties are classified under material factors. Examples are the contact material, hardness, surface finish, frictional polymerization, grain size, metal oxides, coefficient of friction, and electrochemistry. The following sections discuss the effect of each individual factor in detail.

9.1.1.1 Contact Materials

It is important to observe the effect of the contact material on fretting. Gold connectors might evolve differently under fretting compared to palladium connectors under the same working conditions. Contact resistance is observed to assess the performance of the connector material. As already stated in previous chapters, connectors must maintain low and stable contact resistance to function properly.

Table 9.2 Classification of contact resistance behaviors [14].

	Type I (Bad)	Type II (Intermediate)	Type III (Good)
Cycles to attain 10 mΩ	<5000	>5000	>10^5
Contact resistance by 10^5 cycles	>1 Ω	<1 Ω	<10 mΩ

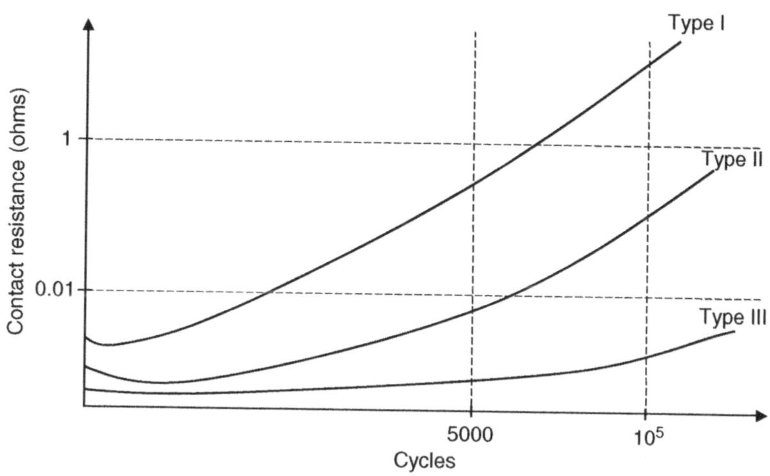

Figure 9.4 Illustration of classification of material behavior.

Antler [14] classified materials into three categories according to their contact resistance behavior. After observing 52 different specimens' results for his fretting study, he concluded that the metals can be compared on two reference observation values: the number of fretting cycles required to reach 10 mΩ resistance and contact resistance reached at the end of 10^5 cycles. According to these values, connector materials were divided into Type I (bad), Type II (intermediate), and Type III (good) behaviors [15]. Table 9.2 specifies the exact conditions of these types. Type I materials showed bad behavior; that is, those materials showed very high contact resistance rise by reaching 10 mΩ within 5000 cycles and exceeding 1 Ω before 10^5 cycles. In Type II materials, contact resistance rose less rapidly than Type I. More than 5000 cycles were required for contact resistance to exceed 10 mΩ and remain below 1 Ω even after 10^5 cycles. Type III material showed low and stable contact resistance for the entire run. Contact resistance remained below 10 mΩ even after completion of 10^5 cycles. Figure 9.4 illustrates these three types of behavior.

Figure 9.5 shows compiled results of the fretting behavior of different materials. Silver contacts have shown almost stable contact resistance over the testing cycle period range. Silver is the most stable noble metal; however, its ability to tarnish over time reduces its applicability. Type III behavior of silver can be attributed to its resistance to oxide and frictional polymer layer formation. Gold is also considered a Type III connector material. Silver and gold layers have comparable behaviors even though gold can form a trace

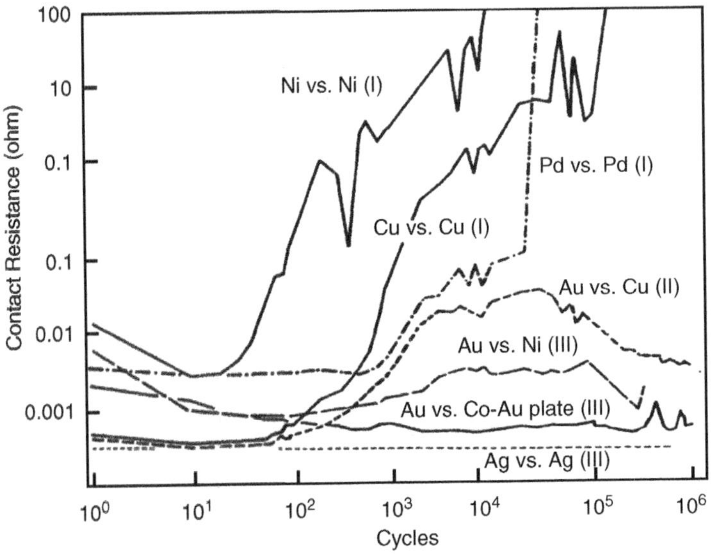

Figure 9.5 Contact resistance behavior of various tested metal combinations [14].

of polymer in benzene vapor or oil. Although polymer formation degrades connector performance, its effect is negligible because the quantity of formation is low.

When contacts made of non-noble metals such as nickel or copper are mated with themselves (nickel versus nickel or copper versus copper), they show similar Type I contact resistance behavior. Initially, contact resistance reduction was observed as the initial metal oxide layer was disrupted within 10 cycles, forming metal-to-metal contact. But as the number of cycles increased, contact resistance rose significantly, reaching 1–10 Ω in 10^3 cycles.

The tin versus tin and Sn-Pb solder versus Sn-Pb solder also fall under Type I, and the contact resistance behavior of solder versus gold is even worse. A few other base metals such as aluminum and copper alloys were studied for fretting and found to fall in the Type I category.

Palladium versus palladium contact mating has shown Type I degradation because of frictional polymer formation. Type III contact behavior has been observed when soft materials such as gold or gold alloys are mated with palladium. This behavior is due to the softer metal transfer on the palladium surface. Materials such as gold, gold alloys, or Au-Ag alloys are softer than palladium. When mated, due to metal transfer, the system becomes all gold, or the softer material gets transferred to the palladium to make it an all gold contact. As discussed previously, the gold versus gold system shows Type III contact behavior. On the other hand, if a harder material such as 75Au-25Cu (248 Hk 25) is mated with palladium, the contact system shows Type I contact behavior because palladium gets transferred to the harder metal. This system forms frictional polymers. Frictional polymers are insulating and are discussed in Section 9.1.1.4. If metals with similar hardness are mated, mostly metal transfer in both directions takes place. Copper and gold have similar hardness. After prolonged fretting, both metals were observed to be deposited on each other's surfaces. This shows some contact resistance instability; hence, it is categorized as Type II behavior.

As previously noted, the nickel versus nickel system shows bad behavior. This performance improves drastically when one contact material is replaced with gold. Similar to the gold palladium system, because gold is a softer material, it is transferred onto the nickel surface to show Type III behavior.

In the case of gold versus electroplated 60Sn-40Pb solder, Type I behavior was observed. Gold versus solder contact behavior was expected to behave better than, if not similar to, solder versus solder contacts because of the noble gold properties. However, gold versus solder contacts performed poorer than solder versus solder contacts. Systems with gold have shown an earlier increase in contact resistance than solder versus solder plates. Antler [14] presents a possible hypothesis for this surprising observation, which corresponds to the fracture mechanism of the oxide film. Solder is transferred over the gold surface because it is softer than gold. The combination becomes gold with thinly plated solder versus solder. Both of the contacts get oxidized because of the solder's affinity to oxidation. An oxidized layer of metal is less plastic than underlying metals. In the case of metals with dissimilar hardness (gold with thin solder plating and solder), the harder material tends to partially support the softer material and delay the oxide layer fracture. Fracture of the oxide layer is necessary to establish metal-to-metal contact, whereas in the case of metals with similar hardness (solder versus solder), the oxide layer fracture is more disruptive.

9.1.1.2 Hardness

The hardness of contact surfaces affects their surface degradation. Harder surfaces are less prone to degradation. Hardness usually influences degradation in two ways. First, higher hardness indicates higher tensile strength and higher fatigue strength. As surface breakdown is caused by a local high-stress fatigue process, it can be inferred that hard surfaces are less prone to surface breakdown. Second, surface oxide debris tends to degrade a surface by abrasive action. The higher the material hardness, the lower the effect of the abrasive action of the debris on the contact surface because it will be more abrasive resistant. Nevertheless, the actual relative hardness between debris and contact material will still be a critical factor. Treatments that improve hardness are expected to be useful in reducing damage by fretting [5].

9.1.1.3 Surface Finish

A lower surface finish is desired for better contact performance. Smooth surfaces have a lower plasticity index than rough surfaces. As a result of a higher plasticity index, asperities of rough surfaces undergo plastic deformation. Still, this deformation is limited further by work hardening, which prevents the surface from becoming completely flat. Work hardening is hardening of metals using plastic deformation. It is also possible that sharper asperities on a rough surface take up some small tangential movement by undergoing elastic deformation. Moreover, preserving surface roughness allows debris to move away from the actual contact surface and accumulate in the valleys of the asperities [5].

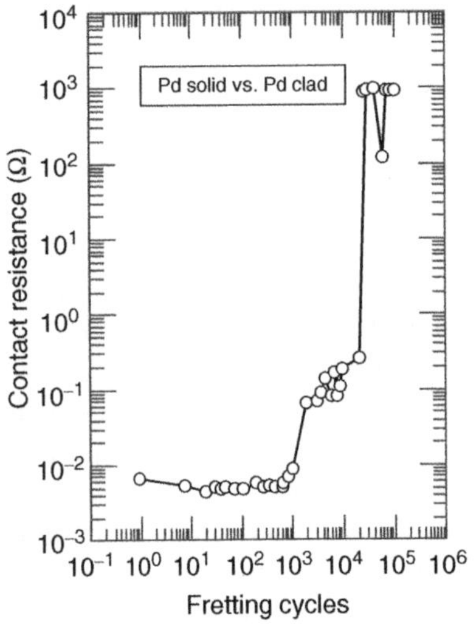

Figure 9.6 Effect on contact resistance of palladium-palladium contacts by frictional polymerization [16].

9.1.1.4 Frictional Polymerization

The formation of organic deposits on contact surfaces due to organic air contaminants is called "frictional polymerization." These high-molecular-weight deposits are insulating and tend to adhere to surfaces, which eventually increases the contact resistance. The necessary conditions for frictional polymerization are the presence of organic air contaminants, catalytically active contact metal, and the occurrence of slide between contacting members [16]. Typically, rhodium, palladium, and other platinum group contact metals and alloys are susceptible to frictional polymerization due to their catalytic activity [17]. The increase in contact resistance in a palladium versus palladium contact can be observed in Figure 9.6. Frictional polymers might help reduce the wear due to fretting but increase contact resistance due to insulation. The effect of frictional polymerization can be reduced by using alloys and lubrication. For example, because silver is a catalytically inactive transition metal, 30% or more silver addition can reduce frictional polymerization in palladium [16]. Lubrication can also reduce frictional polymerization to some extent. The effect depends on the thickness of the lubricant used. As frictional polymers are formed due to organic contaminants, it is certainly possible that organic lubricants can also enhance frictional polymerization. However, surprisingly, excess lubricants disperse the frictional polymers and therefore reduce their overall effect.

9.1.1.5 Grain Size

The grain size of a metal directly affects its hardness. Conventionally, grain sizes are reduced to improve the hardness of a material [18], which is known as "grain size strengthening." As described in Section 9.1.1.2, the hardness of a metal directly affects its fretting behavior. This section can be considered as an extension of Section 9.1.1.2, but because it focuses more on the grain sizes of metals, it is better to study them in a separate section. Noh et al. [19] investigated the copper polycrystals with a grain size of 2–162 μm under fretting motion. Figure 9.7 shows the variation of the critical number of cycles with an increase in grain size. The critical number of cycles is defined as the cycle number at which contact resistance was more than 0.1 Ω for 10 or more

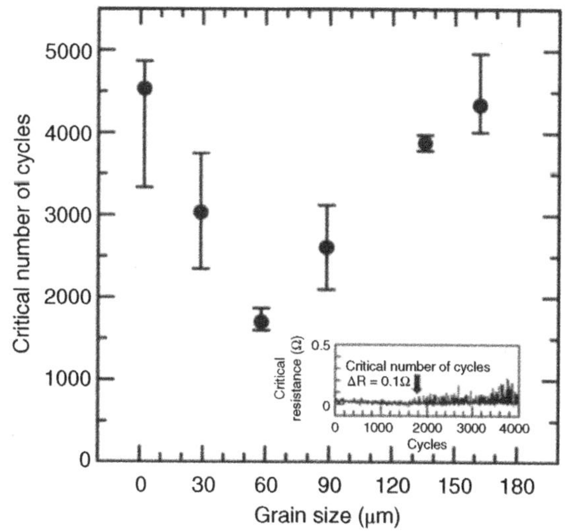

Figure 9.7 Effect of grain size on critical number of cycles for copper connector [19].

subsequent cycles. The critical number of cycles to failure decreased until the grain was 60 μm and again increased for larger grain sizes. Grain size strengthening consists of hardening of a metal as the grain size is reduced. This seems to follow well at grain sizes <60 μm, as the expected life is decreased by increasing the grain size. However, above 60 μm, the mechanism of grain size strengthening does not hold up well. According to the mechanism, the expected life (i.e. the critical number of cycles to failure) should have decreased continuously, showing that a harder material would fail earlier. But the opposite is observed, which indicates the presence of another mechanism causing the life of large-grain copper to improve considerably.

9.1.1.6 Oxides

The properties of metal oxide formed during fretting can further affect the wear rate. Metal oxide gets trapped in between rubbing surfaces; hence, the properties of the oxide are the deciding factor in increasing the wear rate. Metal oxide also mostly increases the coefficient of friction. Metal and metal oxide debris do not necessarily have the same hardness – soft metal can form oxide debris with higher hardness, and harder metal can form soft oxide debris. This relative hardness affects the wear rate. If the softer metal produces harder oxide debris, the wear rate increases, whereas a harder metal producing softer oxide debris may result in less severe wear [11].

Iwabuchi [20] studied the effect of oxide wear debris and noted that there are two ways oxide debris can affect fretting failure – it can either promote or oppose the wear. Iwabuchi reported that the effect depends on the normal load and slip amplitude. The beneficial effect is the result of quick formation of the compacted oxide layer, which acts as a third body in between contacting surfaces, whereas the negative effect is the result of abrasive action of the oxide particles. The nature of the final effect is determined by the balance between the beneficial effect due to formation of a compacted layer and the negative effect due to abrasive action.

9.1.1.7 Coefficient of Friction

In frictional surfaces, the product of normal force and coefficient of friction must be greater than the tangential force for a slip to occur. Slips can be avoided if the amplitude is too short. The effect of the coefficient of friction is straightforward. A high coefficient of friction causes high friction between sliding surfaces, which leads to more plastic deformation and fatigue failure of contacts. High friction causes accumulation of plastic strain, creating plowing by wear particles and thereby increasing the friction to ultimately accelerate the surface damage. Hence, a low coefficient of friction is desired where slip is unavoidable. Thus, elastic sliding can occur instead of plastic deformation. Plastic deformation cannot be completely avoided even in cases of very low friction because there is an elastic-plastic region even for a small frictional force. Reduced friction can decrease the plastic strain and extent of the plastically deformed region [11].

9.1.1.8 Electrochemical Factor

Electrode potential has a significant effect on fretting damage. In the case of steel, the damage rate depends on the electrode potential of the paired metals. Thus, if the mated metals are zinc and cadmium, which are more negative than steel, the steel is protected from wear.

A lower electrode potential metal will degrade more than a higher electrode potential metal. In other words, a higher electrode potential metal is saved by compromising more wear of a lower potential metal.

9.1.2 Operating Factors That Affect Fretting

The factors that define the operating conditions of a connector are external and are not attributed to inherent material properties. Examples are contact load, fretting frequency, fretting amplitude, and electric current.

9.1.2.1 Contact Load

Contact load between surfaces is one of the major factors that affect fretting. The contact area is proportional to the applied contact load. The contact area is related to the applied load by the following formula [21]:

$$A = \frac{P}{P_0} \tag{9.1}$$

where A is the contact area, P is the applied load/contact load, P_0 is the yield pressure $\approx 3Y$, and Y is the yield stress in tension.

Hence, the contact area increases when a higher normal force is applied to the contacts. As a change in the contact area affects contact resistance, a change in contact load has been observed to affect the behavior of fretting damage and contact resistance. Studies have shown that an increase in contact load reduces the deleterious effects of fretting [22–24]. The contact resistance behavior in aluminum versus tin-plated copper has been observed for two loads, 1 N and 10 N, as shown in Figure 9.8. It can be seen from the figure that both loads have distinct curves that are well separated from each other. For the 1 N load, after a certain time, a large number of fluctuations can be observed, leading to open circuit over time. On the other hand, the 10 N load in the graph shows almost stable and unaffected contact resistance over the same period. Another reason

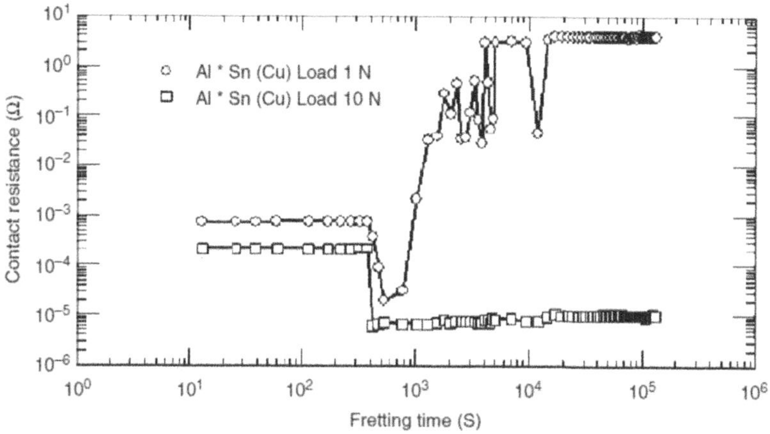

Figure 9.8 Effect of contact load on contact resistance [11].

for the improved performance is the ease in penetration of oxides or frictional polymers because of the increased loads. During fretting motion, higher loads can also be helpful for swiping away the oxide or polymer layer efficiently from the contact area.

Braunovic [23] reasoned the difference of behaviors under low and high contact load as a theory that is explained in this section and in Figure 9.9. In the case of low contact load, when contact is made for the first time, the harder material penetrates the already present oxide layers to establish the metal-to-metal contact. When fretting motions start, these metallic contacts break down, resulting in the formation of wear products. Some of these wear products will get oxidized, but the majority of them remain as metallic particles, which explains the decreased contact resistance in Figure 9.8. Due to prolonged fretting motion, these metallic layers soften and eventually separate. The contacts now have a thick insulating debris layer consisting of oxides and wear debris. Eventually, as fretting continues, all remaining contact area is lost, which causes a sharp increase in contact resistance.

In the case of higher contact loads, when contact is made for the first time, asperities on the surface penetrate the already present oxide layer. The peaks in asperities form contact zones and become practically embedded in the surface. Due to high pressure (10 N), a large number of adhesive bonds are formed, which are called "microwelds." The ruptured oxide film accumulates in the valleys of the surface and seals the contact zone from the atmosphere. When fretting motion starts, the adhesive bonds break, forming a highly concentrated metallic wear product. As the surface is almost sealed off from the atmosphere, oxidation of these wear particles is minimal. This result supports the observation of low contact resistance and low fluctuations in the case of high contact loads. Although higher contact loads show stable contact resistance for a longer time than lower contact loads, the damage due to fretting and ultimate high contact resistance is not completely avoided. In high load cases, the occurrence of very high resistance is just delayed but not prevented.

9.1.2.2 Fretting Frequency

Decreasing oscillation frequency is observed to adversely affect contact resistance. Braunovic [11] compiled the results from studies by Gagnon et al. [25], Antler [22],

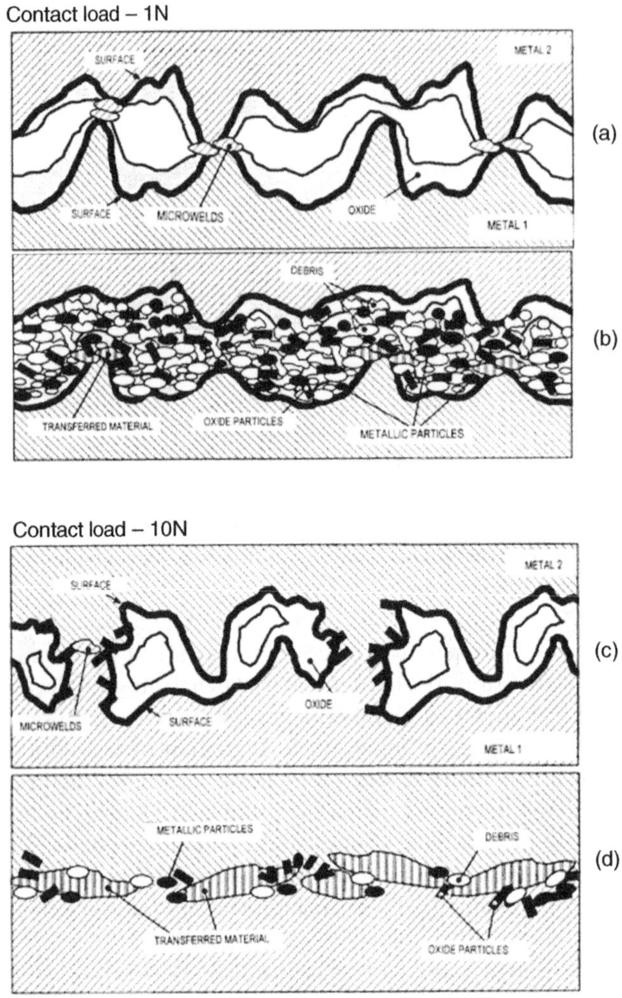

Figure 9.9 Sequence of events in contact zones of contacts with contact load 1 N (a), (b) and 10 N (c), (d) [11].

and Castel et al. [26] about fretting on aluminum versus tin-plated copper and copper versus copper connectors. Figure 9.10 shows the variation of cycles required to reach a predefined contact resistance value as a result of variation in frequency. As the frequency increases, the number of cycles required to attain a predefined value of contact resistance increases. This indicates that higher values of contact resistance can be delayed by increasing the frequency. Hence, increments in frequency can lead to improved connector life span. The results published by Kong and Swingler [27] for gold-plated phosphor bronze connectors, however, were not in accord with this observation. The authors noted worsening of connectors as frequency was increased and hypothesized that the change in contact resistance is due to a change in normal force at different frequencies. The lower frequency results in less serious damage because the effective normal force is lower. As the frequency is increased, the normal force is

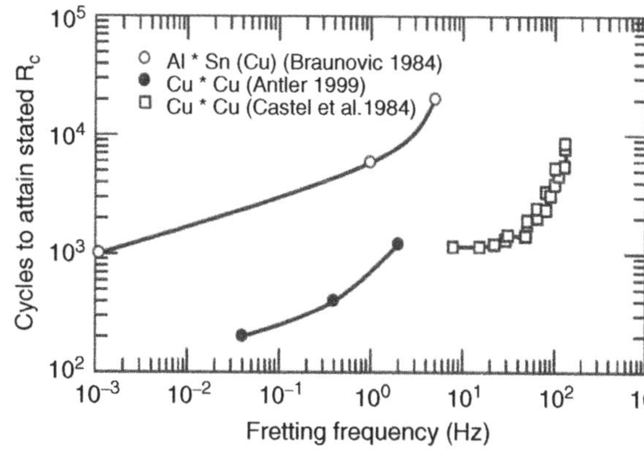

Figure 9.10 Effect of frequency on number of cycles required to attain predefined contact resistance in Al-Sn plated copper and Cu-Cu contacts [11].

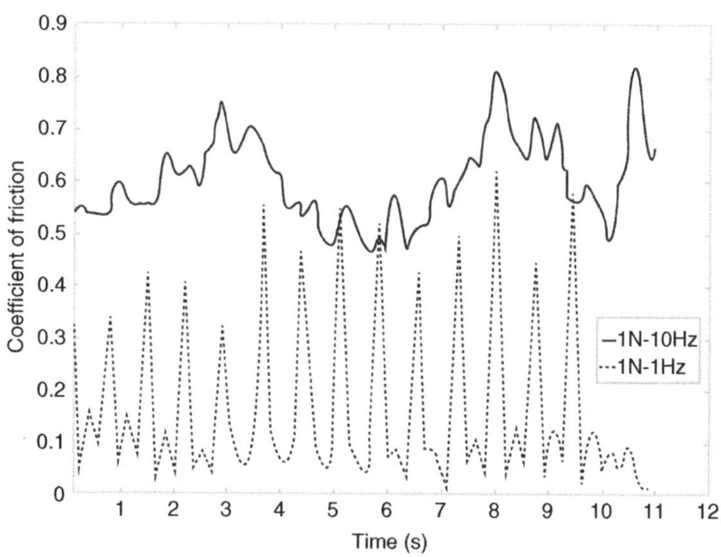

Figure 9.11 Coefficient of friction versus time for different fretting frequencies [28].

also increased owing to stronger vibrations. This action removed the gold plating more rigorously and caused contact resistance to increase quickly in high-frequency cases.

Siddaiah et al. [28] noted the effect of frequency on coefficient of friction and wear. Figures 9.11 and 9.12 show that fluctuation in the coefficient of friction was less for higher frequencies. However, the value of the coefficient of friction was high for higher frequencies.

Both results can be explained by increased exposure time of contact spots to the atmosphere. This gives more time for the exposed metal surface to become oxidized. Because oxidation is a time-dependent process, more oxidation will occur during longer exposure, and the contact spots will be covered with the oxidizing layer. This action builds up the contact resistance significantly.

Figure 9.12 shows that the distance from the initial position (i.e. the wear depth) is high for higher frequencies. The wear depth indicates the amount of degradation due

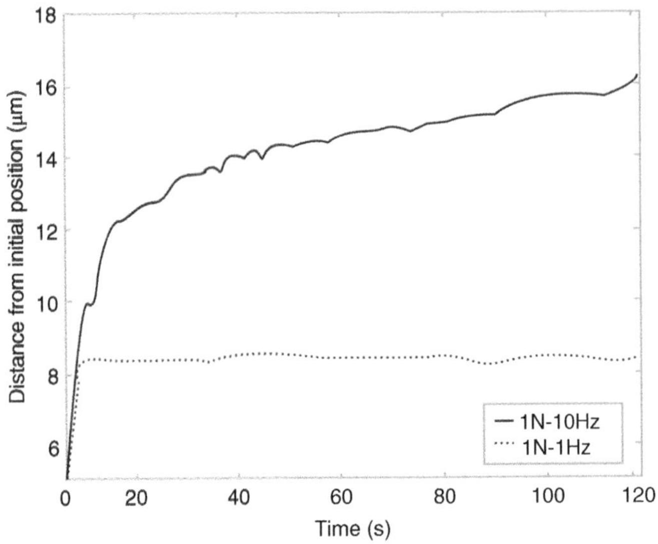

Figure 9.12 Distance from initial position (wear depth) versus time for different fretting frequencies [28].

to wear; hence, it is a good parameter for judging the effect of the considered factor on wear. More wear is observed at higher frequencies.

9.1.2.3 Slip Amplitude

The effect of fretting amplitude is analyzed by observing the number of cycles required to attain a predefined value of contact resistance. Figure 9.13 shows a compiled graph of the results of various studies on different contact materials. Unlike frequency, the increase in slip amplitude causes contacts to take less time to attain the predefined value of contact resistance. This means the predefined contact resistance would be attained more quickly at higher amplitude than it would have been attained at lower amplitudes, and therefore the connector's life span is reduced, causing failure to occur in less time at higher slip amplitude. A lower slip amplitude means fewer contact spots are exposed to atmosphere each cycle, and thereby the area exposed for oxidation is reduced and the contact resistance increase is delayed. On the other hand, for the same connector, a higher slip amplitude means more contact spots are exposed to atmosphere. Therefore, a larger exposed area is vulnerable to oxidation as shown in Figure 9.14. As a result, more oxidation layer is formed on a larger exposed area, which insulates the contact area and increases the contact resistance. This causes failure to occur earlier than the low slip amplitude case.

9.1.2.4 Electric Current

Conducting electric current is a major function of an electric connector, but, as many researchers have reported [7, 29, 30], the variation of electric current load affects the performance of connectors under fretting. Higher electric current has a direct effect on contact resistance. The effects of 10, 100, 500, and 1000 mA electric current were observed in contacts subjected to fretting motion (Figure 9.15). Ren et al. [9] compared

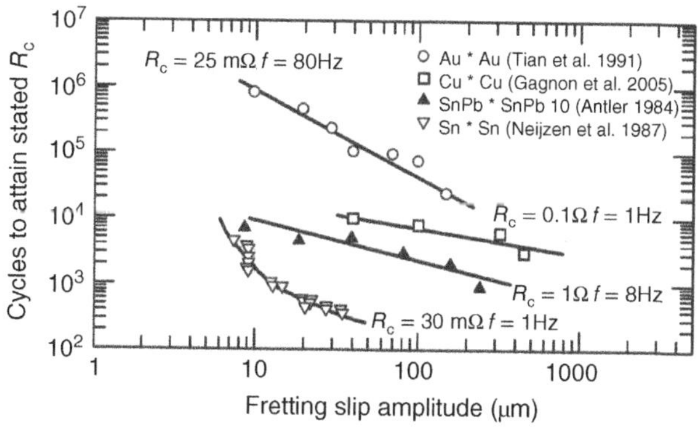

Figure 9.13 Effect of fretting slip amplitude on number of cycles required to attain predefined contact resistance [11].

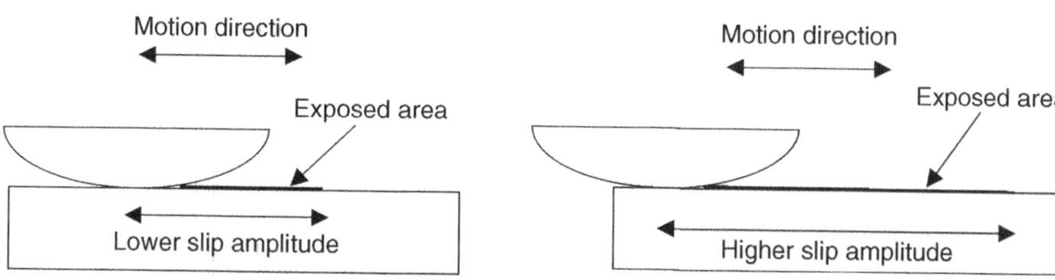

Figure 9.14 Schematic diagram of contact areas with lower (left) and higher (right) slip amplitudes.

the effects of current by maintaining all other parameters constant, such as fretting frequency, normal load, and amplitude. The thick horizontal line shows a threshold contact resistance value of 100 mΩ. It can be easily seen from the figure that the number of cycles required to reach the threshold value of contact resistance is the lowest (8230 cycles) for the lowest current of 10 mA and increases gradually as the current is increased. The highest number of cycles (34 980 cycles) was recorded for 1000 mA. This proves that for given conditions, a connector with 10 mA current fails earlier than a connector with 1000 mA current. Failure due to an increase in contact resistance is delayed as the current is increased. The reduction in contact resistance can be because of higher electric potential in the case of high currents, which helps in penetrating thicker oxide layers easily and electric heating, which causes a larger contact area by softening the metal.

Bowden and Williamson [31] showed that for a given current there is a critical constriction resistance value for every material above which it causes changes in the contact region. Heat generation causes the yield pressure of a material to drop and the contact area to increase. The value of critical constriction resistance is inversely proportional to the current.

9.1.3 Environmental Factors That Affect Fretting

The factors that are induced by the environment and are not part of the system are classified under the third category. Humidity and temperature are major factors in this category.

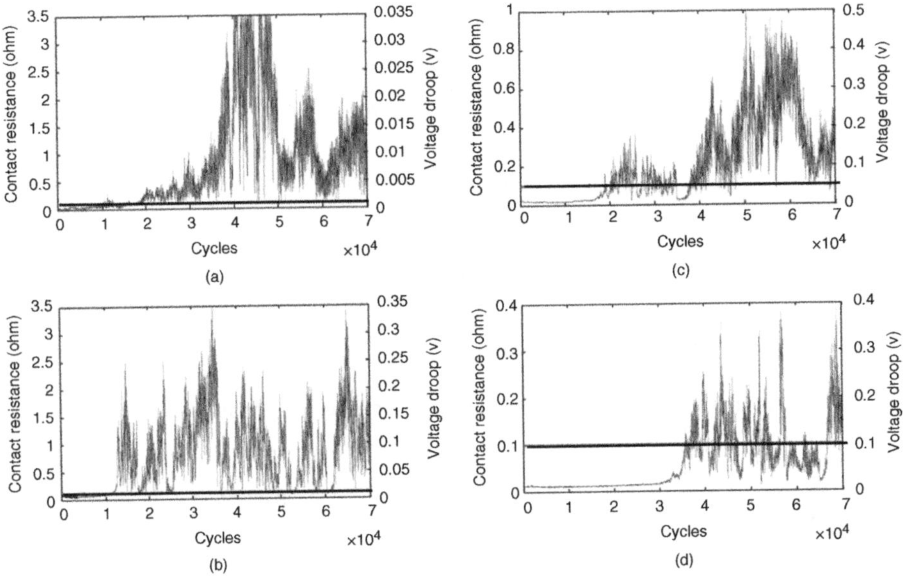

Figure 9.15 Effects of (a) 10, (b) 100, (c) 500, and (d) 1000 mA electric current on contact resistance [9].

9.1.3.1 Humidity

The relative humidity (RH) around a connector affects it significantly because it can change the surface mechanical properties, the interaction between chemical reaction rates, and the physical characteristics of the debris [5]. Since all these phenomena are associated with fretting, RH is one of the most critical factors to control. Fretting damage typically decreases as RH increases, which means that damage due to fretting is less severe in saturated air than in dry air. Wright's [32] investigations showed that damage decreased as RH increased, but at 100% RH, damage increased slightly. Braunovic [11] explained that humidity can affect connectors either favorably or adversely. The water content in a highly humid atmosphere can reduce fretting wear damage by acting as a lubricant between the surfaces and as a result reducing the wear rate, adhesion, and friction. Wright's observations can be attributed to these favorable effects of humidity. On the other hand, humidity can have an adverse effect by inducing anodic corrosion reactions, causing corrosion products to be trapped in surfaces and increase the wear rate.

9.1.3.2 Temperature

Electronic systems can face significant variation in temperature. Connectors operating near a heat source such as an automotive engine can be subjected to very high temperatures, whereas connectors in systems in space can be subjected to very low temperatures. The effect of temperature on fretting failure of connectors should therefore be studied. Temperature variation can affect connectors in two ways. The first way is associated with the change of oxidation or corrosion rates, and the second way is associated with the change in the mechanical properties of connectors [11]. Figure 9.16 shows the variation of cycles to failure of tin-plated copper alloy [30]. Failure was defined as a contact resistance of 20 mΩ. It is evident that the number of cycles to failure gradually decreased as temperature increased from 20 °C to 60 °C. The reduced number of cycles can be

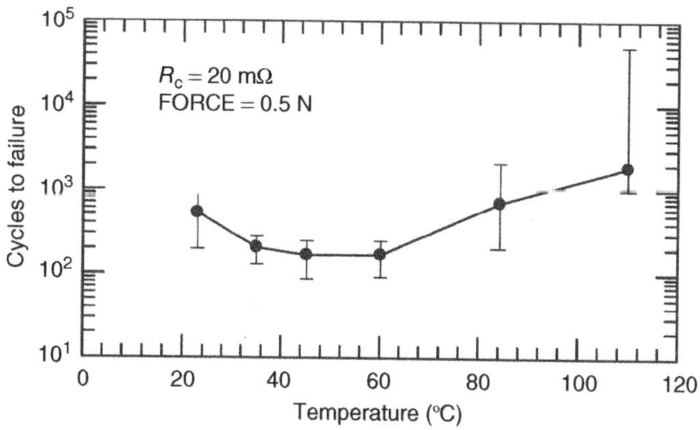

Figure 9.16 Effect of temperature on cycles to attain 20 mΩ by tin-plated copper alloys [30].

attributed to an increase in the oxidation rate because of temperature increase. As previously discussed in Section 9.1, oxidation promotes failure by insulating the contact surfaces and thereby causing contact resistance to rise quickly. At 60 °C, however, the trend of variation reversed, and the cycles to failure increased as temperature increased above 60 °C. This showed that connector performance is better at high temperatures. The latter stage of increased cycles to failure can be associated with the effect of high temperature on the surface of connectors, which softens the tin metal and expands the contact area, resulting in a reduction in the deleterious effect of oxides.

The behavior of all connectors at higher temperature may not be the same because of the difference in softening temperature. In the case of metals with a higher softening temperature, intermetallic compounds can form well before the softening temperature of metal is reached.

9.1.3.3 Dust

Quartz particles are abundant in dust, especially in industrial settings. Figure 9.17 shows the effect of quartz particles on gold-plated phosphor bronze connectors. It compares the performance of connectors with quartz pollutants to the performance of the same connectors without pollutants. Quartz particles are hard, and accumulation of these particles in between contacting surfaces significantly affects the wear rate. Some quartz particles were pushed away during fretting motion, which accumulated around the contact area as shown in Figure 9.18. But other particles were trapped inside and were crushed down to become embedded into the contact surface. Owing to the higher hardness of these embedded particles, the gold layer coating wore out quickly, exposing the inner metal to oxidation. This helps in formation of more debris, which consequently improves the wear rate. Insulating in nature, quartz particles either increase contact resistance by being present in between contacts or they degrade the gold layer by abrasive wear, which catalyzes oxidation of inner metals. Contact resistance in the case of connectors with quartz particles was initially high; however, in the later stages of cycling it almost was the same as or less than connectors without quartz particles. This behavior can be attributed to the extra wear of gold plating due to quartz particles in the initial stages, which increases contact resistance quickly. However, after enough fretting cycles

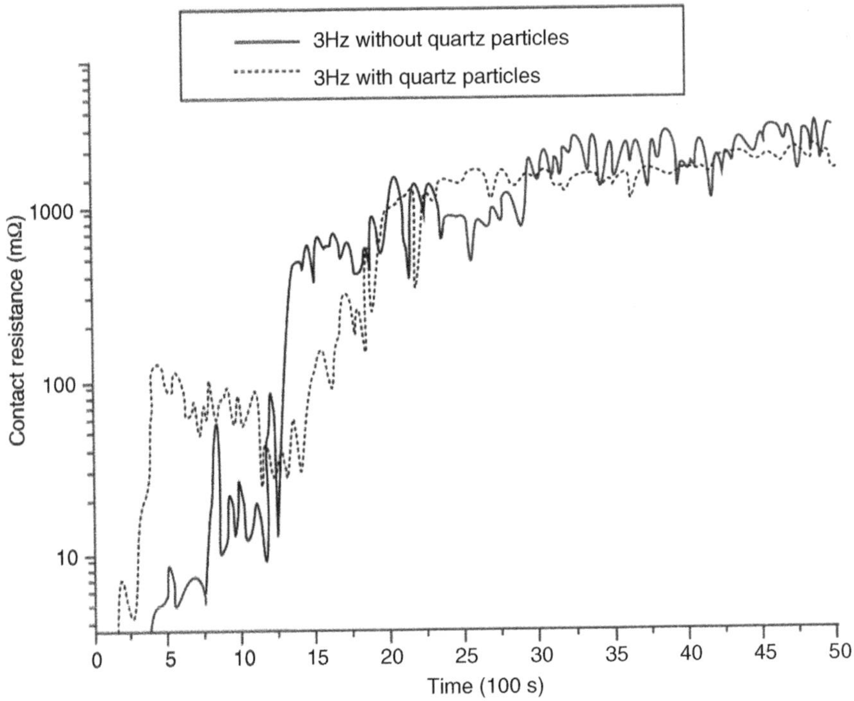

Figure 9.17 Effect of quartz particles on contact resistance [27].

Figure 9.18 Contact surface morphology with quartz particles [27].

the effect of metal oxides due to normal fretting becomes significantly higher so that there is no difference between contacts with and without quartz particles.

Kong and Swingler [27] also studied the combined effect of humidity and pollutant particles. Figure 9.19 shows the effect of quartz particles and high humidity on contact resistance compared with quartz particles in comparatively dry atmosphere. Higher

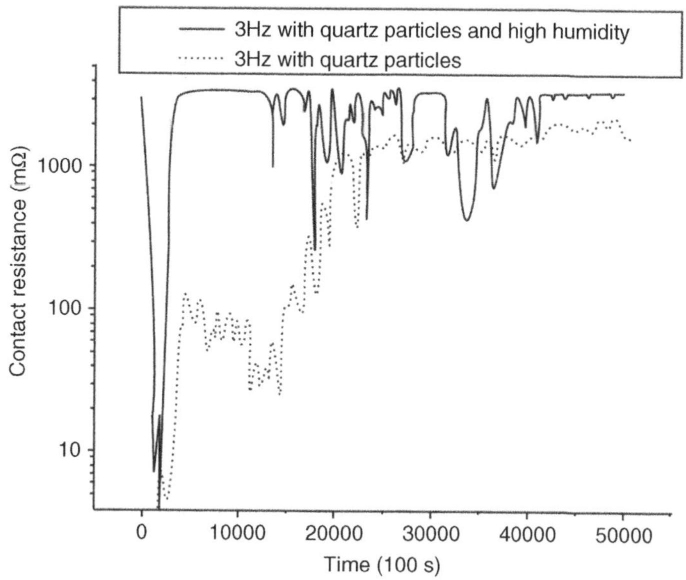

Figure 9.19 Comparison of combined effect of humidity and quartz particles with the effect of quartz particles on contact resistance [27].

contact resistance was observed throughout the cycle life of connectors for the combined effect.

As shown in Figure 9.19, abnormally high contact resistance at the initial phase is due to the mixing of many quartz particles in water on the contact surface, forming a mixture that has high resistance. After the initial stage, the drop in contact resistance can be attributed to removal of this mixture between the surfaces due to fretting motion. This establishes a gold-to-gold contact, which has low contact resistance. After this the contact resistance again rises sharply, reaching its maximum value, after which it fluctuates significantly in the range of 12 500 cycles to 40 000 cycles. The contact resistance remains stable to its maximum value only after 40 000 cycles. During fretting the particles between the surfaces are crushed and swept away. This action continuously varies the contact area available for conduction and creates fluctuations in contact resistance. After tens of thousands of cycles, the gold coating is worn out, exposing the inner metal to form high contact resistance. This explains why the combined effect of humidity and pollutant particles is severe and has a serious impact on connector reliability.

9.2 Reducing the Damage of Fretting

The most effective way to reduce fretting damage is by preventive measures. Micro-motion between surfaces and oxidation of contacting surface are major factors that affect fretting damage. Reduction of micro-motion and oxidation is targeted using various methods. Improvement in connector geometry design can help in avoiding relative motion between connecting surfaces to a certain extent. A noble metal such as gold is commonly used as plating on base metals to avoid oxidation of surfaces. Lubricating the moving surfaces has also found to be beneficial because it reduces friction between moving surfaces and mitigates wear; however, the selection of the right lubricant is a critical factor. If a suitable lubricant is not used, it can affect the contact performance adversely by increasing contact resistance. All three of these factors are discussed in the following sections.

9.2.1 Lubrication

Lubrication is one of the most effective techniques for reducing fretting failure in connectors. Lubrication can improve contact resistance behavior significantly. The type and amount of lubricant required depends on connector failure mechanisms and materials. The physical state of the lubricant can also be an important factor as lubricants in the form of fluids and greases are used more often to reduce the fretting effects than solid lubricants such as wax. The reason is inability of wax to hold between surfaces under fretting motion; it can easily get displaced. The oxides formed during fretting are converted into wear debris. Lubricants can help reduce the rate of formation of oxides and therefore reduce the wear rate. Reducing the wear rate helps stabilize contact resistance. Antler [33] showed the effect of lubricant in gold-plated contacts by comparing the performance of lubricated contacts with that of unlubricated contacts. The lubricant used was polyphenylether with 50 nm thickness. The peaks in contact resistance were significantly reduced, and the occurrence of maximum contact resistance was delayed as well. To analyze the effect of lubricant on surface wear, Antler [33] also studied surface damage degradation of gold-plated cobalt contacts with and without lubrication. In contacts without lubricant, the gold plating was worn away, exposing the nickel underplating, whereas in the case of lubricated contacts, the surface of the gold plating was just burnished, and wear was negligible.

Lubricants can also help stabilize degradation due to formation of frictional polymerization. Here, the effectiveness strongly depends on the thickness of the lubricant. Figure 9.20 shows the dependence of contact resistance behavior on quantity of lubricant. The figure compares palladium contacts submerged into and withdrawn from a solution of lubricants having a concentration of 0.5%, 1%, 2%, and 5% in the volatile solvent. A slight improvement can be seen for 0.5%, but as concentration increases, especially above 2%, the improvement is prominent. The maximum contact resistance of a palladium contact was reduced by three orders of magnitude, and the initial contact resistance rise was also significantly delayed. As discussed in Section 9.1.1.4, organic compounds increase the rate of formation of frictional polymers; hence, off-course organic lubricants will lead to formation of more frictional polymers than contacts without organic lubricants. Even after the presence of excessive frictional polymers, lubrication was still able to dominate their effect by stabilizing contact resistance because excess unreacted lubricants easily disperse the frictional polymers formed. Also, polymers can be less tough if the lubricant incorporated in their matrix produces a gelatinous solid. Lubricants used in the frictional polymerization case should have low migration tendencies and volatility because lubricants should always be present in between contacting surfaces.

9.2.2 Improvement in Design

Fretting arises due to motion between contacts; thus, the elimination of relative motion by design is the ideal solution to prevent fretting. Even this motion cannot be complete

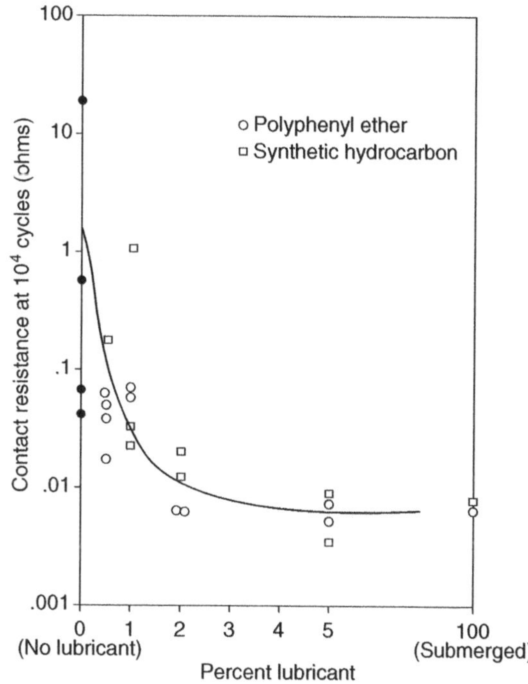

Figure 9.20 Effect of quantity of lubricant on contact resistance [33].

removed, it can be significantly reduced by additional elastic elements. Horn et al. [34] used a spring element, which acted in the direction of actual fretting motion and could significantly reduce the movements by vibration or thermal cycling.

The effects of geometry have also been studied, but the available literature is limited. Dome-shaped and wedge-shaped contacts were studied and compared to flat contacts [35]. Wedge shapes having an angle of 90° and 120° were compared against domed-shaped contacts. For tin-alloy-plated surfaces, the wedge-shaped contacts showed more stability in contact resistance than the domed-shaped contacts. The fretting motion was mechanically or thermally induced along the right angle of line of contact of the wedge.

9.2.3 Coatings

Stable contact performance can be achieved by using noble metals such as platinum, silver, or gold plating on the contact surface. Electronic and electrical industries use protective and wear-resistant coatings abundantly. These coatings not only have economic benefits but also have useful functions such as corrosion and wear protection and diffusion barriers [11]. The effects of silver, gold, and tin coatings on copper alloys can be observed in Figure 9.21b. The contact resistance behavior of silver-plated contacts was the best followed by gold- and tin-plated. The characterization of contact resistance variation behavior is difficult because it depends on multiple factors simultaneously and not only on the coating material used. However, most of the time, contacts coated with protective metals perform better than bare-metal contacts, as observed in Figure 9.21a. Solid palladium fretted against palladium performs the worst. In this case, contact resistance rose quickly and quite early, whereas palladium fretted with

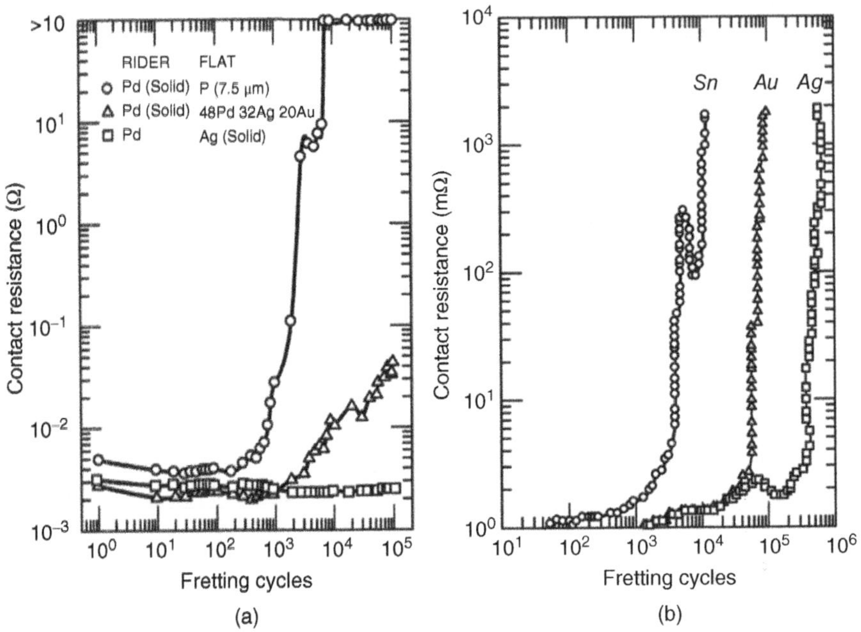

Figure 9.21 Effects of different coating materials on contact resistance: (a) palladium connector against PdAgAu, silver-plated palladium; (b) tin-plated copper against tin-, gold-, and silver-plated tin contact [11].

PdAgAu-plated palladium shows considerably low contact resistance. Palladium fretted against silver-plated palladium showed almost constant resistance behavior over the period of observation.

Despite the advantages of coatings on connectors, there are more influential factors that affect connectors in practical conditions, as described in previous sections. Hence, the sole use of silver or gold does not guarantee improved connector performance.

References

1 Swingler, J. (2000). The automotive connector: The influence of powering and lubricating a fretting contact interface. *Proceedings of the Institution of Mechanical Engineers, Part D: Journal of Automobile Engineering* 214 (6): 615–623. https://doi.org/10.1243/0954407001527484.

2 Timsit, R.S. (March 1990). A possible degeneration mechanism in stationary electrical contacts. *IEEE Transactions on Components, Hybrids, and Manufacturing Technology* 13 (1): 65–68. https://doi.org/10.1109/33.52850.

3 ASTM G40-17. (2017). *Standard Terminology Relating to Wear and Erosion*. ASTM.

4 Ahmad, Z., Institution of Chemical Engineers (Great Britain) (2006). *Principles of Corrosion Engineering and Corrosion Control*, 1e, 242–243. Boston, MA: Butterworth-Heinemann.

5 Braunovic, M., Myshkin, N.K., and Konchits, V.V. (2006). *Electrical Contacts: Fundamentals, Applications and Technology*. Boca Raton, FL: CRC Press.

6 Uhlig, H.H. (1954). Mechanism of fretting corrosion. *Journal of Applied Mechanics* 21: 401–405.
7 Waterhouse, R.B. and Taylor, D.E. (1977). Fretting debris and delamination theory of wear. *Wear* 29: 337–443.
8 Suh, N.P. (1973). The delamination theory of wear. *Wear* 25: 111–124.
9 Ren, W., Wang, P., Song, J., and Zhai, G. (2014). Effects of current load on wear and fretting corrosion of gold-plated electrical contacts. *Tribology International* 70: 75–82.
10 Slade, P.G. (ed.) (1999). *Electrical Contacts: Principles and Applications*, 46–51. New York: Marcel Dekker.
11 Braunovic, M. (2009). Fretting in electrical/electronic connections: A review. *IEICE Transactions on Electronics* 92 (8): 982–991.
12 Park, Y.W., Joo, H.G., and Lee, K.Y. (2010). Effect of intermittent fretting on corrosion behavior in electrical contact. *Wear* 268 (3–4): 353–360.
13 Park, Y.W., Bapu, G.R., and Lee, K.Y. (2009). The influence of current load on fretting of electrical contacts. *Tribology International* 42 (5): 682–689.
14 Antler, M. (1985). Electrical effects of fretting connector contact materials: A review. *Wear* 106 (1–3): 5–33.
15 Antler, M. (1982). Fretting of electrical contacts: An investigation of palladium mated to other materials. *Wear* 81 (1): 159–173.
16 Crossland, W. and Murphy, P. (1974). The formation of insulating organic films on palladium-silver contact alloys. *IEEE Transactions on Parts, Hybrids, and Packaging* 10 (1): 64–73.
17 Antler, M. (1981). Tribology of metal coatings for electrical contacts. *Thin Solid Films* 84 (3): 245–256.
18 Hall, E.O. (1954). Variation of hardness of metals with grain size. *Nature* 173 (4411): 948–949.
19 Noh, H.J., Kim, J.W., Lee, S.M., and Jang, H. (2017). Effect of grain size on the electrical failure of copper contacts in fretting motion. *Tribology International* 111: 39–45.
20 Iwabuchi, A. (1991). The role of oxide particles in the fretting wear of mild steel. *Wear* 151 (2): 301–311.
21 Totten, G.E. (1992). *ASM Handbook, Vol. 18: Friction, Lubrication, and Wear Technology*. ASM International.
22 Antler, M. (1999). Tribology of electronic connectors: contact sliding wear, fretting, and lubrication. In: *Electrical Contacts: Principles and Applications* (ed. P.G. Slade), 309–402. New York: Marcel Dekker.
23 Braunovic, M. (1998). Effect of contact load on the contact resistance behavior of different conductor and contact materials under fretting conditions, *Proceedings of the 19th International Conference on Electrical Contacts*, Nuremberg, Germany (Sept. 21–23, 1998). IEEE. pp. 283–287.
24 Lee, A. (1999). Low power commercial automotive and appliance connections. In: *Electrical Contacts: Principles and Applications* (ed. P.G. Slade), 279–307. New York: Marcel Dekker.
25 Gagnon, D., Braunovic, M., and Masounave, J. (2005). Effect of fretting slip amplitude on the friction behaviour of electrical contact materials. *Proceedings of the 51st IEEE Holm Conference on Electrical Contacts*. IEEE. pp. 186–195.

26 Castel, P., Monet, A., and Caraballeira, A. (1984). Fretting corrosion in low level electrical contacts: A quantitative analysis of the significant variables. *Proceedings of the 12th International Conference on Electrical Contacts*, Chicago. IEEE. pp. 75–81.

27 Kong, Z. and Swingler, J. (2017). Combined effects of fretting and pollutant particles on the contact resistance of the electrical connectors. *Progress in Natural Science: Materials International* 27 (3): 385–390.

28 Siddaiah, A., Kasar, A.K., Khosla, V., and Menezes, P.L. (2019). In-situ fretting wear analysis of electrical connectors for real system applications. *Journal of Manufacturing and Material Processing* 3 (2): 47.

29 Bock, E. M., and Whitely, J. H. (1974). Tunnel film resistance utilizing non-linear constriction resistance measurements. *Proceedings of the 20th IEEE Holm Conference on Electrical Contacts*. IEEE. pp. 45–49.

30 Lee, A., Mao, A., and Mamrick, M. S. (1988). Fretting corrosion of tin at elevated temperatures. *Electrical Contacts, Proceedings of the 34th Meeting of the IEEE Holm Conference on Electrical Contacts*. IEEE. pp. 87–91.

31 Bowden, F.P. and Williamson, J.B.P. (1958). Electrical conduction in solids I. Influence of the passage of current on the contact between solids. *Proceedings of the Royal Society of London, Series A, Mathematical and Physical Sciences* 246 (1244): 1–12.

32 Wright, K.H.R. (1953). An investigation of fretting corrosion. *Proceedings of the Institution of Mechanical Engineers* 167 (1b): 556–574.

33 Antler, M. (1985). Survey of contact fretting in electrical connectors. *IEEE Transactions on Components, Hybrids, And Manufacturing Technology* 8 (1): 87–104.

34 Horn, J., Kourimsky, F., Baderschneider, K., and Lutsch, H. (1995). Avoiding fretting corrosion by design. *AMP Journal of Technology* 4: 4–7.

35 Garte, S. M. (1976). The effect of design on contact fretting. *Proceedings of the Holm Seminar on Electrical Contacts*, Illinois Institute of Technology, Chicago (Sept. 21–23). IEEE. pp. 65–70.

10

Testing

Bhanu Sood[1] and Michael G. Pecht[2]

[1] *NASA Goddard Space Flight Center, Greenbelt, MD, USA*
[2] *Center for Advanced Life Cycle Engineering (CALCE), University of Maryland, College Park, MD, USA*

Testing is conducted to verify that the connectors will properly perform over time (reliability) per the application requirements. Testing by the manufacturer usually involves performance testing per the datasheet requirements. However, the customer may have applications that require performance under unique operational and environmental conditions. In general, electrical tests for connectors include dielectric withstanding voltage, low-signal-level contact resistance, insulation resistance, contact resistance, and standing wave ratio. Environmental tests include humidity, temperature, and contaminating conditions, as well as mechanical stress environments including vibration, mechanical shock, and durability cycling exposures. To assure the performance and reliability of connectors, manufacturers usually do qualification testing, which includes a series of tests that specify connectors to certain requirements. Table 10.1 shows an example of qualification testing.

10.1 Dielectric Withstanding Voltage Testing

The purpose of the dielectric withstanding voltage test is to verify that the connector can operate at its rated voltage and will be able to withstand momentary over-potentials due to switching and/or surges. Also known as a high-potential, over-potential, or dielectric-strength test, this test differs from a dielectric-breakdown test. In cases where the applied voltage causes a sudden breakdown of the insulation material that provides a flow of current, the insulation is determined to be dielectric-insufficient. While the dielectric withstanding voltage test is widely used, the real objective of the test is often misunderstood, which may lead to incomplete testing or misleading test results. Simply stated, the test voltage breaks down the insulating properties of the material [1, 2]. The dielectric breakdown of an insulating material is a complex physical phenomenon but it may be generally characterized as a sudden change in the resistance of the insulation under test due to the applied voltage. The mechanism of dielectric breakdown begins with the application of a strong electric field to the insulating material by a high voltage, which in this case would equate to the dielectric withstanding voltage set in the test.

Electrical Connectors: Design, Manufacture, Test, and Selection, First Edition. Edited by San Kyeong and Michael G. Pecht.

Different materials require different levels of electric field for dielectric breakdown to occur. Metals and other conductors have free electrons without the application of any electric field, but insulating materials typically require a high electric field to allow electrical current to flow. When the electric field is sufficient, it energizes the insulator's electrons until they gain enough energy to cross the bandgap and move into the conduction band, dramatically increasing the conductivity of the material. This transformation is called "dielectric breakdown," and the electric field necessary to start the breakdown is called the "dielectric strength" or "breakdown strength." Thus, dielectric breakdown is a dramatic and sudden increase in the conductivity of a material due to an applied voltage.

The specimens required for the dielectric withstanding voltage test are usually a plug, receptacle, or mated combination as specified in the individual connector specification. The equipment is a voltage source adjustable to within a certain tolerance of the required test voltage (DC or root-mean-square [RMS]) and capable of delivering a minimum current of approximately 1 mA. The applied voltage can be either alternating current (AC) or direct current (DC). The voltage should be applied across the two closely spaced contacts as well as between the connector shell and the contacts closest to the shell. In the case of coaxial connectors, the test voltage is applied between the inner and outer conductors. The voltage is usually ramped up uniformly from zero to the rated value. The test can be performed at various pressures depending on the application for which the connector is going to be used. For example, in undersea applications, the pressure on the connectors will be high [3].

All insulation will break down at some specific voltage. Above this critical voltage, the current flow will increase catastrophically. Per methods such as the MIL-STD-1344A, Method 3001.1, the dielectric withstanding voltage is 75% of the minimum breakdown voltage; it is suggested that the operating rated voltage should be one-third of the dielectric withstanding voltage. During measurement, usually an alternating potential is applied between the adjacent contacts, the voltage is increased from zero to the specified value as uniformly as possible at a specified rate and, the test voltage is maintained at the specified value for one minute to see whether the material breaks down. This method is often called the "step-by-step test." In another method, the short-term dielectric withstanding voltage is obtained by steadily increasing the test voltage from zero to breakdown. The magnitude of the test voltage is expressed as its RMS value. Since the barometric pressure greatly affects the withstanding voltage characteristics of the connector, the dielectric withstanding voltage is usually specified for sea-level applications. The breakdown voltage is influenced by the dielectric strength of the insulator, duration of the applied voltage, thickness of the sample, temperature, surrounding medium, and frequency of the applied voltage. The dielectric strength is a property of an insulator, expressed as the maximum voltage gradient that causes insulator breakdown. The dielectric strength of insulators can be obtained from many books.

10.2 Insulation Resistance Testing

The purpose of the insulation resistance test is to assess the resistance offered by various insulation materials and seals of a connector or coaxial contacts to the DC voltage that tends to produce a leakage current through or on the surface of these contacts [3].

The test is designed to assess the quality of insulation material used in the connector by measuring the amounts of leakage current between the conductors of a connector. Electrical insulation begins to age as soon as it is installed, and this deterioration can affect the performance of the connector. Insulation resistance testing is also used as quality control during production of the connector and preventive maintenance tasks and as a troubleshooting tool.

The insulation resistance is measured using a mega-ohm bridge or a mega-ohm meter. Voltage is applied to the conductors of the connector, – the voltage is always lower than the voltage required for performing the dielectric testing. The current flowing through the conductors of the connector is then measured. Typical values of insulation resistance, as measured by a high-resistance meter, such as the Keysight 4339B DC High-Resistance or the B2980A Series Femto/Picoammeter and Electrometer/High Resistance Meter[1] for the insulator of a connector, range from mega-ohms to tera-ohms. Temperature and humidity can affect the value of the insulation resistance and secondarily the value of the leakage current. These factors may significantly affect the measurement result. Temperature causes the insulation resistance value to vary quasi-exponentially. If the end-use condition of the connector is going to include application conditions with elevated temperature or humidity, then those environmental requirements need to be incorporated in the test protocol.[2]

The insulation resistance test is performed on a sample of the connector receptacle or mated connector as specified in the individual connector specification. In some cases, the connector or receptacle is cleaned prior to connections or the test. The cleaning process ensures that the contacts are free from contaminants that can influence the insulation resistance; contaminants can include moisture, excess dust, oil, or other surface contaminants. If the receptacle or the connector has been subjected to environmental conditions, the tests should be performed rapidly to prevent the equalization of the samples with the ambient conditions in the laboratory. Typically, a period of a half-hour to three hours after removal from the chamber is recommended.[3] The resistance is typically measured either between the individual pairs of immediately adjacent contacts pins and the shell or between the hardware to which the connector is attached and the closest contact. Once the connections are made, a test potential, typically 500 VDC, is applied for a two-minute period. The insulation resistance is measured immediately after the electrification period.

Some standards for measuring the dielectric breakdown constant and insulation resistance are listed here for reference:

American Society for Testing and Materials (ASTM)-D-149: Standard test method for dielectric breakdown voltage and dielectric strength of solid electrical insulating materials at commercial power frequencies

Electronic Industries Association (EIA)-364-TP20: Withstanding voltage test procedure for electrical connectors

MIL-STD-1344, Method 3001: Dielectric withstanding voltage test method for electrical connectors

1 B2980A Series Femto/Picoammeter and Electrometer/High Resistance Meter datasheet.
2 TP-29C. *Contact Retention Test Procedures for Electrical Connectors*, EIA/ECA-364-29C.
3 IPC Test Document. *IPC-TM-650 3.6, Insulation Resistance, Connectors.*

Table 10.1 Qualification testing sequence for slot connectors required by Intel.

	Test Group									
Test	**1**	**2**	**3**	**4**	**5**	**6**	**7**	**8**	**9**	**10**
Visual inspection	1, 11	1, 6	1, 6	1, 9	1, 8	1, 5	1, 4	1	1	1
Dimensional verification	2	2	2	2	2	2	2	2	2	2
Contact resistance	4, 6, 9	3, 5	3, 5		3, 5, 7					
Insulation resistance				3, 7						
Dielectric withstand voltage				4, 8						
Vibration	8									
Shock	7									
Durability	5									
Mating force	3									
Un-mating force	10									
Thermal shock										
Temp. cycling				5	4					
Temp. life		4		6	6					
Mixed flowing gas			4							
Resistance to solder heat							3			
Porosity								3		
Plating thickness								4		
Solvent resistance									4	
Normal force								5		
Solderability								6		
Contact retention									3	
Max. force on connector						3				
Contact back out wipe						4				
Current rating										
Substrate visual inspection	12									3
Sample size	16	8	4	8	8	4	4	4	4	4

ASTM-D-257: Standard test method for DC resistance or conductance of insulating materials
EIA-364-TP20: Insulation resistance test procedure for electrical connectors
MIL-STD-1344, Method 3003: Insulation resistance test method for electrical connectors

Table 10.1 shows an example of qualification testing.

10.3 Contact Resistance Testing

The contact resistance test is used to identify problems with loose connections, eroded contact surfaces, and contaminated or corroded contacts. It is also used to measure the contact resistance between the mated connectors contact attached to the wire. The test is performed by measuring the milli-voltage drop across the connector when it is carrying a rated current. The resistance is calculated from the voltage and current data [3].

A connector introduces extra contact interfaces between electronic devices. To maintain a consistent and reliable contact interface, a contact force should be applied. The applied contact normal force depends on the contact resistance that can be achieved. When two contacts are mated, the work of an external force, such as the insertion force, causes contact deflection. Consequently, a contact normal force is exerted on the contacts, and a contact interface is created. This interface is usually not as tight as it seems to be, from a microscopic view. The surface roughness, surface insulation film, contamination, and dust in the contact interface prevent real metallic contact. The effective contact area is usually a fraction of the total contact area; this fraction is determined by the contact manufacturing process, contact finish, and contact cleanness. The surface roughness is usually described by "asperities." Asperities are the protruding spots on a surface; during mating, only asperities actually come into contact. Due to their small size (the radii are measurable in micrometers), the asperities deform plastically even at low applied loads. With increased loads, the asperities deform further, the contact area is enlarged constantly, and more asperities touch each other. The number of asperities mainly depends on the surface roughness, material hardness, and magnitude of contact normal force. The limited contact area results in a contact resistance called "restriction resistance": the current flow is restricted to flowing through the asperities. Figure 10.1 shows a schematic of the contact interface, asperities, and restricted current flow.

The constriction resistance as a function of the number, combined area, and distribution of multiple asperities is described by the following equation:

$$R_C = \frac{\rho}{na} + \frac{\rho}{D} \tag{10.1}$$

where ρ is the resistivity of the contact material (assuming the same materials), n is the number of asperities, a is the diameter of the asperity, and D is the diameter of the area over which the contacts are distributed.

As mentioned, the area of the asperities is determined by the applied load; thus, the constriction resistance can be expressed in terms of the contact normal force [4]:

$$R_C = k\rho\left(\frac{H}{F_n}\right)^{1/2} \tag{10.2}$$

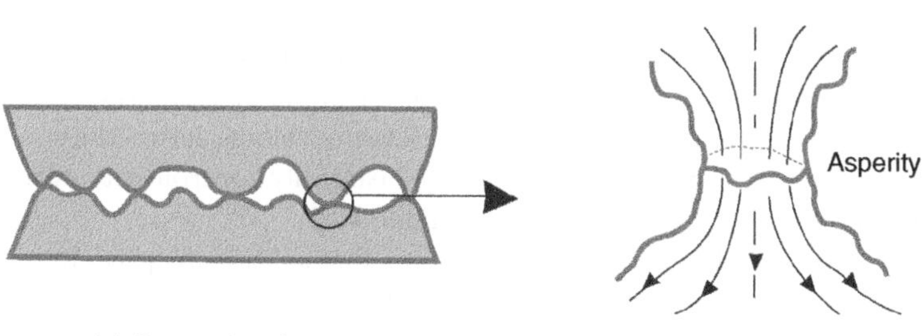

Figure 10.1 Schematic of (a) contact interface and (b) asperity.

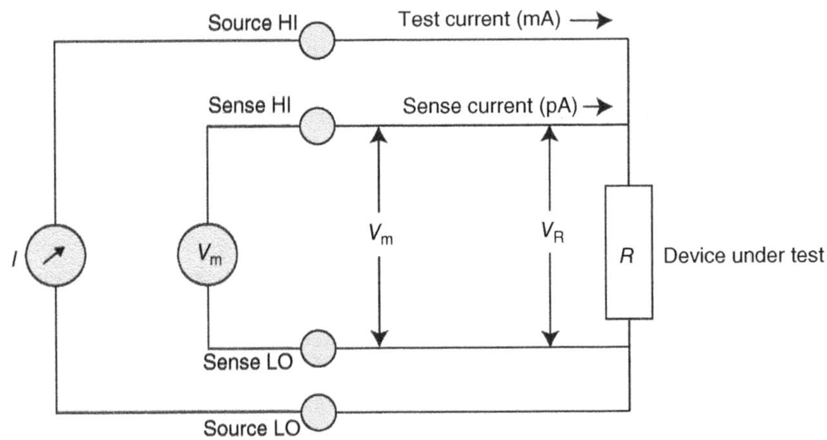

Figure 10.2 Schematic of four-wire measurements.

where k is a coefficient including the effects of surface roughness, contact geometry, and elastic/plastic deformation, which can be determined experimentally, H is the hardness of contact material, and F_n is the contact normal force.

In Eqs. (10.1) and (10.2), a pure metallic surface is assumed; however, in most applications, the conditions of the contact surfaces are not perfect; surface films may grow initially or develop gradually during connector application. Surface films may be displaced or disrupted completely or partially or remain intact, depending on the applied contact force, and applied bias, as well as film composition, structure, and thickness. Applied bias may cause the electrical breakdown of surface films. Applying a normal force may mechanically disrupt the oxide layers and expose the metallic contacts. The film composition, structure, and thickness depend on the contact finish and application environment. In principle, the overall contact resistance can be regarded as a combination of constriction resistance, due to asperity contact, and film resistance, due to the oxide or corrosion film accumulated on the contact surfaces.

The four-wire method is commonly used to measure contact resistance. A sourcing current is applied to flow through the contact interface, and the voltage across the interface is measured (Figure 10.2). The measurement can be made using a micro-ohmmeter, or a separate current source and voltmeter. With this configuration, the test current (I) is forced through the contact resistance through one set of test leads, while the voltage across the device under test is measured through a second set of leads called "sense" leads (the voltage is usually negligible, typically in the order of pA or less). Therefore, the voltage measured by the meter is essentially the same as the voltage across the contact resistance. The sourcing current and voltage must be carefully controlled to avoid the breakdown of interface insulation films (shown in Table 10.2).

The test condition is often called a "dry-circuit" condition; that is, with an open-circuit voltage less than 20 mV (the breakdown voltages of metal surface films are usually in the range of 30–100 mV) and a short-circuit current less than 100 mA, devices are tested in a manner that will not puncture any oxide film that may have built up on the contacts and changed other physical properties.

Table 10.2 Breakdown and melting voltage of surface insulation films.

Contact Material	Softening Voltage (mV)	Melting Voltage (mV)
Au	80	430
Ni	220	650
Cu	120	430
Ag	90	370
Sn	70	130

Some standards for testing contact resistance are listed here for reference:

EIA 364-TP04: Normal force test procedure for electrical connectors

ASTM B-539: Test methods for measuring contact resistance of electrical connections (static contacts)

EIA 364-TP06: Contact resistance test procedure for electrical connectors

MIL-STD 1344, Method 3004: Contact resistance test method for electrical connectors

International Electrotechnical Commission (IEC) 60512-2: Electromechanical components for electronic components: Basic testing procedures and measuring methods – Part 2: General examination, electrical continuity and contact resistance tests, insulation tests and voltage stress tests

The test involves a four-terminal resistance measuring technique wherein a measured and controlled test current is introduced into the sample using two "terminals" or connecting points, and two other points are selected on the sample across which a voltage drop is measured. This voltage drop, divided by the test current, is the effective overall resistance of the sample included between the voltage probes. The voltage-measuring points are chosen so as to measure as closely as possible the voltage drop due only to the contact resistance of the sample and to eliminate from the measurement as much as possible the resistance of the metal pieces comprising the contact and the resistance of the wires and connections used to introduce the test current into the sample. Usually, the test is run with two levels of test current. The choice of which level to use is governed by the application and requirements of the electrical connection being tested.

10.4 Current Rating Testing

Current rating, also called current carrying capacity, specifies the maximum current that a conductor can carry safely. Due to current flow, Joule heat is generated in the conductor, and the local temperature increases. The local temperature rise, compared with the ambient temperature, depends on the balance between Joule heat and heat dissipation to the neighboring regions. If the flowing current is too large, excessive heat will be generated and accumulated, and the local temperature will rise so high that it may surpass the maximum operating temperature of the insulators that separate the connector contacts. The maximum operating temperature of the connector housing determines the maximum current flow through a connector contact.

Although the current rating can be specified in terms of the transient current rating or overload current rating, the continuous current rating has generally been adopted by the connector industry. This current rating is based on the local temperature rise above the ambient, as induced by the current flow. It is commonly taken as the current that produces a 30 °C temperature rise, though other criteria can be used, such as a 10 °C T-rise. The criterion can be applied to both AC and DC current.

The current rating depends on the contact size, contact pitch, contact type, and heat-sinking capability. A large contact size assures a high current rating. The current rating for a connector may be 10 times larger than the current rating for an integrated circuit (IC) socket. The contact pitch for the socket contacts also limits the applicable current rating. A high-conductivity contact may be adopted to compensate for the reduction of contact size and pitch. High conductivity connector contacts not only generate less heat but also dissipate heat more effectively. The current rating performance will be greatly improved if a heat sink is attached, as a heat sink greatly enlarges the area of heat dissipation.

Some standards for current rating are listed here for reference:

EIA 364-TP70A: Temperature rise versus current test procedure for electrical connectors and sockets

IEC 60512-3: Electromechanical components for electronic equipment; Basic testing procedures and measuring methods – Part 3: Current-carrying capacity tests

IEC 60512-10-4: Electromechanical components for electronic components: Basic testing procedures and measuring methods – Part 10: Impact tests, static load tests, endurance tests and overload tests, Section 4: Test 10d: Electrical overload

The test is usually run with an electrical overload current flowing through contacts for a limited period of time ranging between 100 ms and 20 seconds. The test procedure is based on measuring the increase of temperature during the specified period of time when the electrical overload is applied to the contacts. Temperature increase over time is measured for an electrical overload current, specified as an integer multiple of the rated current, and an overload current-over-time diagram is plotted. Practice shows that, for limited periods of time, contacts can conduct a multiple of the maximum permissible current without damage to the contact area. At least three mated connectors are wired with the maximum wire size for the contacts with the shortest possible wire lengths compatible with the contact arrangement. The overload current is applied for a period of time and switched off as soon as the temperature has reached the upper temperature limit defined in the specification. The samples are then allowed to recover to room temperature before the next cycle is carried out. The test is repeated with different overload currents to plot temperature–time curves.

10.5 Electromagnetic Interference and Electromagnetic Compatibility Testing

Electromagnetic interference (EMI) is the effect of an electromagnetic phenomenon degrading the performance of a device, equipment or system. Related to EMI, electromagnetic compatibility (EMC) is the ability of a device, equipment or system to function satisfactorily in its electromagnetic environment without introducing intolerable electromagnetic disturbances to that environment. Products must not be susceptible, or, in other words, must be immune, to EMI, such as electrical fast transients (EFTs) and electrostatic discharge (ESD). Furthermore, systems may be required to operate in severe electromagnetic environments, such as during lightning strikes, and withstand threats such as electromagnetic pulses (EMPs). Because connectors act as a transmitting or receiving antenna, the grounding, filtering, and shielding on the connectors plays an important role in optimizing EMC. Connectors must comply with several domestic and international standards governing EMC prior to commercial release. Connector in a fully assembled configuration or in the complete terminated assembly are subjected to EMI testing. There are several techniques to help achieve EMC including grounding, shielding, balanced lines, and filtering. All have their place as well as their limitations, depending on the applications and the severity of the EMI problem. EMI and EMC testing are performed per the following standards:

IEC 61000-4-3: Electromagnetic compatibility (EMC) – Part 4-3: Testing and measurement techniques – Radiated, radio-frequency, electromagnetic field immunity test

IEC 61000-4-4: Electromagnetic compatibility (EMC) – Part 4-4: Testing and measurement techniques – Electrical fast transient/burst immunity test

IEC 61000-4-5: Electromagnetic compatibility (EMC) – Part 4-5: Testing and measurement techniques – Surge immunity test

IEC 61000-4-6: Electromagnetic compatibility (EMC) – Part 4-6: Testing and measurement techniques – Immunity to conducted disturbances, induced by radio-frequency fields

IEC 61000-4-7: Electromagnetic compatibility (EMC) – Part 4-7: Testing and measurement techniques – General guide on harmonics and interharmonics measurements and instrumentation, for power supply systems and equipment connected thereto

IEC 61000-4-8: Electromagnetic compatibility (EMC) – Part 4-8: Testing and measurement techniques – Power frequency magnetic field immunity test

IEC 61000-4-9: Electromagnetic compatibility (EMC) – Part 4-9: Testing and measurement techniques – Pulse magnetic field immunity test

IEC 61000-4-10: Electromagnetic compatibility (EMC) – Part 4-10: Testing and measurement techniques – Damped oscillatory magnetic field immunity test

IEC 61000-4-11: Electromagnetic compatibility (EMC) – Part 4-11: Testing and measurement techniques – Voltage dips, short interruptions and voltage variations immunity test

IEC 61000-4-12: Electromagnetic compatibility (EMC) – Part 4-12: Testing and measurement techniques – Oscillatory waves immunity test

IEC 61000-4-13: Electromagnetic compatibility (EMC) – Part 4-13: Testing and measurement techniques – Harmonics and interharmonics including mains signalling at a.c. power port, low frequency immunity tests

IEC 61000-4-14: Electromagnetic compatibility (EMC) – Part 4-14: Testing and measurement techniques – Voltage fluctuation immunity test

IEC 61000-4-15: Electromagnetic compatibility (EMC) – Part 4-15: Testing and measurement techniques – Flickermeter – Functional and design specifications 1

IEC 61000-4-16: Electromagnetic compatibility (EMC – Part 4-16: Testing and measurement techniques – Test for immunity to conducted common mode disturbances in the frequency range 0 Hz to 150 kHz immunity test

10.6 Temperature Life Testing

For the temperature life test, the connector is exposed to constant elevated temperature. This temperature could be either the rated temperature of the connector or the maximum ambient temperature at which the connector is expected to work. Electrical loading can be performed during the operation of the connectors. In the case of signal connectors, electrical loading will not cause any significant increase in temperature and can be ignored. However, in the case of power connectors, electrical loading will cause a significant rise in temperature. Therefore, the test temperature should be brought down in case the connector is operating at maximum rated conditions. In the case of the ambient temperature test, the temperature of the surroundings should be brought down if the sum of the ambient temperature and temperature rise due to electrical loading goes beyond the rated temperature of the connector.

The temperature tests evaluate potential degradation mechanisms such as dry oxidation mechanisms; diffusion, migration, or intermetallic compound formation; potential creep tendencies of plastics under stress with the possible loss of contact retention and normal force; and stress relaxation of contacts, which may result in a loss of normal force and hence a loss of mechanical stability over time [5].

When the connector is to be used in ambient temperature conditions, a temperature rating of 65 °C is often used. When high-power devices are to be operated, a temperature rating of 85 °C is often used. When tin and its alloys are used in the connector, a temperature rating of 105 °C may be used. Cobalt-hardened gold connectors are tested at 125 °C. A 240-hour test is usually sufficient to establish the magnitude of stress relaxation that may occur (assuming that the trend can be determined). A 1000-hour test is required to evaluate all the potential degradation factors described earlier [6]. Per IEC 60512-11-9, the connector specimens are subjected to a 125 °C temperature for a period of 168 hours (1 week).

10.7 Temperature Cycling with Humidity Testing

Temperature cycling with humidity is a dynamic test wherein the temperature is cycled between two extremes while at high humidity. The purpose of this test is to: (i) determine the sensitivity of the housing material to swelling, moisture absorption, and dimensional stability; (ii) determine the susceptibility of the connector system to the wet oxidation process; (iii) check the susceptibility of the contacting surfaces to fretting corrosion due to the thermal cycling environment; and (iv) verify whether the wet oxidation process happens to the materials in contact as well as any wear debris or particulates that may be present on these surfaces [8].

10.8 Thermal Cycling Testing

In a thermal cycling test, the temperature is varied between two extremes with a specific ramp time. This test is the same as temperature cycling with humidity except that the humidity is absent. It evaluates the same factors as mentioned earlier with the addition of fretting degradation due to thermal excursions and the associated thermal expansion mismatches in the connector system [6].

10.9 Thermal Shock Testing

The thermal shock test is similar to the thermal cycling test except that the ramp-up and ramp-down times are extreme, because the item under test is moved from one temperature extreme to the other. Thermal shock is used to test the ability of the connector to withstand the shock of alternate exposures to extreme temperature, which could arise in some storage, transportation, and application conditions (e.g. going from an outside environment to an inside environment). This test is used to assess and explore the connector's susceptibility to fretting corrosion, features involving different materials, and significantly different masses (e.g. the effects of thermal dissipation) of the materials [6]. IEC 60068-2-14 provides the details of the test that involves rapid cycling from cold to hot. This test determines the ability of connectors to withstand rapid changes of ambient temperature. The exposure times adequate to accomplish this are selected based on the nature of the connector specimen. The connector specimen is exposed to rapid changes of temperature in air, or in a suitable inert gas, by alternate exposure to low temperature and to high temperature. Two separate chambers or one rapid temperature change rate chamber may be used. If two chambers are used, one for the low temperature and one for the high temperature, the location allows transfer of the specimen from one chamber to the other within the prescribed time. Either manual or automatic transfer methods may be used. The severity of the test is defined by the combination of the two temperatures, the transfer time, the exposure time of the specimen, and the number of cycles. The lower and higher temperatures are selected from the test temperatures of IEC 60068-2-1 and IEC 60068-2-2. The exposure time of each of the two temperatures depends upon the heat capacity of the specimen. It may be 3 hours, 2 hours, 1 hour, 30 minutes, or 10 minutes, or as specified in the relevant specification. Where no exposure period is specified in the relevant specification, it is understood to be 3 hours. The preferred number of test cycles is five, unless otherwise specified in the relevant specification. The two-fluid-bath method determines the ability of connectors to withstand rapid changes of temperature and is applicable to specimens with glass-metal seals. In this test, the specimen is immersed alternately in two baths, one filled with liquid at low temperature and one filled with liquid at a high temperature. The liquids used for the test are compatible with the materials and finishes used in the manufacture of the connector specimens. The rate of heat transfer will depend upon the liquids used and will affect the severity of the test for a given temperature range. In special cases, the relevant specification should specify the liquids to be used.

10.10 Humidity Testing

Steady-state humidity testing is performed at a constant temperature and relative humidity (RH)., [6].[4] Since this test is conducted in a static environment, its effectiveness is limited. In a situation when the test chamber humidity is cycled instead of

4 MIL-STD 1344, Method 1002.2: *Test Methods for Electrical Connectors: Humidity.*

being held at a constant state, the insulating materials are subjected to creep loads and the materials exhibit higher creep rates than in a constantly humid state. This higher creep is a more general phenomenon consistent with sorption-induced stress-gradient explanations.

The purpose of this test is to: (i) determine the sensitivity of the plastic material to swelling, moisture absorption, and dimensional stability; (ii) determine the susceptibility of the connector system to the wet oxidation process; and (iii) determine the potential surface degradation of a plastic housing. The two common test condition combinations used are 40 °C and 95% RH, and 85 °C and 85% RH [7].

Some standards for humidity testing are listed here for reference:

EIA-364-31F: Humidity Test Procedure for Electrical Connectors and Sockets
MIL-STD 1344, Method 1002.2: Test methods for electrical connectors: Humidity

10.11 Corrosion

At the heart of every electrical contact is the interface. For an electrical contact to adequately perform its function over a normal operating life, the contact interface must establish and maintain a low, stable contact resistance. The conductive, metallic contact area essentially determines the contact resistance over the life of the electrical contact. Due to surface roughness, contact between two surfaces is achieved only at higher isolated points where the asperities on the two mating surfaces touch. The surfaces are not in contact away from the asperities [8].

Corrosion affects contact resistance in two ways. First, corrosive attack can occur at the periphery of any or all of the contact spots, and the constriction resistance increases as the contact spot area and the spot distributions are reduced in size. Second, corrosion products can occur in the spaces between the contact spots. The corrosion occurring in this region can affect the contact resistance in three ways. First, a contact spot might occur in the proximity of a corrosion product and then the contact spot returns to its original position, resulting in a circuit intermittent. Alternatively, the contact might ride up on the corrosion product and remain there. Finally, when the contact spot moves back, it might carry the corrosion product back. In all of these cases, the end result is the same: contact resistance is increased.

Electrical connectors typically fall into one of three categories: pin-in-hole, edge and socket, and pressure connectors. Pin-in-hole connectors normally use wire ribbons and require attaching either the male or female part of the connector to the board. The pin-in-hole connection is limited by the increase in insertion force and the number of connections. Edge connections are achieved by designing a row of fingers on the rigid board. The edge of the board is inserted into a socket that has spring-loaded contact fingers. This design is seen predominately in motherboard–daughterboard interconnections. The third connection scheme involves plating rows of contact fingers on the printed wiring board and forming rows of gold bumps on a flexible printed wiring ribbon. The third connection scheme is more costly than the other two but allows for the highest density of interconnection.

Flexible printed wiring has become an increasingly attractive alternative for connecting electronic equipment. In a growing number of applications, flexible connectors have been used to replace the traditional wire ribbon connectors. One particular advantage over wire cable connectors is a higher level of interconnection and better surface contact. In addition, flexible printed wiring connectors have better control of electrical characteristics, as well as aesthetic appeal.

In all connection designs, the electrical contact resistance at the connection is a primary concern. Contact resistance is typically reported either as electrical resistance versus force or as electrical resistance versus pressure. The contact resistance is a function of the materials, geometries, and contact area. Changes in the contact resistance can be attributed to a change in contact area and materials due to wear and corrosion. Copper, which is used in most printed wiring board designs, is extremely susceptible to corrosion from chlorides and sulfides. In addition, copper oxidizes relatively quickly. Oxide films can result in up to 1000 times increase in the contact resistances. Gold surface finishes are used in high-reliability applications due to good electrical characteristics and resistance to corrosion. The contact resistance for clean gold surfaces is typically below 1 mΩ. The gold-to-gold contact resistance under minimum load conditions starts about seven to eight times lower than the gold-on-copper contact resistance. As the contact area increases with increased load, the copper-to-gold contact resistance approaches the equivalent gold-to-gold contact resistance.

In pressure contact connections, the contact load is a design parameter that must be considered as well as movement at the contact interface due to temperature expansion mismatches. Movement produces surface wear of the noble plating finish, which can expose the non-noble base material to corrosive agents. In addition, particles produced during wear can cause shorts and opens.

Previous work [9–13] on common contact metals such as aluminum, copper, and nickel have shown that surface films are one of the key factors affecting the conductive area in contact between two metallic surfaces. When electrically insulating films cover a portion of the asperities in contact, the total area of electrical contact is reduced, thus increasing the contact resistance. As the corrosion films cover the contact surfaces, many of the original metal-to-metal asperity electrical contacts have been changed into, or replaced by, insulating asperities. Several researchers [12, 14, 15] have shown that the electrical contact area is decreased with the ingress of surface films such as oxides onto the asperities.

Usually a copper alloy is used as the base metal for contact pins and springs. They are plated with precious metal to increase corrosion resistance. The gold contact plating typically will be in the range of 0.4–1.3 mm thick. However, despite having noble metal plating, corrosion is still a problem due to porosity in the plating. Porosity is a function of the plating process used, and for a given process, the porosity is directly proportional to the thickness of the noble metal plating – the thicker the plating, the lower the porosity. Porosity exposes the non-noble under plate or base metal to the environment, which leads to its corrosion. These corrosion products creep out of the pores and spread over the noble metal plating.

Testing must be conducted to assess the potential connector failure mechanisms of pore corrosion, edge creep, corrosion due to particulates or contaminants on the contact surfaces, and corrosion of non-noble material systems. The key environments that can lead to such degradation are temperature, humidity, gaseousness, and dust.

Corrosion can be categorized as four types: dry, creep, moist, and fretting. The following sections discuss these types.

10.11.1 Dry Corrosion

Dry corrosion occurs when an oxidizing gas is present. It does not require the presence of a liquid electrolyte. The copper oxides are both ionic and electronic conductors; thus, the oxide layer serves as the electrolyte. The oxide layer also forms the electrode where oxygen is reduced and forms the diffusion layer through which the ions and electrons must migrate. The more metal-rich oxides exist at the metal interface, while the more oxygen-rich oxides exist at the atmospheric interface.

A surface cross section of copper would reveal the following layers: Cu, CuO, CuO_2, and O_2. The film generated by dry corrosion on copper is porous, and is electrically insulating but not self-limiting. In addition to the noble metal plating on the contact surface, nickel underplating is often employed as a two-way barrier against corrosion. The migration of copper to the contact surface is drastically slowed down because copper diffuses through nickel approximately 1000 times slower than through gold. Nickel exposed at the base of the pores in the gold plating forms a self-limiting film that does not grow out of the pore defect and covers the gold surface [16].

10.11.2 Creep Corrosion

Corrosion products can be generated at pores and any other places (cracks, edges, scratches) where the base metal is exposed to the environment and can spread over the protective precious metal coating without reacting chemically with the plating. This phenomenon, known as "creep corrosion," could be regarded as an extension of the pore corrosion process, especially when corrosion products from adjacent pores begin to merge with each other [17]. However, it is believed that pore corrosion is driven primarily by chlorine ions, while creep corrosion is usually a sulfur-dominant process [18]. In mixed flowing gas (MFG) testing, an established accelerating corrosion test for qualification of connectors [19], creep corrosion is regarded more as a separate failure signature than pore corrosion.

Sulfide products, especially silver sulfide, creep at the highest rate [20]; other reacting products, such as chloride, do creep but at orders of magnitude lower than those of sulfides. Creep of all products is highest across pure gold. A study on creep corrosion phenomena of IC packages [21] found that creep corrosion may also occur on a plastic surface. In this case, creep corrosion products were found to be a mixture of chloride and sulfide of the base metal (copper) and the diffusion-barrier plating layer (nickel).

The physics behind the phenomenon of corrosion products creep is not yet apparent. However, two theories are being considered. The first, a surface diffusion theory, states that creep corrosion is driven by concentration gradients of chemical species of the corrosion product [22]. A surface diffusion coefficient can be used to quantify the mobility of corrosion products over a surface under given environmental conditions. The second, a galvanic corrosion theory, provides another interpretation [23]. Metallic ions move from anode (base metal) toward cathode (noble plating) and are deposited near the cathode, combining with other anions in the electrolyte. In this way, corrosion products propagate over a noble metal plating surface and have a tendency to cover more area of the plating surface.

The effect of creep corrosion on contacts is the degradation of performance by increasing the contact resistance. Once corrosion products begin to spread over a plating surface, the contact resistance of the surface increases, owing to the poor electrical conductivity of corrosion products. Noise in the electrical signals or current leakage may occur [18].

In engineering design of contacts with noble metal plating, the resistance to creep corrosion is usually considered as a critical characteristic. MFG tests are often conducted in the laboratory to reproduce creep corrosion phenomena to simulate field use, and contact resistance can be measured to evaluate the effect of creep corrosion [18].

10.11.3 Moist Corrosion

Moist corrosion is caused by the humidity in the air. A thin layer of moisture from the ambient RH combines with gaseous corrosion drivers to form an electrolyte. Either a galvanic cell or a differential cell is created with the metals of the electrical contact forming the anode and cathode, while monolayers of water contaminated with such impurities as chlorine and sulfur form the electrolyte [17].

A contact may be susceptible to creep corrosion if the intended contact area has been selectively gold-plated. This phenomenon occurs when sulfides cause tarnish films to form on exposed copper and migrate across the gold-plated surfaces, electrically insulating them. In addition, pores on the gold plating surface (called "pores") are a source of the creep corrosion. Copper exposed at the base of the pores in the gold plating forms the corrosion products that grow out of the pore and creep over the gold surface. An electrical contact surface affected by pore corrosion develops discrete "mounds" of corrosion on an otherwise uncorroded (native) surface. Discontinuities in the plated surface and a corrosive environment are required for pore corrosion to occur [17].

Pore corrosion can also cause a mechanical disruption in the plating. The corrosive products formed at the base of the pore have a larger volume because they are less dense than base copper. As the new corrosive products form, they can exert a force on the plating layers to break them apart. This disruption exposes more of the corrosion-prone underplate and intensifies the corrosion problem [17].

It is not always possible to differentiate between these two types of atmospheric corrosion because, depending on the corrosion conditions, there may be a gradual transition from one form to another. Structures initially corroded in air by the dry corrosion mechanism can begin to corrode by the moist mechanism as a result of increase in moisture or formation of hygroscopic corrosion products [17].

10.11.4 Fretting Corrosion

Owing to the increasing number of low-power connections in electronic packaging, fretting corrosion of electrical contacts is becoming more of a concern. Thermal cycling, as well as ordinary micro-vibration, provides the driving mechanism for the micromotion that induces fretting corrosion, thereby decreasing the reliability of the electrical system [17].

Fretting occurs when two contacts are brought together with a given contact normal force and then moved with respect to each other cyclically. This motion is typically 10–100 μm in amplitude at a frequency dependent on the driving mechanism. The degree of fretting corrosion is highly dependent on environmental conditions, frequency, normal force, contact materials, and presence of lubricants. However, the increased resistance associated with fretting corrosion is only one part of the problem. Electrical intermittences associated with the fretting process can degrade digital circuit signals [17].

Gold-plating technology has been widely used to decrease fretting corrosion in electrical contacts. However, formation of an insulating surface film and further fretting corrosion in the gold-plated contacts as a result of the gold-plate damage and diffusion of substrate metal through gold is still one of the major concerns in producing reliable electrical contacts with low contact force application [17].

10.12 Mixed Flowing Gas Testing

One of the most widely used environment tests is the MFG test. This test is considered to be a realistic accelerated test that is designed to simulate the kinetics and degradation mechanisms found in indoor environments. Battelle Laboratories in Columbus, Ohio, conducted a study on electrical contacts and developed four classes, each associated with a distinct pollutant. It found that while the environment class depends on the pollutants in the environment, it is best defined by how the degradation of the contact materials occurs.

There are four environment classes. Class I represents a benign, nonindustrial business office environment with good atmospheric control. Class II represents typical conditions in business offices, control rooms, and telephone exchanges that are associated with a light industrial area. Class III represents poorly controlled industrial and related locations where moderate amounts of pollutants are present. Class IV represents extremely corrosive, highly polluted, and heavily industrial locations. Table 10.3 outlines the predominant degradation mechanism and the pollutants that are present in the various environment classes.

Tests may be conducted on plated metallic coupons rather than on actual connector samples if the similarity is appropriate. The coupons are placed in the accelerated chambers for varying periods of time. They are then taken out, and their contact resistance is measured using a contact resistance probe. The probe measures contact resistance at varied normal forces.

An MFG test is a laboratory test in which the temperature, relative humidity, concentration of gaseous pollutants, and other critical variables (such as volume exchange rate and airflow rate) are defined, monitored, and controlled. The purpose of this test is to simulate corrosion phenomenon due to atmospheric exposure [20]. Test samples that have been exposed to MFG testing range from bare metal surfaces to electrical

Table 10.3 Predominant degradation mechanism and pollutants in the environment classes.

Class	Degradation Mechanism	Pollutants
I	Pore corrosion	Humidity, low chlorides
II	Surface and pore corrosion	Chlorides
III	Pore corrosion and edge corrosion creep	Chlorides, sulfides
IV	Edge corrosion creep – chloride-enhanced or sulfide-dominant	Chlorides, sulfides

connectors to complete assemblies. MFG testing has been widely accepted as a qualification test method to evaluate the performance of connectors.

In the 1980s, researchers at Battelle Labs (Columbus, OH), Telcordia (previously Bellcore), and IBM carried out tests on the use of MFG to accelerate atmospheric corrosion and its effect on electronic applications. In the early 1990s, professional organizations, including the ASTM, EIA, IEC), and Telcordia, began to standardize these test methods and published corresponding documents as guidelines. Among them, ASTM provided a comprehensive list of documents covering the methods for a well-controlled MFG test. These documents include the following:

ASTM B827-97: Standard practice for conducting mixed flowing gas environmental tests

ASTM B845-97: Standard guide for mixed flowing gas tests for electrical contacts

ASTM B810-01a: Standard method for calibration of atmospheric corrosion test chambers by change in mass of copper coupons

ASTM B825-97: Standard test method for coulometric reduction of surface films on metallic test samples

ASTM B826-97: Standard test method for monitoring corrosion tests by electrical resistance probes

ASTM B808-97: Standard test method for monitoring of atmospheric corrosion chambers by quartz crystal microbalances

ASTM International, which grows from US industry, is a nonprofit organization that provides a global forum for the development and publication of voluntary consensus standards for materials, products, systems, and services.[5] ASTM publishes voluntary consensus standards for materials, products, systems, and services. Therefore, ASTM standards are more likely a review of existing MFG practices, rather than a mandatory procedure for individual situations. The EIA is a US national trade organization that includes the full spectrum of US electronic product manufacturers.[6] The IEC, primarily based on the European electronics industry, is the international standards and conformity assessment body for all electrical, electronic, and related technologies.[7] For industrial applications, the following test methods are more specific and application-oriented. Each MFG test method is reviewed in the following sections.

IEC 68-2-60 (second edition): Environmental testing – Part 2: Tests – Flowing mixed gas corrosion test

EIA 364-TP-65A: Mixed flowing gas

Telcordia GR-63-CORE Issue 2, Section 5.5: Airborne contaminants test methods

Each MFG test method is reviewed next.

10.12.1 Battelle Labs MFG Test Methods

Table 10.4 lists the classifications and parameters for the Battelle Labs MFG test methods. The operational environments for electronic equipment in atmosphere are divided into four classes, from least corrosive (Class I) to most corrosive (Class IV).

5 American Society for Testing and Materials (ASTM): www.astm.org

6 Electronic Industries Alliance (EIA): www.eia.org

7 International Electrotechnical Commission (IEC): www.iec.ch

Table 10.4 MFG test methods developed by Battelle Labs [15].

Class	Temp (°C)	RH (%)	H_2S (ppb)	Cl_2 (ppb)	NO_2 (ppb)
I	—	—	—	—	—
II	30 ± 2	70 ± 2	$10 + 0/-4$	$10 + 0/-2$	200 ± 25
III	30 ± 2	75 ± 2	100 ± 10	20 ± 5	200 ± 25
IV	50 ± 2	75 ± 2	200 ± 10	50 ± 5	200 ± 25

Class I means a well-controlled office environment with continuous adjustment. Class II means a light industrial environment, such as business offices without effective or continuous environment control. Class III means a moderate industrial environment, such as storage areas with poor environment control. Class IV means a heavy industrial environment, such as locations adjacent to primary sources of atmospheric pollutant gases. Since available data for Class I indicates no precedent for environmental effects on reliability, there is no accelerated testing for Class I. The other three classes use a combination of three corrosive gases, NO_2, H_2S, and Cl_2, to accelerate corrosion. Most other standards also use a fourth gas, SO_2. The reason is that some researchers believe that H_2S and SO_2 have the same synergistic effects on metal corrosion and SO_2 is necessary to stress nickel in corrosive environments [24].

Since an MFG environment is an accelerated testing method, the determination of acceleration factor is required to understand the durability or reliability of the "device-under-test." In other words, if a sample has a certain degree of corrosion after *N* days in the testing chamber, how long would it take to have the same amount of corrosion in the fielded conditions? However, there is currently no model for this, although some authors suggest an acceleration factor of two days in the chamber for one year in the field for Battelle accelerated test conditions [25].

10.12.2 EIA MFG Test Methods: EIA 364-TP65A

EIA published its own specifications for MFG testing, as shown in Table 10.5. The latest version was approved on November 6, 1997. Class II, III, and IV parameters come directly from Battelle research. Classes IIA and IIIA are adapted from Classes II and III by adding SO_2 along with the other three corrosive gases.

Table 10.5 MFG test methods developed by EIA.

Class	Temp (°C)	RH (%)	H_2S (ppb)	Cl_2 (ppb)	NO_2 (ppb)	SO_2 (ppb)
I	—	—	—	—	—	—
II	30 ± 2	70 ± 2	10 ± 5	10 ± 3	200 ± 50	—
IIA	30 ± 1	70 ± 2	10 ± 5	10 ± 3	200 ± 50	100 ± 20
III	30 ± 2	75 ± 2	100 ± 20	20 ± 5	200 ± 50	—
IIIA	30 ± 1	70 ± 2	100 ± 20	20 ± 5	200 ± 50	200 ± 50
IV	40 ± 2	75 ± 2	200 ± 20	30 ± 5	200 ± 50	—

Table 10.6 MFG test methods developed by IEC.

Method	Temp (°C)	RH (%)	H_2S (ppb)	Cl_2 (ppb)	NO_2 (ppb)	SO_2 (ppb)
1	25 ± 1	75 ± 3	100 ± 20	—	—	500 ± 100
2	$30 + 1$	70 ± 3	10 ± 5	10 ± 5	200 ± 50	—
3	30 ± 1	75 ± 3	100 ± 20	20 ± 5	200 ± 50	—
4	25 ± 1	75 ± 3	10 ± 5	10 ± 5	200 ± 20	200 ± 20

10.12.3 IEC MFG Test Methods: IEC 68-2-60 Part 2

IEC 68-2-60 Part 2 about MFG testing was published in December 1995 but is now withdrawn. Table 10.6 shows the parameters for MFG testing by IEC 68-2-60. Test method 1 can be used as a pore corrosion test on gold coatings. Test method 1 is for testing of contacts with gold-plated surfaces to be used in mild environments. Test methods 2 and 4 are appropriate for electronic products to be used in moderate corrosive environments. Such environments may be found in telecommunication centers, most office environments, and some industrial instrument rooms. Test method 3 is appropriate for more corrosive environments. Such environments may be found in some industrial locations.

10.12.4 Telcordia MFG Test Methods: Telcordia GR-63-CORE Section 5.5

Telcordia, previously known as Bellcore, is a center that provides standardization within the telecommunication industry. MFG test methods developed by Telcordia focus on electronic equipment in telecommunication applications. Since these kinds of equipment may operate indoors or outdoors, two MFG test methods are available from Telcordia, which are known as indoor and outdoor tests. Table 10.7 lists the parameters for these two methods.

10.12.5 IBM MFG Test Methods: G1(T)

In contrast to Battelle and Telcordia, IBM divided the working conditions for electrical equipment into three classes, which are G1 (business office), G2 (industrial), and G3 (harsh industrial). To simulate the accelerated corrosive effect of equipment in the G1 environment, IBM designed and verified the G1 (T) MFG test environment, where it used four corrosive gases. IBM's recommended gas concentrations are very different from that of Battelle (see Table 10.8) [26]. Interestingly, the IBM MFG test method has not gained much popularity in the industry, perhaps because of the costs of using four gases.

Table 10.7 MFG test methods developed by Telcordia.

Method	Temp (°C)	RH (%)	H_2S (ppb)	Cl_2 (ppb)	NO_2 (ppb)	SO_2 (ppb)
Indoor	30 ± 1	70 ± 2	10 ± 1.5	10 ± 1.5	200 ± 30	100 ± 15
Outdoor	30 ± 1	70 ± 2	100 ± 15	20 ± 3	200 ± 30	200 ± 30

Table 10.8 G1 (T) MFG test method developed by IBM.

Temp (°C)	RH (%)	H_2S (ppb)	Cl_2 (ppb)	NO_2 (ppb)	SO_2 (ppb)
30 ± 0.5	70 ± 2	40 ± 5%	3 ± 15%	610 ± 5%	350 ± 5%

Figure 10.3 MFG chamber located at CALCE at the University of Maryland.

Table 10.9 CALCE MFG chamber capability.

		Corrosive Gas Concentrations (ppb)				
Temp. Range (°C)	RH (%)	NO_x	SO_2	H_2S	Cl_2	Interior Dimensions (inches)
25 ~ 50	20 ~ 95	10 ~ 1000	10 ~ 1000	10 ~ 1000	10 ~ 1000	29 × 30 × 38

10.12.6 CALCE MFG Chamber Capability

The Center for Advanced Life Cycle Engineering (CALCE) at the University of Maryland provides quality services to electronic applications that require testing using an MFG chamber. Figure 10.3 shows this MFG testing system. Table 10.9 lists the capability of the MFG chamber, including up to four corrosive-gas-testing.

Mechanisms of electrical connector failures include film growth that influences contact force and voltage drop. Contact force can suppress surface film growth either through a pore in a gold layer or ingress at the contact area, which contributes to the resistance stability of electrical contacts [17].

10.13 Vibration

The purpose of the vibration test is to establish the mechanical integrity of the connector as part of a product or system. The test can be used to determine whether the electrical system is stable under vibration conditions, to determine discontinuities that may exist in the electrical system, and to check the susceptibility of the connector(s) in the system to fretting wear and corrosion. Too often, vibration is included in connector testing without any real thought as to what purpose the vibration serves. Mechanical vibration is typically used as a test environment to establish or verify the mechanical integrity of connector, to determine whether electrical discontinuities exist, the impact on electrical stability, and to determine whether the connector system is susceptible to fretting wear or corrosion. Typical vibration testing is either monitored or unmonitored. The monitored testing typically is looking for interruptions in contact resistance of 1 μs or longer. More recent test requirements for high-speed systems are looking at interruptions of 10 Ω or more in a time frame of 10 ns (known as low nanosecond events). Unmonitored vibration testing relies upon low-level contact resistance (LLCR) measurements to detect the presence of contact degradation.

It is recognized that test sequencing must be considered; i.e. a monitored vibration test cannot proceed with an LLCR measurement due to current-induced oxide breakdown, but traditional vibration testing does not fully consider the role of vibration in precipitating and detecting potential connector failures. The connector failure mechanisms that are driven by vibration include acceleration-driven contact separation, contact wear, and contact fretting.

The vibration test is typically performed on the three axes of the system under test. The most common types of vibration tests are sine vibration, random vibration, and vibration scans. Sine vibration tests involve sweeping through various frequencies at a defined amplitude and seep duration. In the random vibration test, the frequencies and amplitude appear randomly in a given frequency spectrum. The vibration scan test is performed to identify the dominant resonances in the connector. Then the connector can be vibrated at the dominant resonant frequencies for a specified period of time.

The effects of vibration on connector performance can be monitored by measuring interruptions in resistance and the variation in the LLCR.

Some standards for vibration testing are listed here for reference:

EIA 364-28C: Vibration test procedure for electrical connectors and sockets
IEC 60512-6-5: Dynamic stress test – section 5: test 6a: random vibration
MIL-STD 1344, Method 2005.1: Test methods for electrical connectors: vibration

10.13.1 Mechanical Shock

The mechanical shock test is meant to simulate conditions a connector may be subjected to during assembly, transportation, and other forms of handling. The mechanical shocks should be associated with the range of conditions expected and often vary from 30 to 300 G with specified durations. When this test is used to simulate transportation conditions, it is often performed as a precursor to vibration tests.

10.13.2 Mating Durability

Mating durability is a test to simulate the wear of a connector finish system normally expected during the anticipated life of the connector. The test is normally performed using automatic cycling equipment with an adjustable speed control and counter system. The rate of cycling has to be controlled to prevent frictional heating.

Two basic attributes need to be monitored during this test: mating/unmating forces on the first and last cycles and LLCR and contact resistance at rated current (CRRC) upon completion of the test. These attributes will determine whether the force requirements are met and the electrical stability and mechanical integrity are maintained.

10.14 Highly Accelerated Life Testing

Highly accelerated life testing (HALT) can uncover many of the weak links of a new product. These accelerated tests are used for finding weaknesses at various steps of a manufacturing process. Design weaknesses are easier to correct during product development, resulting in reduced product development costs and compressed time-to-market. When accelerated tests are used after the products are introduced into the market, they can determine product reliability due to changes in components, suppliers, and manufacturing processes. Because of their accelerated nature, these tests are typically faster and less expensive compared to normal tests [27].

In highly accelerated life tests, environmental stresses are applied to the system. These stresses eventually lead to product failure when the stress applied becomes more than that expected during use. The stresses applied are typically cold and hot temperatures, thermal cycling, power cycling, random vibrations, and power margining. When stress-introduced failures occur, the causes for the failure should be identified and corrected. The tests are then repeated, the second test can also identify additional weaknesses, in addition to verifying the effectiveness of the correction.

The typical HALT chamber should be capable of applying a pseudo-random vibration with frequency ranging to 10 000 Hz in 6° of freedom, with the ability for rapid changes in temperature. Often, high-power resistive heating elements are used for heating and liquid nitrogen (LN_2) for cooling purposes [28].

The product under test must be monitored continuously to detect degradation and failure. The test equipment usually includes thermocouples, performance measuring sensors, accelerometers, data loggers, and multimeters. The most common causes of failure in these tests are weak product design, bad workmanship, and poor manufacturing techniques [28].

10.15 Environmental Stress Screening

Environmental stress screening (ESS) refers to the process of exposing a newly manufactured or repaired product or component to stresses such as thermal cycling and vibration to force the latent defects to manifest themselves by permanent or catastrophic failures. The remaining population is assumed to be more reliable than a similar unscreened population.

ESS tests are mainly used for military and aerospace applications. The tests usually involve temperature variations, vibration tests, and pressure and flexibility tests. ESS can be performed as part of the manufacturing process, or it can be used in new product qualification testing.

The various steps involved in manufacturing processes are as follows: Visual inspection is done first. Testing of the product is done to check whether the unit under test

(UUT) is within design guidelines. Then the physical environment conditions that the product will face in its lifetime, including during the transportation, storage, handling, and operational use phases, are simulated and applied. This includes vibration tests while transporting the product and thermal cycling tests while using it. Qualification is done to ensure that the design, manufacturing, assembly, and repair processes have resulted in a product that conforms to the specifications. Typically, qualification samples are derived from several production lots to ensure that the assessment covers the extent of the variability of the process. The qualification procedures used are: MIL-STD 810H for line replaceable unit (LRUs), MIL-STD 202G for electronic piece parts, and MIL-STD 883H for microelectronic devices. EMI/RFI testing is usually included in ESS qualification testing and requires application of MIL-STD 461E. The qualification tests generally involve aggravated ESS testing, which means subjecting a UUT to harsher conditions than it will experience in its lifetime. These are destructive tests. There is then acceptance ESS where the products are tested to specifications given by the procurement agency. Passing this test means the unit is accepted by the procurement agency.

References

1 Underwriters Laboratories, The Dielectric Voltage Withstand Test – Benefits and Limitations.library.ul.com/wp-content/uploads/sites/40/2015/02/UL_WP_Final_The-Dielectric-Voltage-Withstand-Test_v5_HR.pdf (accessed Dec. 2, 2019).

2 IPC-TM-650 3.13 (1975). *Test methods manual for withstanding voltage, connectors.* Institute for Interconnecting and Packaging Electronic Circuits.

3 NASA (1977). Military Standard – Test Methods for Electrical Connectors. nepp.nasa.gov/docuploads/A2C0D7A5-1F9E-4CA1-89040C1F55D2A974/MIL-STD-1344.pdf (accessed Dec. 2, 2019).

4 McReynolds, D., “Optimize Your RF/MW Coaxial Connections.” www.rfcoaxconnectors.com/Technical_wp-OptimizeConnections.htm (accessed Nov. 26, 2019).

5 Sun, M., Pecht, M.G., Natishan, M.A.E., and Martens, R.I. (1999). Lifetime resistance model of bare metal electrical contacts. *IEEE Transactions on Advanced Packaging* 22 (1): 60–67.

6 Holm, R. (1967). *Electric Contact*, 4e. New York: Springer-Verlag.

7 Kulwanoski, G., Gaynes, M., Smith, A., and Darrow, B. (1991). Electrical contact failure mechanisms relevant to electronic package. *Proceedings of the 37th IEEE Holm Conference on Electrical Contacts*. IEEE. pp. 184–192.

8 Timsit, R.S. (1990). A possible degeneration mechanism in stationary electrical contacts. *IEEE Transactions on Components, Hybrids, and Manufacturing Technology* 13 (1): 65–68.

9 Takano, E. and Mano, K. (1968). The failure mode and lifetime of static contacts. *IEEE Transactions, Parts, Materials and Packaging* PMP-4 (2): 51–55.

10 Antler, M. (1987). Contact resistance of oxidized metals: dependence on the mating materials. *IEEE Transactions on Components, Hybrids, and Manufacturing Technology* 10 (3): 420–424.

11 Bryant, M.D. and Jin, M. (1991). Time-wise increases in contact resistance due to surface roughness and corrosion. *IEEE Transactions on Components, Hybrids, and Manufacturing Technology* 14 (1): 79–89.

12 Malucci, R.D. (1992). Dynamic model of stationary contacts based on random variations of surface features. *IEEE Transactions on Components, Hybrids, and Manufacturing Technology* 15 (3): 339–347.

13 Ming, S., Pecht, M., and Natishan, M.A.E. (1999). A comparative assessment of gold plating thickness required for stationary electrical contacts. *Microelectronics Journal* 30 (3): 217–222.

14 Liu, W. and Pecht, M. (2004). Failure modes and mechanisms. In: *IC Component Sockets* (eds. W. Liu and M. Pecht), 138–154. Wiley-Interscience.

15 Abbott, W.H. (1988). The development and performance characteristics of flowing mixed gas test environments. *IEEE Transactions on Components, Hybrids, and Manufacturing Technology* 11 (1): 22–35.

16 ASTM B845-97 (1997). *Standard Guide for Mixed Flowing Gas (MFG) Tests for Electrical Contacts*. American Society for Testing and Materials.

17 Slade, P.G. (1999). *Electrical Contacts: Principles and Applications*. New York: Marcel Dekker.

18 Zhao, P. and Pecht, M. (2003). Field failure due to creep corrosion on components with palladium pre-plated leadframes. *Microelectronics Reliability* 43: 775–783.

19 Xie, J.S. and Pecht, M. (2001). Palladium-plated packages: creep corrosion and its impact on reliability. *Advanced Packaging* 10 (2): 39–42.

20 Leygraf, C. and Graedel, T.E. (2000). *Atmospheric Corrosion*. New York: Wiley.

21 Bhagath, S. and Pecht, M. (1993). Modeling the effects of mixed flowing gas (MFG) corrosion and stress relaxation on contact interface resistance. *Journal of Electronic Packaging* 115: 404–409.

22 Mroczkowski, R.S. (1998). *Electronic Connector Handbook*. New York: McGraw-Hill.

23 EIA 364-31F (2019). *Humidity Test Procedure for Electrical Connectors and Sockets*. Herndon, VA: Electrical Components Industry Association.

24 Gore, R.R., Witska, R., Kirby, J.R., and Chao, J.L. (1990). Corrosive gas environmental testing for electrical contacts. *IEEE Transactions on Components, Packaging and Manufacturing Technology* 13 (1): 27–32.

25 Williams, D.W. (1988). The effect of test environment on the creep of base metal surface films over precious metal inlays. *IEEE Transactions on Components, Hybrids and Manufacturing Technology* 11 (1): 36–42.

26 Chao, J. G., and Gore, R. R. (1991). Evaluation of a mixed flowing gas test. *Proceedings of the 37th IEEE Holm Conference on Electrical Contacts*, Piscataway, NJ. IEEE. pp. 216–228.

27 Lin, X.-Y., and Zhang, J.-G. (2004). Dust corrosion. *Electrical Contacts, 2004. Proceedings of the 50th IEEE Holm Conference on Electrical Contacts and the 22nd International Conference on Electrical Contacts*. IEEE.

28 Levin, M.A. and Kalal, T.T. (2003). The reliability toolbox. In: *Improving Product Reliability: Strategies and Implementation* (eds. M.A. Levin and T.T. Kalal), 64–126. John Wiley.

11

Supplier Selection

Reliability Capability Assessment

Michael H. Azarian, Diganta Das and Michael G. Pecht

Center for Advanced Life Cycle Engineering (CALCE), University of Maryland, College Park, MD, USA

Rapid globalization of several industries has taken place in the last decade of the twentieth century and the first decade of the twenty-first century. Competitive and regulatory pressures have driven companies to low-cost manufacturing and to the evolution of a worldwide supply chain. Today, external sourcing of components and contract manufacturing is widespread. Companies are dependent upon worldwide suppliers that provide them with materials, parts, and subassemblies. Therefore, for any product design, it is essential that the reliability requirements be applied to all the incoming subcontracted elements so that reliability can be managed across all the tiers of the supply chain. The ultimate goal is that each supplier's reliability practices will be adequate to satisfy the end-product requirements of their customers.

System integrators, which are at the top of the supply chain, generally set the requirements for system reliability. Parts and manufacturing processes purchased on the market as commodities are selected based on information provided by suppliers. However, system integrators cannot wait until they receive purchased parts or subassemblies to assess whether they are reliable. This would lead to an expensive iterative process of part delivery, product assembly, and reliability testing followed by part re-specification. An up-front evaluation of suppliers based on their ability to meet reliability requirements can provide a valuable competitive advantage. A company's capability to design for reliability and to implement a reliable design through manufacturing and testing can yield important information about the likelihood that the company will provide a reliable product.

11.1 Connector Reliability

Connector reliability depends on the requirements the connector has to meet and the degradation mechanisms it is subjected to. In simple terms, connector reliability is the probability that the connector meets a specified range of connector resistance under the operating conditions it was designed for during its specified lifetime. The application of the connectors determines their resistance requirements and operating conditions, which in turn determine the degradation mechanism that might be active. The degradation mechanisms might be intrinsic (i.e. related to the design and materials

Electrical Connectors: Design, Manufacture, Test, and Selection, First Edition.
Edited by San Kyeong and Michael G. Pecht.

of construction of the connector) or extrinsic (i.e. related to the application). Some examples of intrinsic degradation are corrosion, loss of normal force through stress relaxation, and joule heating, which leads to temperature-related degradation. Examples of extrinsic degradation include contamination and fretting corrosion, and both of these are dependent on the application of the connector. The connector manufacturers have control over intrinsic degradation but not extrinsic factors [1].

To determine connector reliability, the following issues must be addressed: the active degradation mechanisms must be identified and categorized with respect to their importance in the application of interest; appropriate tests, acceleration factors, and exposures must be known, defined, or determined for these degradation mechanisms; failure criteria appropriate to the application of interest must be established; and the statistical approach to determining or calculating reliability values must be agreed upon [1].

11.2 Capability Maturity Models

The maturity approach to determining organizational abilities has roots in quality management. Crosby's Quality Management Maturity Grid describes the typical behavior of a company, which evolves through five phases (uncertainty, regression, awakening, enlightenment, and certainty) in its ascent to quality management excellence. Consequently, maturity models have been proposed for a wide range of activities, including software development, supplier relationships, research and development effectiveness, product development, innovation, collaboration, product design, and reliability information flows. This leads to the following metric for reliability capability:

> A reliability capability maturity metric is a measure of the practices within an organization that contribute to the reliability of the final product, and the effectiveness of these practices in meeting the reliability requirements of customers.

11.3 Key Reliability Practices

The Institute of Electrical and Electronics Engineers (IEEE) Reliability Program Standard 1332 defines broad guidelines for the development of a reliability program, based on three objectives:

1. The supplier, working with the customer, should determine and understand the customer's requirements and product needs so that a comprehensive design specification can be generated.
2. The supplier should structure and follow a series of engineering activities so that the resulting product satisfies the customer's requirements and product needs with regard to product reliability.
3. The supplier should include activities that adequately verify that the customer's reliability requirements and product needs have been satisfied [2].

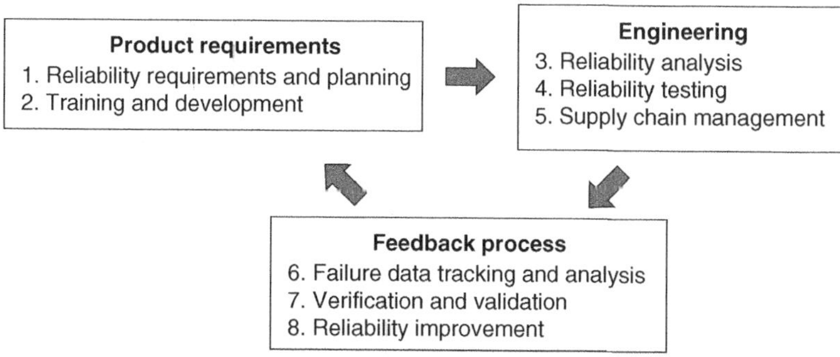

Figure 11.1 Key reliability practices [3].

For each of these reliability objectives, key practices for evaluating reliability capability can be assigned. Figure 11.1 presents eight key practices identified from a study of reliability standards from the industry and reliability literature. Each of the eight key reliability practices is described in the following sections.

11.3.1 Reliability Requirements and Planning

During product development, the customer's needs and operational conditions for all phases of the product life cycle must be understood to arrive at a set of customer reliability requirements. The different considerations for establishing reliability requirements for a product include the design and operational specifications (information about the manner in which the product will be used), regulatory and mandatory requirements, definition of failure, expected field life, criticality of application, cost and schedule limitations, and business constraints, such as potential market size.

Establishing reliability requirements and planning early incorporates activities needed to understand customers' requirements, generates reliability goals for products, and plans reliability activities to meet those goals. The inputs for generating reliability requirements for products include customer needs, reliability data specifications for competitive products, and lessons learned from the reliability experience with previous products, including test and field failure data.

Reliability planning is needed to establish and maintain plans that define reliability activities and manage the defined activities. The planning activity starts with identifying available resources, such as materials, human resources, and equipment, and determining the need for additional resources. Reliability analysis and testing needed for the product and the logistics to obtain feedback on the implementation of these activities can be identified.

The output from this key practice is a reliability plan. The reliability plan identifies and ties together all the reliability activities. The plan should allocate resources and responsibilities and include a schedule to follow. Decision criteria for altering reliability plans can also be included.

11.3.2 Training and Development

Training and development enhances the specialized skills and knowledge of people so that they can perform their roles in the development of a reliable product effectively and efficiently. The aim is to ensure that employees understand the reliability plans and goals for products and have sufficient expertise in the methods required to achieve those goals. This includes the development of innovative technologies or methods to support business objectives.

Training and educating employees enhances the possibility of obtaining a better, more reliable product. Reliability managers must be aware of how specific reliability activities can impact or improve reliability, and business managers should appreciate the importance of reliability to ensure successful implementation of reliability practices within a company. The implementation of regular training programs indicates the willingness of business managers to spend time, effort, and money on training employees.

Effective training requires assessment of needs, planning, instructional design, and appropriate training media. The main components of employee training include a training development program with documented plans and means for measuring its effectiveness. The main activity for this key practice is the development of a training plan, including training needs for individual personnel, with a schedule. The implementation of the plan requires the procurement of a training infrastructure, including training instructors and training material.

The different modes of imparting training include in-class training, mentoring, web-based training, guided self-study, or a formal on-the-job training program. Employees must be trained in the life cycle reliability management of products, including specific areas such as failure analysis, root-cause analysis, and corrective action system. The training should incorporate an understanding of reliability concepts and statistical methods.

11.3.3 Reliability Analysis

Reliability analysis incorporates activities to identify potential failure modes and mechanisms, to make reliability predictions, and to quantify risks for critical components to optimize the life cycle costs for a product. Prior experience and history can be helpful in this analysis. The data used to make reliability predictions may be historical, from previous testing of similar products or from the reported field failures of similar products.

Reliability analysis activities include conducting failure modes, mechanisms, and effects analysis (FMMEA) to identify potential single points of failure, failure modes, and failure mechanisms for a product. The next step is to identify the criticality of these failure modes and mechanisms. Criticality may be based on complexity, application of emerging technologies, demand for maintenance and logistics support, and the impact of potential failure on overall product success. Reliability analysis also includes identification of reliability logic for products as a system, and creating reliability models at the component and product levels to make reliability predictions. Assessing adherence to design rules, including derating, electrical, mechanical, and other guidelines, is also part of reliability analysis.

The outputs from this analysis include an assessment of the reliability of the product, expected failure modes, and identification of design weaknesses to determine the suitability of the existing design for avoiding early-life failures and the product's susceptibility to wear-out failures. The information from reliability analysis can be used to create a list of reliability critical materials, parts, subassemblies, or processes and to design reliability tests. Predictions regarding expected warranty costs and logistics support, including spares provisioning, can also be made.

11.3.4 Reliability Testing

Reliability testing can be used to explore the limits of a product, to screen products for design flaws, and to demonstrate (or qualify) the reliability of products. The tests may be conducted according to some industry standards or to required customer specifications. The reliability testing procedures may be generic – that is, common for all products – or the tests may be custom-designed for specific products. The tests may or may not be used for the verification of known failure modes and mechanisms. Detailed reliability testing plans can include the sample size for tests and the corresponding confidence level specifications.

Important considerations for any type of reliability testing include establishing the nature of the test (failure- or time-terminated), the definition of failure, and the correct interpretation of the test results, and then correlating the test results with the reliability requirements for the product. The information required for designing product-specific reliability tests includes the expected life cycle conditions, the reliability plans and goals for a product, and the failure modes and mechanisms identified during reliability analysis. The different types of reliability tests that can be conducted include tests for design marginality, destruct limits determination, design verification testing before mass production, ongoing reliability testing, and accelerated testing.

The output from this key practice is the knowledge obtained from different types of tests. Test data analysis can be used as a basis for design changes prior to mass production, for identifying the failure models and model parameters, and for modification of reliability predictions for the product. Test data can also be used to create guidelines for manufacturing tests, including screens, and to create test requirements for materials, parts, and subassemblies obtained from suppliers.

11.3.5 Supply-Chain Management

Supply-chain management activities include monitoring a list of potential suppliers, conducting supplier assessment and audits, and selecting vendors or subcontractors for parts or processes. Other activities include part and process qualification through review of process, quality, reliability testing, or accelerated test data from the suppliers. Activities such as tracking product change notices, changes in the part traceability markings, and management of part obsolescence are also included under this key practice. These activities are essential for sustaining product reliability throughout its life cycle.

The decision criteria for supplier selection include their ability to provide reliable components effectively and their demonstrated ability to control their own supply chain. Possible control over the supplier's reliability practices through exchanging technological expertise and sharing information also increases the possibility of achieving and maintaining product reliability. In some cases, multisourcing of parts may be necessary due to considerations of product manufacturing schedules, supplier capacity, or anticipated supply fluctuations.

Key outputs from this practice are a list of preferred/qualified/approved parts, vendors, and subcontractors, and a system for supplier rating. Other outputs include component qualification reports, supplier audit reports, and development of supply contracts that include contractual quality and reliability requirements.

11.3.6 Failure Data Tracking and Analysis

Failure tracking activities are used to collect manufacturing, test, and field-failed components as well as related failure information. Failures must then be analyzed to identify the root causes of manufacturing defects and test or field failures and to generate failure analysis reports. These records can include the date and lot code of the returned product, the failure point (quality testing, reliability testing, or field), the return date, the failure site, the failure mode and mechanism, and recommendations for avoiding the failure mode in existing and future products. For each product category, a Pareto chart of failure causes can be created and continually updated.

The failure sources that initiate failure analysis of a product include manufacturing, production testing, reliability testing, pre- and post-warranty field returns, and customer complaints. Failure analysis includes statistical analyses of failure data and analysis of the cause of failure at various levels down to the identification of the root cause of failure.

The outputs for this key practice are a failure summary report arranged in groups of similar functional failures, forward and backward traceability of failed components through date and lot code information, actual times to failure of components based on time-specific part returns, and a documented summary of corrective action implementation and effectiveness. All the lessons learned from failure analysis can be included in a corrective actions database for future reference. This database can help save the considerable cost in fault isolation and rework associated with problems that may be encountered.

11.3.7 Verification and Validation

Verification and validation through an internal review/audit of reliability planning, testing, and analysis activities helps to ensure that planned reliability activities are implemented so that the product fulfills the specified reliability requirements. Benchmarking can be used to study the best internal practices that produce superior reliability performance and to ensure that noncompliance is addressed. Part of the process is to understand how some practices are better than others and to find ways to improve others by pushing for improved facilities, equipment, and methodologies.

The inputs for this key practice are the outputs from previous practices such as planning, analysis, testing, and failure data tracking. The inputs include reliability plans and goals for products, potential failure modes and mechanisms identified during reliability analysis, information on failure mechanisms from reliability testing, specific reliability test plans and specifications, and the corrective actions database.

Verification and validation activities include comparison of identified potential problems against those experienced in the field. This includes comparison of expected and field failure modes and mechanisms and of reliability prediction models for a product against field failure distributions.

The outputs from this key practice include an updated failure modes and mechanisms database, modification of reliability predictions and failure models for a product, and modification of warranty costs and spares provisioning. Reliability test conditions may also be modified based on field information on products.

11.3.8 Reliability Improvement

Reliability improvement is concerned with applying lessons learned from testing, reported field failures, technological improvements, and any additional information from previous tests or experiences. This key practice primarily involves implementing corrective actions based on failure analysis. It also involves initiating design changes in products or processes due to changes in reliability requirements or in life cycle application conditions (operating and nonoperating).

Reliability improvements can be affected either by making design changes in products or by using alternative parts, processes, or suppliers. Design changes can include an improved design using an established technology or implementing developing technologies within an older design. New modeling and analysis techniques and trends that could improve reliability can also be used.

The inputs required to initiate reliability improvement also come from previous key practices. Such information includes Pareto charts for field failure modes and mechanisms, recommendations from the corrective actions database, and documented anomalies from verification and validation. Other factors that can initiate a reliability improvement process are changes in life cycle usage conditions for a product or changes in reliability requirements due to business or other considerations.

11.4 Reliability Capability of an Organization

Reliability is the ability of a product or system to perform as intended (i.e. without failure and within specified performance limits) for a specified time, in its life cycle application environment. For any electronics business, time-to-profit is a key metric for establishing product design, product operation, and high-level management goals, including cost, schedule, and social responsibility. Since reliability is associated with preventing or minimizing the likelihood of failure occurrences, it is a risk factor associated with profit making. Failures lead to costs that extend the time-to-profit for a product. Failures can stain the reputation of a company and cause financial losses. Financial losses can be in the form of loss of market share due to damaged consumer confidence, increase in insurance rates, costs to replace parts, claims for damages resulting from personal injury, and maintenance of a service infrastructure to handle failures. Legally, most states in the United States operate on the theory of strict liability. Under this law, a company can be liable for damages resulting from a defect for no reason other than that one exists, and a plaintiff does not need to prove any form of negligence to win a case. A history or reputation of poor reliability can also prevent potential future customers from buying a product, even if the causes of past failures have been corrected. Therefore, to be competitive, electronics manufacturers need to know how things fail as well as how they work.

11.5 The Evaluation Process

This topic presents a procedure for evaluating and benchmarking the reliability capability of electronics companies.

The reliability capability evaluation process of an organization is comprised of three phases. In the first phase, initial information about the process is sent to the company being evaluated. A reliability capability evaluation questionnaire is included for the company to answer and collect evidence supporting the answers. In the second phase, evaluators visit the facility and verify the responses to the questions with the supporting evidence. The third phase involves the compilation of an evaluation report.

The first phase is initiated by sending information about the concept of reliability capability and maturity evaluation to the company being evaluated. This helps the personnel within the company to appreciate the benefits of such evaluation and enables them to answer the questions asked during the evaluation with a positive frame of mind. A questionnaire for the evaluation is sent at least 20 days before the evaluators visit the company. The evaluation consists of nine subsections: eight sections pertaining to each of the key practices essential to reliability achievement, and one section on background information about the company. A schedule for the second phase involving physical evaluation is also included. The respondents are required to provide objective evidence in support of their responses. The evidence may be in the form of data, reports, policy drafts, or other documents.

In the second phase of the evaluation, evaluators visit the facility. The evaluation team usually includes one representative from the company. The company presents an overview of their reliability objectives and practices. The evaluation team then reviews the responses to the questionnaire and the supporting evidence. Additional follow-up questions are asked, and additional supporting information is identified to clarify some responses and obtain the correct information. Evidence is sought and judgments are made based on the following: commitment to perform (leadership, resources), ability to perform (experience, training, tools), methodology used to perform (logic, framework, planning), tasks and activities performed (tasks, activities), and results of product performance used (integration at organizational level).

In the third phase, the company is provided a draft evaluation report summarizing the evaluation team's observations and recommendations for reliability improvement. The company is typically given one week to review the draft report and provide comments. A final report incorporating the feedback comments and clarifications is sent to the company, usually within four weeks after the evaluation. Based on the documented information and responses received, a reliability capability level is assigned to the company [2].

References

1 Mroczkowski, R.S. (1998). *Electronic Connector Handbook*. New York: McGraw-Hill.

2 Tiku, S., Pecht, M., and Strutt, J. E. (2003). Organizational reliability capability. *Proceedings of Canadian Reliability and Maintainability Symposium*, Ottawa, Canada (Oct. 16–17, 2003). IEEE.

3 Gullo, L.J., Azarian, M.H., Das, D., Schenkelberg, F., and Tiku, S. (2006). Assessment of organizational reliability capability, *IEEE Transactions on Components and Packaging Technologies*, 29 (2): 425–428.

12

Selecting the Right Connector

Ilknur Baylakoglu[1] *and San Kyeong*[2]

[1] *SATURN Engineering, TUBITAK, Ankara, Turkey*
[2] *Center for Advanced Life Cycle Engineering (CALCE), University of Maryland, College Park, MD, USA*

There are many factors to consider when selecting a connector for a targeted application, including function, performance, and reliability. Severalproperties of a connector should be considered before selecting a candidate part. From the design perspective, the connector with the best performance and functional specifications is a key factor. From the parts management perspective, the most generic industry-standard part with the best quality rating and that is least likely to become obsolete is important. From the procurement perspective, lead time and the availability of multiple sources are important. From the business perspective, the cost of the part is important. All of these aspects should be considered before a candidate connector is selected.

The design engineer bases the initial parts selection decisions on a combination of functional and environmental requirements. Besides previous experience and professional preferences, designers often use available databases, websites, and promotional materials and brochures from parts manufacturers to search for parts that conform to the required electrical and mechanical specifications.

There are several parts databases available from parts distributors and other information service providers [1]. However, these parts databases, although convenient, can be outdated, incomplete, or just plain wrong [2]. In some cases, the bill of materials (BOM) for a previous product or a previous version of the product is the starting point for a new BOM. Selection of a connector must be conducted at the beginning of the system design process to get the optimal design. The following guidelines are presented for the optimal choice of a connector.

12.1 Connector Requirements Based on Design and Targeted Application

Generally, connectors should be specified that directly meet the requirements of the targeted application. However, some connectors, known as "hybrid" connectors, serve more than one purpose and may be a combination of data and power, data and signal, or power, signal, coax, fiber, liquid, and/or gas. Hybrid connectors simplify the user interface by eliminating the need for multiple connections. A simpler interface can solve the

Electrical Connectors: Design, Manufacture, Test, and Selection, First Edition.
Edited by San Kyeong and Michael G. Pecht.

need for several combinations of different connections at the same time, especially in medical applications such as surgical procedures, dentistry, cosmetic skin treatments, and eye surgery [3].

Regardless of connection functions, there are key factors/requirements that must be considered when selecting a connector. These considerations include the physical application, electrical requirements, mechanical requirements, application environment (e.g. space, submarine, harsh environments, etc.), mating force, number of mating cycles, current and power ratings, termination types, environmental conditions (temperature, vacuum, chemicals, radiation, vibration, etc.), materials, and reliability. The mating or unmating force of connectors depends on the connector type and pitch [4]. The contact geometry, the contact spring rate, and the coefficient of friction determine the mating force, which occurs in two stages – a spreading stage and a sliding stage. Additionally, insulator binding or bottoming out affect the mating/unmating forces and should be avoided and controlled by the connector user.

For low-mating-force interconnects, new technology brush contacts with multiple-contact interfaces (strands of high-tensile wire bundled together to form brush-like contacts) are available. By intermeshing two multistrand wire bundles, an electrical connection is made [5]. For example, the proven durability of brush contacts for Amphenol line replaceable module (LRM) surface mount connectors is 20 000 cycles of quad-lock connector (QLC) mating/unmating [6]. The Amphenol high-density brush (HDB) connector mating cycle is 100 000 as compared to a pin-and-socket connector mating cycle of 500 [7].

Horn and Egenolf [8] demonstrated that the optimization of the shape at the entry range of a pin-and-socket spring reduces insertion force and wear. The insertion force of a connector with wedge-shaped pins has been determined as about 50 N, and the pin shape with optimized geometry reduced the insertion force to about 20 N.

When connectors mate, friction enables good contact to be made and the contacts to be cleaned. For connector types with a large number of pins (e.g. 60, 100, 360 … 500–1000 or more signal layouts), the force per contact is multiplied by the total number of pins used, which results in the need for a significant amount of force to connect and disconnect the connectors. For some applications, the required amount of force may be a problem and determines whether a certain type of connector can be considered. In some specialized applications where large numbers of pins are needed (e.g. portable medical applications, robotics, stage lighting and sound systems, and avionics) zero-insertion force (ZIF) connectors may be used. ZIF connectors do not allow the male pins to contact the female mating pins, considerably reducing the force. A lever on the side of the female connector brings the two halves into contact once they have mated and reduces the contact retention to a small value [9].

12.2 Mating Cycles

Because the process of connecting and disconnecting wears out connectors, their lifetime is finite and is presented in terms of numbers of mating cycles on manufacturer datasheets. This number varies widely depending on the technology. For example, a Universal Serial Bus (USB) connector's lifetime in terms of mating cycles may be in the thousands or tens of thousands, while a board-to-board connector designed for use inside of consumer electronics may be limited to tens of mating cycles. After the number of mating cycles has been determined according to the usage expectation, the appropriate connector can be identified that provides a suitable lifetime for the application. During this selection process, the connector's mating style and electrical and mechanical requirements are also taken into consideration. Connector mating/unmating cycles simulating the expected life can be determined according to IPC-TM-650 Test Method 3.4, Durability, Connectors.

In addition, ASTM B794 – 97(2015), "Standard test method for durability wear testing of separable electrical connector systems using electrical resistance measurements," and IEC 60512-9-4:2011, "Connectors for electronic equipment – Tests and measurements – Part 9-4: Endurance tests – Test 9d: Durability of contact retention system and seals (maintenance, ageing)" are also test methods to assess the ability of a connector to withstand stresses caused by repeated extraction and insertion of contacts.

The actual "mating points" of a contact in electrical connectors occur at the asperities. As asperities become compressed and wear out, the resistance can decrease due to increased contact area; but the resistance can also increase as new high spots make contact and the plating becomes worn. In most cases, the resistance will gradually increase, then drop back down, and then start increasing again as new high spots are created and worn away [10].

12.3 Current and Power Ratings

A connector's mated contact pair's current carrying capacity depends mainly on contact material, geometry, and spring force between the pin (male) and receptacle (female) contacts. The contact current rating is defined as the current stage that creates a certain level of temperature rise in the contact spring (usually 20 or 30 °C) [11]. This rating is determined by the diameter of the connector as well as the cable. The voltage rating of the connector is determined by the insulating material between the contact points.

For voltage rating determination, dielectric withstanding voltage (DWV) requirements are assessed. The purpose is to ensure that there are no failures that occur as disruptive discharge or current leakage due to a dielectric (insulation) inadequacy or breakdown. EIA-364-20 and MIL-STD-1344 both specify that the DWV should be 75% of the minimum breakdown voltage of the connector (breakdown voltage is defined as the voltage at which a connector failure is observed). In addition, if the connector is exposed to repeated DWV tests, the insulation may degrade; therefore, the DWV test voltage should never be used as an operating/working voltage. To prevent degradation, EIA-364-20 and MIL-STD-1344 suggest that the operating rated voltage be established as 1/3 of the withstanding voltage (see Table 12.1). Current derating curves are constructed as shown in Figure 12.1b and are used in the connector selection according to the electrical needs of the designed product by using the maximum rated current and voltage information, which is usually provided in the datasheet. The maximum current that the contact can carry depends on the maximum safe operating temperature of the spring material, the ambient temperature, the heat transfer coefficient, and the contact resistance. The current-carrying capacity is limited by the thermal properties of the materials that are used for the contacts, terminals, and insulating materials of housing. Therefore, it is a function of the self-generated heat and the ambient temperature at

Table 12.1 Dielectric withstanding voltage (DWV) and suggested rated operating voltage [12].

Military Connector Specification	AirBorn Connector Series	DWV Sea-level (rms 60 Hz)	DWV 70 000 ft. (rms 60 Hz)	Rated Operating Voltage Sea-level (rms 60 Hz)	Rated Operating Voltage 70 000 ft. (rms 60 Hz)
MIL-DTL-83513 All slash sheets	M	600	150	200	50
MIL-DTL-55302 /55 through /66, /138, /139	W	1000	300	333	100
MIL-DTL-55302 /190 through /193	R	750	250	250	83
MIL-DTL-32139 All slash sheets	N	250	100	83	33

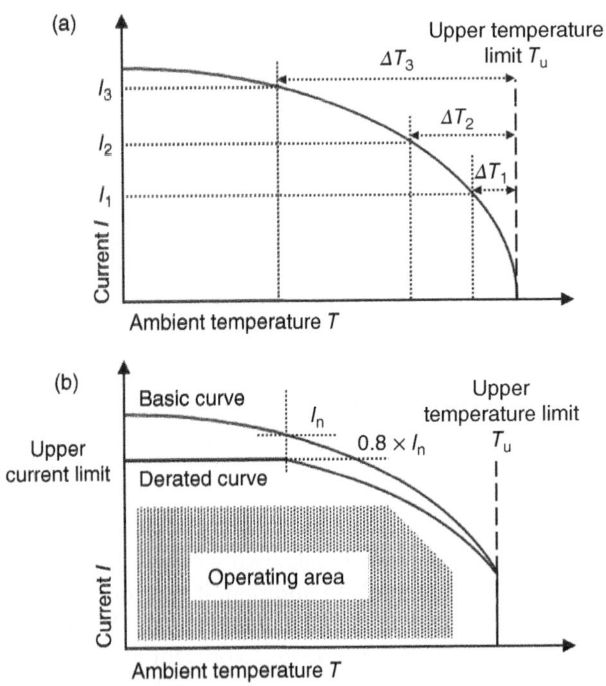

Figure 12.1 (a) Construction of the basic current-carrying curve and (b) derating curve derived from the basic curve.

which a device operates. The relation between the current, the temperature rise, and the ambient temperature of the component is represented by a curve, as shown in Figure 12.1a. Unless otherwise specified in the detail specification, the temperature rise is based on the mean current of three specimens. The mean value derived from the measured values of these three specimens serves as the basic curve. At least three points of the basic curve shall be established. The permissible upper-limit temperature of the materials employed is plotted as a vertical line on the graphs shown in Figure 12.1 with current I as the ordinate and temperature T as the abscissa. The temperature rise T (mean value of three specimens), determined at current I_n, is deducted. From this, the maximum permissible ambient temperature T_u for the load current I_n is obtained, since the sum of the ambient temperature T_u and the temperature rise t shall not exceed the upper temperature limit of the materials [13].

The dotted area shown in Figure 12.1b indicates the permissible operating range. This derating curve is obtained by applying a reduction factor of 0.8 on the current value of the basic current-carrying curve, unless otherwise clearly specified on the derating curve.

The connector current carrying capacity is primarily affected by the ambient temperature and the number of contacts that carry current simultaneously. The current carrying ability of any connector contact is limited mainly by the maximum allowable operating temperatures of the socket contact material(s) and the plastic connector housing material. For example, a significant amount of heat is generated in a size 30 contact when it carries a current of more than 1 A. This internal heat generation causes the contact to become hotter than its surroundings, and this causes rise in the temperature of the contact above the ambient temperature until it reaches an "equilibrium temperature." The equilibrium temperature is the temperature at which the rate of heat transfer from the hot contact to the surroundings is exactly equal to the rate at which heat is generated internally. When the heat output balances the heat input, the contact temperature will stabilize. At stabilization the difference between the contact temperature and the ambient temperature is referred to as the "temperature rise" of the contact. If the equilibrium temperature is within the safe operating temperature range of the pin spring material, then the contact is capable of carrying this level of current continuously, for long periods of time (years) [14].

The electrical conductivity of the spring contact is also an important design variable; for example, MIL-DTL-55302 specifies the use of several different contact materials with a significant difference in electrical conductivities. As the conductivity increases, the contact interface resistance decreases, and for power applications, minimizing both contact interface and bulk resistance is critical to avoid heating. A high-conductivity alloy also provides a size advantage and makes it possible for smaller contacts to support high-current requirements.

For a higher heat transfer rate (with a larger coefficient), high-velocity airflow around the connector is better, whereas stagnant air reduces it. In space applications, operation in a vacuum where there is no air and airflow is a separate issue. Connecter selection for a space environment is discussed in Section 12.17. By mounting a connector to a large, metal-core circuit board or other thermally conductive surface, heat can be removed more efficiently, and the connector will be able to carry a higher current rate than a connector suspended in mid-air and connected only to small-gauge wires in a cable.

Heat transfer is affected by the number of contacts that carry current simultaneously. When contacts are clustered together in a small area of the connector, the contacts in the center of the cluster must dissipate the heat they generate through the hotter surrounding contacts. That is, the apparent ambient temperature of the contacts in the center of a cluster is higher than the actual ambient temperature and can limit their ability to dissipate heat. Therefore, if only one contact in a connector is required to carry high current, this single contact will be able to carry more current without overheating than if several closely spaced contacts are used to carry large currents simultaneously.

The heat transfer coefficient can be easily measured for a given set of mounting, airflow, and simultaneously powered contact conditions by running a constant current through the contact(s) while monitoring the socket contact spring temperature using a very fine-gauge thermocouple [15]. The contact temperature is measured after it reaches equilibrium, and the ambient air temperature and the contact resistance are also measured. These values can then be used in Eq. 12.1 to calculate the value of the heat transfer coefficient (H). The test is repeated several times with different numbers of current carrying contacts clustered together, and a new H (or K) value is calculated for each combination. Equation 12.2 defines how the contact temperature T_c is affected by the following variables: current (I), ambient temperature (T_a), contact resistance (R_c), and heat transfer:

$$I^2 R_c = H(T_c - T_a) \tag{12.1}$$

which, when solved for T_c, yields the following:

$$T_c = \frac{I^2 R_c}{H} + T_a \tag{12.2}$$

Once the current rating and voltage rating are chosen, one needs to look at the connector datasheet to make sure the connector meets the given specifications. In the datasheet, one should look at the test procedure used to arrive at the specifications and check whether it meets the test standard because not every supplier uses the same testing criteria. A noncompatible conductor could cause overheating issues, leading to premature connector failure.

12.4 Environmental Conditions

Environmental conditions that connectors are exposed to include shock and vibration, crushing, pulling, bending, twisting, dust and dirt, water, oils, chemicals, corrosive gasses, temperature extremes, metallic shavings and filings, humidity, solar radiation, electrostatic discharge, radiated radio frequency (RF), electromagnetic interference (EMI), and electrical transients [16]. These environmental conditions must be identified and considered when selecting a connector. For example, if the connector is to be used in extreme weather conditions, one should check the manufacturer's ingress protection (IP) rating for sealing out dust and water at various depths and operating timeframes. A connector with an IP rating that satisfies the requirements of expected weather conditions should be chosen [17]. An IP67 code rating indicates these cables and connectors are protected against dust ingress as well as up to 30 minutes of total underwater immersion up to 1 m deep. In IP67, the 6 means that the connector prevents dust from entering the enclosure, and the 7 indicates that the connector prevents ingress when the enclosure is immersed in 1 m of water for 30 minutes. In IP68, the 8 indicates the protection from immersion in water with a depth of more than 1 m (manufacturer must specify exact depth). International IEC 60529 [18], "Ingress Protection," defines the ratings and classifies the degrees of protection provided against the intrusion of solid objects (including body parts such as hands and fingers), dust, accidental contact, and water in electrical enclosures.

Sections 12.18 and 12.20 describe the harsh environment and high-power subsea environmental considerations for connector selection.

12.5 Termination Types

The connector assembly process and the ability to seal the connector are both dependent on the termination type, which is based on how the connector will be used in the system. Connectors with solder contacts are easier to seal and thus prevent moisture ingress, whereas crimp contacts may be easier to maintain and repair. There is a trade-off between the two termination types, so the final decision is often made after discussions with manufacturing and design groups. These groups must know exactly how and where the connector will be used and whether field reparability is a requirement, as this decision has a significant impact on the assembly equipment and processes used in manufacturing [19]. Miniature connectors fit in small places, but only a few can carry power and signal. These small connectors are extremely difficult to terminate; therefore, miniature plugs and receptacles are often sold prewired to maintain better reliability.

12.6 Materials

A connector is intended to transmit power or signals between two subunits of an electronic system without any unacceptable power or signal distortion. The resistance introduced into the system by the connector must be acceptably low and stable. So, the material selection process for connectors is dedicated to ensure the resistance stability. Material selection can be dependent on the material's electrical, mechanical, and environmental capabilities and the ease with which connectors can be attached to other materials through processes such as crimping and soldering. Connector materials are chosen that have minimal resistance, proper electrical conductivity, suitability for machinability, good stability over time, and enough hardness such that there is minimal wear and performance deterioration even with repeated coupling cycles.

Brass connectors with nickel/chrome plating are more wear-resistant and have longer life cycles than many other materials. Aluminum connectors are an option if lighter material is needed. Plastic requires adequate testing to verify that it will continue to perform well per the end-use application.

It is also necessary to review the operating temperature of the insulating materials used in the connectors. In humid conditions, insulation will absorb the humidity in the atmosphere and can cause insulation resistance (IR) failures. When insulation absorbs water or water vapor, electricity begins to leak (water is a conductor) between conductive parts. The more water absorbed, the worse the problem.

Insulation may absorb water in high relative humidity conditions. In addition, insulation made of materials other than Teflon/TFE or polyvinyl chloride (PVC), such as nylon, rubber, neoprene, or PVC with special additives, tends to respond differently in humidity by absorbing water. Residues that contaminate the materials, such as no-clean

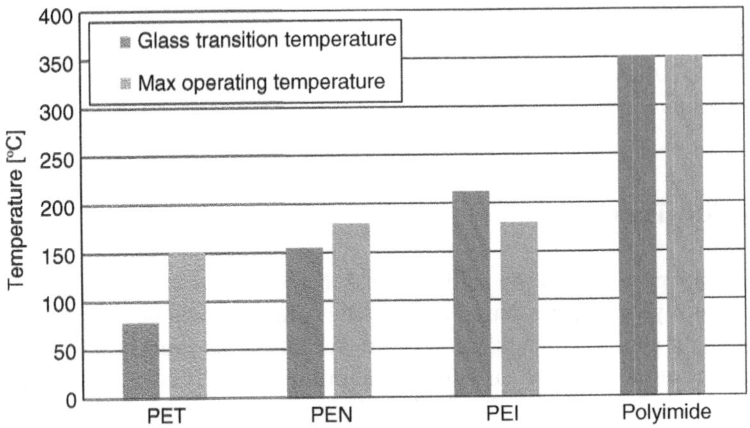

Figure 12.2 Thermal properties of electrical insulating materials [21]. Graphically renewed.

flux, cleaning alcohol, or even oils from the operator's hands, can also cause humidity absorption.

The selection of the optimal electrical insulating material, whether it is polyethylene terephthalate (PET), polyethylene naphthalate(PEN), polyetherimide(PEI), or polyimide, is a function of the anticipated operational and environmental requirements. For mission-critical applications, the correct choice of electrical insulation materials and processes starts when the engineer addresses these requirements and balances the various design and application risks against expected cost constraints. Because of the polymer transitions from a hard, glassy material to a soft, rubbery material at the glass transition temperature (T_g), insulation materials for connectors are decided considering the T_g (see Figure 12.2) [21].

The hydroscopic nature of insulating materials is evaluated by IPC-TM-650 Test Method 3.5, "Humidity, Connectors," which looks for evidence of deteriorated physical properties (e.g. dimensions and mechanical strength) or degraded electrical properties (e.g. insulation resistance). Zhang et al. [17] reported that 22.3% of connector failures were caused by deteriorated insulation, which leads to reduction in the resistance between the different contacts or the contact and the shell or possibly a short circuit.

For example, maritime electronic equipment often operates in an environment of high temperature, high humidity, and high salt fog. Although the connector is sealed, the protection degrades with time because the moisture and salt particles are absorbed in the connector. The moisture and its corrosion effects diminish the resistance of the insulation. For effective insulation materials, the surface resistance coefficient of the materials is often sufficient to prevent current leakage – the salt and moisture are combined in the polluted insulator and a leakage passage for the current forms, causing a reduction of the insulation resistance and conduction of the insulator surface.

Zhang et al. [17] were primarily focusing on modeling the failure of the insulation and the effects of the transmission signals caused by this failure. As a result of deterioration of insulation materials, the insulation resistance was reduced. Because the failure of the insulation material and the change of insulation resistance are gradual, the scanning parameters are customized for this gradual change process and are used in their analysis. Zhang et al. concluded that the voltage drop across the load resistor changes with the continued decrease of insulation resistance and the smaller the insulation resistance is the greater the signals transmitted on the contact changes. Thus, the voltage of one of the contacts is affected not only by the signals transmitted on it but also by the adjacent interference signals coming through the insulation resistance.

High-grade jacket materials, such as polyurethane (PUR), are resistant to harsh conditions such as oils, fire, abrasions, and chemicals. Lower-grade jacket materials, such as PVC, while less expensive cannot handle exposure to some oils and chemicals and can degrade. PVC also becomes brittle and cracks at low temperatures.

Stainless steel is a steel alloy with a small amount of chromium content that does not rust or corrode when exposed to moisture. It is a hard material that is extremely durable and is often used for connector housings, but not for contact parts because of its hardness and relatively low electrical conductivity. Stainless steel costs more than bronze or brass, but its high stability, high durability, and good corrosion resistance make it reliable and safe in different extreme operating environments.

Brass, a copper–zinc alloy, is used for connector housings as well as for connector contacts owing to its lower cost. However, it is less durable and softer than stainless steel. It is an excellent conductor of both heat and electricity, and it provides good durability in most industrial and marine environments because it is not subject to corrosion from contaminants found in those environments.

Connector parts are plated with different metals to improve the electrical and thermal conductivity, the contact between conductors, and the weldability and solderability. Noble metals, such as gold and silver, are excellent conductors and are resistant to corrosion, but these are expensive materials. Thus, only thin layers of these noble metals may be used on top of other metals during manufacturing. Plating allows manufacturers to take advantage of the electrical and thermal properties of these metals while limiting the actual amount used. Gold, for example, is an excellent conductor – it features good oxidation resistance, and it enhances the electrical conductivity of connector parts made from copper or brass. Owing to its high cost, gold is plated in thin layers, which can sometimes suffer from diffusion or wearing away of the gold finish. To minimize gold diffusion, nickel is used as underplating beneath the gold layer.

For coaxial connection fabrication, the metals used are stainless steel, brass, nickel, gold, silver, or white bronze. SMA and Type N connectors are available with either stainless steel or brass housings with various finish types. SMA connectors can have passivated, nickel, or gold finishes, while Type N connectors can be passivated or plated with nickel, silver, or white bronze. A white bronze finish can be an alternative to silver without the tarnish problems and without the passive intermodulation (PIM)–inducing tendencies of nickel plating. The 7/16 DIN connectors have brass bodies with silver or white bronze finish, QMA connectors use brass with white bronze finish, and 3.5-mm connectors are constructed of 303 stainless steel with passivated finish to deter corrosion. In cases where there are choices, it is often a clear tradeoff between lower cost (brass) and higher performance and operating lifetime (stainless steel) [22, 23].

The choice of material impacts the reliability, weight, and cost of the system. Brass connectors with nickel/chrome plating are used in applications where there is a lot of wear and tear involved. For light-weight requirements, aluminum connectors can be used. Plastic connectors can be used in harsh, humid, high-temperature, and high salt fog conditions. If the application requires operating in a highly corrosive environment, then stainless steel may be required.

12.6.1 Connector Housing Materials

The connector housing electrically isolates the individual contacts from one another, maintains the dimensional spacing of the contacts, and facilitates mating of the connector by providing mechanical protection and support for the contact springs. It also shields the contacts from the environment. Connector housing materials are selected primarily on the basis of the material properties, which determine the isolation, spacing of contacts, and protection requirements. Table 12.2 lists some of the relevant properties for a few typical connector housing materials. The stability of the property with respect to manufacturing and application requirements also influences material selection. In general, more than one material will be acceptable in the selection.

Electrical isolation depends on dielectric strength and surface and volume resistivity. The dimensional stability of a housing depends on material selection and molding, assembly, and exposure conditions, so appropriate material selection will involve trade-offs based on application and manufacturing requirements and material characteristics. Connector housings are typically molded from crystalline thermoplastic resins. The advantages and disadvantages of the materials listed in Table 12.2 are as follows [24]:

Polyamides (nylons) (PA): Nylon 66 possesses a good balance of mechanical, electrical, and temperature characteristics, good chemical resistance, and stability, but it has a tendency to absorb moisture, which has a significant negative effect on its dimensional and electrical stability.

Polybutylene terephthalate (PBT) and PET: PBT is the most used connector housing material due to its processability and functional characteristics. PET shares the structural and performance characteristics of PBT with an improvement in temperature capability, but the creep and warpage of PET are greater than PBT. Both PBT and PET are used in glass-reinforced grades (PET resins with 30–50% glass).

Polyphenylene sulfide (PPS): PPS has a major advantage because of its high temperature capability. Its mechanical properties are good, but it has a tendency toward brittleness.

Liquid-crystal polymers (LCPs): LCPs are good candidates for surface-mount applications owing to their stiffness, mechanical strength, thermal expansion characteristics, mold flow characteristics that provide good capability to fill thin-wall sections, and near zero shrinkage.

Table 12.2 Connector housing material properties [24].

Property	PA	PBT	PET	PPS	LCP
Dielectric strength ($kV\,mm^{-1}$)	18	24	26	18	44
Tensile strength (MPa)	133	119	154	126	168
Flexural modulus (GPa)	7.0	7.7	10.5	11.9	14.7
Notched, I_{ZOD} 75 °C ($J\,m^{-1}$)	54.9	37.6	46.2	37.6	63.6
Heat deflection test, 1.8 MPa (°C)	241	204	224	260	240
Temperature index (°C)	130	135	150	210	130

Table 12.3 Contact spring material characteristics [24].

Material	Young's Modulus (10^9 Pa)	0.2%Yield Strength (10^6 Pa)	Stress Relaxation	Formability	%IACS[a)] conductivity
C17200 (Beryllium Copper)	130	560–875	Excellent	Good–excellent	21–26
C26000 (Brass)	112	420–630	Poor–fair	Fail–good	26–28
C51100 (Phosphor Bronze)	112	490–770	Fair–good	Good–very good	18–20

a) IACS: International Annealed Copper Standard.

12.6.2 Connector Spring Materials

Connector spring materials must have sufficient mechanical strength to generate the forces required to make and maintain the contact interface. Spring materials must have sufficient formability to tolerate the deformation of the crimping process and the stamping and formation of the contact geometry. Copper alloys are the most common contact spring materials. Contact springs provide the normal force that produces and maintains the contact from the separable interface to the point of permanent connectivity. This function is easily met by copper alloys [25]. In signal applications, the stability of resistance is more important than conductivity, but for power applications, conductivity is significant in material selection. Selection criteria for connector spring materials are based on Young's modulus, yield strength, and stress relaxation resistance to provide the required contact normal force with deflection characteristics of the spring. Spring materials should have sufficient formability under the manufacturing conditions. Table 12.3 lists the pertinent material characteristics for common connector spring materials.

Commonly used connector spring alloys are as follows. Brass (C26000) is used in consumer goods in high-force systems because its limitations of poor stress relaxation hinder its temperature capability and formability and it is not suitable for many contact geometries. Phosphor bronze (C51000, C51100, and C52100) is the most commonly used alloy because of its good strength, formability, and stress relaxation conductivity. Beryllium copper (C17200) provides superior strength/formability performance with excellent stress relaxation characteristics, but its use is limited by its high cost.

12.7 Contact Finishes

To optimize the contact interface, the contact finish must be free of surface films (which is a nonconductive film that isolates the mechanical interface of the contact); the contact area must be adequate; and there must be sufficient electrical conductivity. These requirements determine the type of metal used – noble or precious metals such as gold or non-noble metals such as tin. Maintaining the contact interface requires corrosion protection for the contact springs, a diffusion barrier against the base metal spring constituents, and an acceptable level of durability. Contact spring materials are subject to corrosion in typical environments. A diffusion barrier is a corrosion-related

Table 12.4 Contact finish characteristics [24].

Finish	Contact Resistance (mΩ at 0.1 kg)	Knoop Hardness	Coefficient of Friction	Durability
Gold	1.1	190–240	0.2	Good–very good
Palladium	1.4	150–350	0.3–0.5	Good–very good
Palladium–silver (60 : 40)	1.7	100	0.7	Very good
Palladium–nickel (80 : 20)	8.0	400–500	0.3–0.5	Very good
Silver	0.5	50–125	0.5–0.8	Fair
Tin	0.8	10	0.6–1.0	Poor–fair
Tin–lead (93 : 7)	0.7	10	0.5–0.8	Poor–fair

requirement – adding corrosion resistance to the finish ensures that the finish is not worn away during the specific mating cycle lifetime of the connector.

Gold over nickel is an example of a noble metal contact finish. Gold is an ideal contact finish because of its electrical and thermal characteristics and corrosion resistance. Palladium is also a noble metal, but it is a catalytic material and organic deposits can form on its surface. In many cases, palladium–nickel and palladium–silver are used with a gold surface for better corrosion resistance. Non-noble metal finishes always have some amount of surface film. Tin is the most commonly used non-noble metal finish. Tin has tin-oxide on its surface, and fretting corrosion is its major degradation mechanism.

Silver reacts with sulfur and chlorine, resulting in surface films. However, silver does not exhibit fretting corrosion. Silver finishes form surface tarnish and silver sulfide and have a tendency to electromigrate. Silver can be used for high-current contacts due to its high electrical and thermal conductivity and resistance to welding. Table 12.4 summarizes the connector finish properties of noble and non-noble metals that relate to connector performance.

12.8 Reliability

There are two aspects to ensuring connector reliability: building reliability into the connector through materials and design choices, and assessing whether reliable performance has, in fact, been realized. A connector is an electromechanical system that provides a separable electrical connection between two subsystems without unacceptable signal distortion or power loss in relation to the designed application/lifetime performance. On the microscale of the contact interface, all surfaces are rough, and this roughness determines the performance characteristics of the contact interface; the performance of the contact interface, in turn, determines the connector reliability [26].

"Reliability needs" refers to how much repeated wear and tear the connector can handle. It includes repeated mating cycles (i.e. connection and disconnection of the device over its lifetime) and externally applied stress and strain. If a high number of mating cycles is a requirement, then appropriate connectors with proper features should be chosen, such as a connector with 5000–10 000+ mating cycles. Another requirement to consider is whether the connector will be able to remain stable in harsh and extreme environments [27].

12.9 Raw Cables and Assemblies

After choosing the connectors, the cables to go with the connector should be considered. There should be good compatibility between the connector and the cables. When determining the cable types, it is important to know exactly how and where the connector will be used and whether field reparability is a requirement. This decision has a significant impact on the assembly equipment and processes used in manufacturing [27]. Most connector manufacturers provide assembly services. The manufacturing facility's ability to provide termination-only capabilities or complete over-molding, sealing, and testing capabilities needs to be considered.

The connector assembly process takes into account the crimp type (crimp contacts are assembled to conductors outside of the connector and are subsequently installed in the connector body), the solder type (the stress relief of soldered connectors needs to be sufficient), whether the connector is hermetic or environmental sealing type, whether it is an assembly of adapters and cable clamps (adapters, cable clamps, and connector back shell tongs are torqued to connector bodies using calibrated tooling), and whether it is an RF connector and uses coaxial contacts (assembly procedures vary from normal shop practice for cables operating at less than 1 gigahertz [GHz] to precision techniques for those operating up to 18 GHz) [25]. The assembly process controls, inspection, and verification testing should take place. Verification testing consists of tensile testing, push testing, pull testing, and visual examination depending on the connector type.

12.10 Supplier Reliability Capability Maturity

System integrators purchase the parts and subassemblies by assessing their reliability, which is provided by evaluating the supplier company's design for reliability and implementing a reliable design through manufacturing and testing capability. An up-front evaluation of suppliers based on their ability to meet reliability requirements can provide a valuable competitive advantage for system integrators.

The supplier should be selected based on its ability to provide a high-quality and reliable product. A reliability capability maturity metric is a scale to measure the procedures within an organization that affect the reliability of the final product and the effectiveness of these practices in meeting the reliability expectations of customers. The maturity metric to determine organizational capabilities has roots in quality management. Crosby's Quality Management Maturity Grid [28] describes the typical behavior of a company, which evolves through five phases (uncertainty, regression, awakening, enlightenment, and certainty) in its ascent to quality management excellence. The grid is a simple 5×6 matrix that shows different stages of maturity of the company's quality management for six different quality management categories (management understanding of quality, problem handling, and cost of quality, quality improvement actions, and summary of company quality posture). The minimum score is 6 (all categories are at "Uncertainty"), and the maximum is 30 (all are at "Certainty") [29]. Once the supplier has been evaluated and the risk is acceptable, then connectors from that supplier can be considered.

12.11 Connector Selection Team

When a customer problem is tied to a specific connector, one often asks the question, "Who selected that connector in the first place?" The answer is usually the person or group that selected it in the *last* place rather than in the *first* place. The selection process often starts with the design stage as an R&D function, where the primary expectation is to prove system functionality regardless of cost and manufacturability. Connector selections at the early design stage are passed to the development team, which is responsible for refining and developing the system further and often for maintaining the original connectors specified in the design. The design then reaches the purchasing/procurement/supply chain management department, which tries to reduce cost and assure availability rather than make the correct technical selection. Problems may arise when the selected connectors are changed based on equivalent form, fit, and function. Whereas an alternate connector may appear to be the same as the original selection, it may not pass an adequate qualification process.

The recommended connector selection team is composed of an electrical design engineer, a mechanical design engineer, a system engineer, and a reliability engineer. Some recommendations for creating a connector selection team are as follows [30]:

- Apply a project team approach in product development, making sure the manufacturing and purchasing departments participate in the development process and are aware of its importance.
- Assign detailed responsibilities to each team and make their role and task clear in connector selection and approval throughout the complete design and manufacturing cycle.
- Recognize that the function or team accountable for product performance must control any outsourcing of parts for the product design. The design of the mechanical packaging and component selection is as important as the design of the electrical circuit and the firmware.
- Consider connectors as critical-to-design-and-function components, and include them in the design review process.
- Avoid the use of commercial item part numbers, which are often used for generic hardware items.
- Include qualification of connectors in the product design verification and validation steps of the product development process.
- Document approved connector suppliers and specific part numbers.
- Require original equipment manufacturer (OEM) engineering approval and traceable documentation for connector substitutions and/or changes.
- Require those upstream (connector manufacturers) and downstream (manufacturing assembly) to document and requalify all connector changes and modifications.
- Consider adopting an organizational "RACI" plan, which is basically a matrix outlining who in the chain of events is to be **R**esponsible, **A**ccountable, **C**onsulted, or **I**nformed regarding changes to connector selection.

12.12 Selection of Candidate Parts from a Preferred Parts Database

Connector manufacturers provide selection guides for their products depending on different parameters according to the user's needs. The parts database contains archived information gathered on parts so that designers and procurement specialists do not have to repeat previously performed part selection and qualification processes. A preferred (or approved) parts lists is a subset of the complete parts database containing only parts that have been approved for use in the company's products. Some companies also have a guaranteed parts database containing a subset of the preferred parts list that is guaranteed to be in stock (or procurable).

The information that should be contained in the database includes baseline data gathered from part manufacturers and other sources on the design, function, structure, and properties of the connector (e.g. part life cycle information, qualification and "reliability" monitoring conditions and results, and test and manufacturing yield data); temporal data for a part that varies from lot to lot of the product, such as that which comes from product change notices (PCNs); usage data about the use of the part within the organization; and failure data about the failure history of the part within the organization (documentation of rejected parts, field return data, and lessons learned).

12.13 Electronic Product Manufacturers' Parts Databases

Many electronic product manufacturers have internal parts databases to archive information gathered on parts so that engineers and procurement specialists do not have to "reinvent the wheel" every time they consider a part. Many of these databases have evolved into "preferred" or "approved" parts lists containing only parts that have been approved for use in the company's products. Ideally, the selection of a part from an approved parts list is a guarantee to the design engineer that the part has been qualified for use within the company and can be successfully procured. For the parts database to perform this role, several key elements must be present and managed in the database:

- A formal method should be used for tracking and linking product change notifications (PCNs) to the parts in the database [31] so that the PCNs received by different functional groups (e.g. purchasing, component engineering) within a company are input to the database [1]. When a new PCN is received, a flag should be raised in the database to alert future potential users of the part changes.
- Expiration dates should be assigned to each part to indicate when the "approved" status of a part expires. The information gathered and analyzed for making parts selection decisions is time specific. For example, the "reliability" test results that are used for parts family assessment are periodically (e.g. quarterly, biannually) updated by parts manufacturers [32].
- Criteria for parts reassessment must be defined. Examples of possible criteria include a fixed period, expiration of the part status, issuance of certain types of part changes (e.g. change in package type, site of manufacture), or unexpected or unexplained failures.

- The parts database should link to the latest parts datasheet source on which assessment was made. The parts datasheet can change with the parts life cycle [33], and there is no guarantee that the original datasheet will be available from outside sources.
- The parts database should be linked to other company databases such as those on assembly yield, test yield, warranty, and other field return analyses so that detailed information can be accessed by the parts team when necessary.
- Parts that have been considered for use in the company's products but rejected for any reason should be documented, along with the reasons for their rejection.
- Part life cycle information should be placed in the database to prevent design engineers from choosing parts that are designated for discontinuance or that are already obsolete.
- If the part assessment resulted in qualification only for a limited environment and application, that should be clearly noted.
- The availability and accessibility of the part by multiple company sites and by contract assemblers should be in the database.
- Each part in the approved parts list should have a distinct part number (even if it is the same part from different manufacturers). Manufacturers are assessed [34, 35] separately from parts in the parts selection and management process, and a part is not uniquely identified unless a manufacturer is specified.
- Besides the part family assessment information, additional test data, which may be obtained by the manufacturer, by an independent testing facility, or by the company, should be linked to the database.

A part on the "approved" list should not mean that no further assessment is necessary; it means only that a part on that list can be a candidate part for further assessment. The process of adding a part to the approved parts list must provide design engineering with feedback as quickly as possible. Different companies have different policies regarding the use of parts during the approval process; however, in most organizations, design engineering can proceed with product development (using the part in their designs) while waiting for component engineering approval. If component engineering returns a "reject" decision, significant cost and schedule issues can occur if the approval process is lengthy.

Based on benchmarks at numerous companies, three weeks is a reasonable amount of time to turn around 90% of part approval requests. More lengthy approval processes may force design engineers to continue the design assuming that the part will be approved. If the part is eventually rejected by component engineering, it may be impractical, due to the solidification of the rest of the design, to reject the part. Regardless of where a product is in the design phase, if component engineering rejects a part, component engineering and the design engineers must have a process to resolve disputed rejections in a timely manner (in less than a week in most cases).

Other types of parts databases that may be useful include the following:

- The *demonstration database* contains parts for use in demonstrator products, i.e. those products designed and made for demonstration of a new technology or new system that is not intended for sale. Parts used in the demonstrator products are reviewed

and signed off by purchasing, but other approvals are not generally necessary to put the part in the demonstration database.
- The *guaranteed parts database* contains all the information included on an approved parts list, but only for a subset of the most common parts used in the company's products. The additional attribute that sets this type of database apart from the approved parts list is that the parts in this database are guaranteed to be in stock at the company or at the contract assembler so that parts procurement (availability) problems are not an issue.

12.14 Parts Procurement

The procurement process for most companies in the electronics industry consists of several steps, including selecting parts, finding the right vendor (parts manufacturer or distributor), negotiating the price, getting approvals, ordering the part, receiving the part, and paying the invoice.

The time, labor, and resources associated with procurement can be costly. Errors or delays in shipment add to procurement costs, slow down the design process, and leave design engineers waiting for parts they need to complete their jobs on schedule.

12.15 Parts Availability

Parts availability is measured in terms of lead time and the volume of available parts and is usually addressed by the procurement process in the company. During the candidate parts selection process, engineers provide procurement specialists with bills of materials. The procurement specialist performs several checks on each part in the BOM, including the following:

- It must be determined whether the part can be purchased from the company's existing qualified distributors and is manufactured by one of the company's qualified manufacturers. If not, then the procurement specialist must have engineering initiate a manufacturer and/or distributor assessment process.
- Depending on the quantity needed for the company's products and the short-term availability of the part, the procurement specialist may search for alternative sources of the specified part.
- If the part can be procured from an approved source, the cost of the part must be determined either from long-term strategic pricing agreements that the company already has in place, via catalog prices (for small quantities), or through quotes/negotiations with the supplier.
- The deliveries schedule must be determined (i.e. the number of parts the manufacturer/distributor can provide and when).
- The long-term availability of the part must be assessed in order to consider the necessity and viability of making a lifetime buy of parts with the initial purchase.
- Obsolescence of the part (i.e. whether it is already obsolete or soon to be obsolete) may be assessed by the procurement specialist.

12.16 High-Speed Connector Selection

High-speed digital design needs require connector manufacturers to meet the demand for multiple high-speed low-loss connections in a small connector shape. Connector design, therefore, is becoming increasingly difficult. High-speed digital design engineers can use the RF knowledge of connector suppliers, which is available in datasheets, in application notes, and on the Internet.

In selecting a connector, first the connector type should be chosen – backplane, board-to-board, board-to-cable, or mezzanine. Second, the manufacturers that supply connectors with the right physical parameters should be evaluated. The manufacturer's electrical specifications, test reports, and other published references should be examined. During the connector selection, the EMI parameters (translated to series inductance), cross talk (translated to mutual inductance), and signal propagation (known as "parasitic capacitance") should also be evaluated [36].

The most basic effect that a connector causes to a circuit is series inductance, of which the primary factor is the pin length of the connector. Together with the series inductance of each connector pin, the pin layout of the connector determines the radiated EMI. Signals traveling through a connector need a current return path (ground). Even if no return path is provided through the connector, large inductive loops can be created. This will result in substantial EMI emission. This effect can be minimized where there are enough ground pins in the connector.

Parasitic capacitance exists between the parts of an electronic component or circuit simply because of their proximity to each other. Parasitic capacitance is unwanted and usually unavoidable like another effect caused by a connector – mutual and shunt (pin-to-pin) capacitance. Capacitance slows down the system edge rate. In multidrop backplane applications, parasitic capacitance places a greater burden on connectors than in point-to-point applications. The cumulative effect of parasitic capacitance caused by many connectors has a larger impact than series inductance in a multidrop bus. Therefore, the chosen connector needs to have low parasitic capacitance, even at the expense of higher inductance [37].

To provide high-quality high-speed connectors, manufacturers need to control and manage in the design stage the previous parameters as well as many more. Engineers are now able to reach an extensive amount of data measured and calculated by connector manufacturers.

12.17 NASA Connector Selection

Materials are a primary consideration in selecting connectors for space flight applications. The requirements for NASA connectors are described in NASA EEE-INST-002, "Instructions for EEE Parts Selection, Screening, Qualification, and Derating" [38]. Figure 12.3 shows a schematic of the NASA connector selection and application process.

A connector is a complex system having many failure modes and safe, and reliable use of a connector requires close attention to its design and implementation in each application and environment to avoid each of the many failure modes.

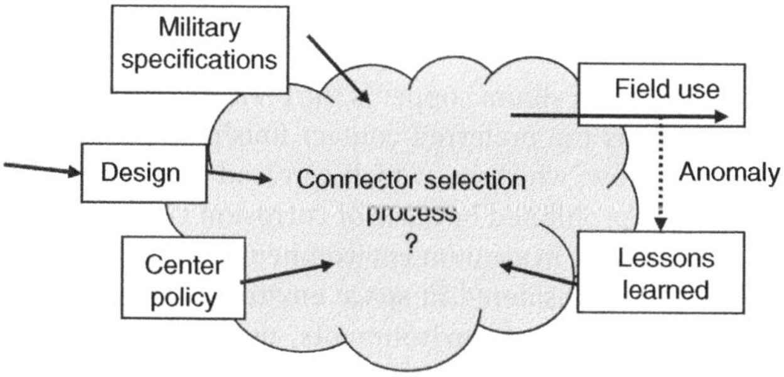

Figure 12.3 NASA connector selection process [39].

Materials are a primary consideration in selecting connectors for space flight application, and the following requirements and guidelines for housing are provided in NASA EEE-INST-002:

Base materials for metal shell connectors: "For metal shell connectors, base metals should demonstrate low permeability (i.e. resist the establishment of magnetism in the material). Machined aluminum alloy, corrosion-resistant steel, and brass are the preferred metals. For connector contacts, copper, beryllium copper, or half hard brass are the preferred metals. In some applications, it may be necessary to screen connectors and contacts for specific levels of residual magnetism."

Preferred finishes for metal shell connectors and contacts: "Electroless nickel plating is the preferred finish for circular, general-purpose D-subminiature, and microminiature metal shell connectors. Gold over copper flash is the required finish for D-subminiature connectors when residual magnetism is a consideration. Passivated stainless steel or gold is the required finish for coaxial connectors. Gold plating in accordance with ASTM-B488, Type II, Grade C, Class 1.25 (50 µin) (formerly MIL-C-45204, Type II, Grade C, Class 1) is the preferred finish over the entire contact, including the engagement area."

Prohibited finishes for metal shell connectors, contacts, terminals, and terminal lugs: "Cadmium, zinc, chemically coated cadmium or zinc, or silver should not be used as a connector or contact finish. Silver should also not be used as an underplate or as a finish due to corrosion concerns when exposed to atomic oxygen in lower earth orbits. Due to the risk of tin whisker growth, which can lead to short-circuit conditions, pure tin finishes are prohibited."

Preferred dielectric materials: "The preferred dielectric insulating materials for multi-contact connectors (color is optional) are as follows:

- Molding compound diallyl phthalate (DAP), type SDG-F or type GDI-30F of ASTM-D5948 Thermoplastics Polyester, glass-reinforced, MIL-M-24519 type GPT-30F or type TPES013G-30A0000 (per ASTM-D5927).
- Polyphenylene sulfide (PPS), glass-reinforced, MIL-M-24519 type GST-40F or type PPS000G40A00000 (per ASTM-D4067).
- Liquid-crystalline polyester, glass-reinforced, MIL-M-24519 type GLCP-30F.

Diallyl phthalate is the preferred material for solder applications. PTFE Teflon is the preferred dielectric material for radio frequency and data bus connectors. In general, the nonmetallic materials used in connectors should be noncombustible or self-extinguishing.

As a summary for connector materials: Beryllium copper is the preferred material for contacts; 50 μin minimum gold plating is the preferred contact finish; 100% tin plating should not be used; pure tin can grow 'whiskers,' which can lead to catastrophic electrical short circuits; silver plating is prohibited because of corrosion concerns; and cadmium is prohibited because it is unstable in vacuum environments.

Outgassing properties also need to be considered in space environment connector selection. Because outgassing occurs in vacuum environments, unreacted additives, contaminants, absorbed gasses, or moisture can evaporate from molding materials, and these outgassed materials can condense on cold surfaces and cause performance degradation. Outgassed materials can also become more rigid or brittle. Nonmetallic materials should not exceed 1% total mass loss (TML) or 0.1% collected volatile condensable material (CVCM) when tested in accordance with ASTM-E595, "Test Method, Outgassing."

Acceptable materials should be selected from NASA Reference Publication 1124, "Outgassing Data for Selecting Spacecraft Materials," at outgassing.nasa.gov or Materials and Process Technical Information Service (MAPTIS) at mpm.msfc.nasa.gov/materialdb.html. However, materials listed as acceptable in these documents may have been baked out prior to evaluation to reduce outgassing, and the user may have to perform similar processing to achieve acceptable levels of outgassing. Testing should be performed per ASTM-E595 for materials that are not traceable to the previous references. Processing generally consists of a bakeout at 125 °C and 10^{-6} Torr for 24 hour. The TML cannot exceed 1.00% of the total initial mass. During the test, outgassed matter condenses on a cooled collector plate. The quantity of outgassed matter is calculated to determine the CVCM. The CVCM cannot exceed 0.10% of the original specimen mass.

NASA states, "A bakeout for outgassing control is driven by the application and may be required where tight contamination control must be maintained" [38]. NASA generally recommends that military circular connectors undergo outgassing processing. An inexpensive oven bakeout has better results than the more costly thermal vacuum outgassing. Higher-temperature oven bakeouts are more effective at removing volatile materials. However, both methods ensure compliance with outgassing limits.

The recommended NASA outgassing processes are as follows:

Circular connectors: "The silicon rubber environmental seals and lubricants applied to keyways are outgassing concerns. Additional processing for contamination control due to outgassing is available by special request from some military qualified parts list (QPL) manufacturers. Processing may also be performed in-house by replacing normal lubricant with low-outgassing lubricant followed by a bakeout. A selection list of low-outgassing lubricants and greases is found in NASA Reference Publication 1124, sections 8 and 13, or NASA-MSFC Handbook 527. Removal and replacement of lubricant requires some disassembly of the connectors and is usually labor-intensive. A bakeout of military connectors is helpful in controlling outgassing, but is not guaranteed to resolve all outgassing concerns."

MIL-DTL-83513 microminiature connectors: "Certain receptacles contain a thin press-on fluorosilicone–rubber or silicone–rubber interface. Silicone may represent an outgassing concern. If necessary, this seal should be removed with small tweezers and discarded. If a seal must be maintained, it may be necessary to bake out the connector or the interface seal in order to prevent outgassing."

MIL-PRF-39012 SMA RF connectors: "Certain plugs contain a gasket interface seal inside the coupling nut, inside the cable mounting nut, or on the mounting flange, which may represent an outgassing concern. Additional processing may be required for outgassing control."

MIL-DTL-83513 microminiature and DSCC 94031 nanominiature connectors: "Certain plugs contain an epoxy potting material that must be traceable to acceptable outgassing test results."

D-Subminiature connectors: "NASA GSFC 311 connectors have acceptable outgassing properties, and no further processing is normally required. However, some commercial versions contain an optional rear silicone rubber sealing grommet behind the connector that provides wire support and seals the rear of the connector. The silicone rubber and the bonding agent represent an outgassing concern, and additional processing is recommended for outgassing control. Other military and commercial connectors may contain a cadmium or zinc finish that is unacceptable for use in high-temperature vacuum environments due to sublimation problems. Right-angle printed wiring board connectors may have a nylon mounting bracket that may require a bakeout for outgassing control."

MIL-DTL-55 302-type printed wiring board connectors: "Military PWB connectors are made from molding materials with known good outgassing stability, and no further processing for outgassing control is normally required. However, certain commercial connectors may be molded with thermoplastic polyamides, such as nylon, that represent an outgassing concern. Certain nylons can contain additives that can outgas. Nylon is also hygroscopic and can absorb moisture, even after a bakeout. The use of nylon is not recommended."

When the application requires special processing for outgassing control, color bands shall be omitted from the contacts, or low outgassing epoxy inks shall be used. For coaxial and high-voltage contacts, the insulator material should meet 1% TML and 0.1% CVCM requirements when tested in accordance with ASTM-E595.

12.18 Harsh Environment Connector Selection

A harsh environment can generally be described as a setting in which survival is difficult or impossible. The extreme cold of Siberia, −70 °C, or the extreme heat of the Sahara Desert, 57 °C, may be considered harsh environments for humans [40].

For properly functioning connectors in different environments, manufacturers use all sorts of metals, plastics, rubbers, and ceramics throughout the connector to meet both interconnect and harsh environment requirements. The materials in internal components vary due to performance and cost. External components, connector shells, and inserts are often metal and can be aluminum, stainless steel, brass, titanium, or even composite to meet the demanding harsh environment conditions [41].

Table 12.5 compares the various base materials and finishes common in the connector industry. These are relative comparisons among the materials for cost, weight, electrical conductivity, and corrosion resistance. Also included are temperature and salt spray corrosion resistance performance characteristics of the base material and finish. Finally,

Table 12.5 Harsh environment connector properties for connector selection.

Base Material	Finish Material	Cost (1–6, 6 Most Expensive)	Temperature Range	Corrosion Resistance	Weight (1–6, 6 Heaviest)	Conductivity (1–3, 3 least Conductive)	RoHS-Compliant	Color Availability
6061-T6 Aluminum	Cadmium	2	−55 to 175 °C	2000 hr	2	1	No	Olive drab
	Zn-Ni	3	−55 to 175 °C	500 hr	2	2	No	Olive drab, black
	Zn-Co	3	−55 to 175 °C	500 hr	2	2	Olive drab (no), black (yes)	Olive drab, black
	Anodized	2	−55 to 175 °C	300 hr	2	3	Yes	Red, green, blue, gray, tan
	Hard anodized	2	−55 to 175 °C	>2000 hr	2	3	Yes	Natural (gray/green), black
	Nickel PTFE	3	−55 to 175 °C	1000 hr	2	1	Yes	Silver/gray
303 CRES		4	−55 to 175 °C	Very good	5	1	Yes	Natural (silver), black
304 CRES		5	−55 to 175 °C	Excellent	5	1	Yes	Natural (silver)
316 CRES	Unplated	5	−65 to 225 °C	Excellent	5	1	Yes	Natural (silver)
	Ni-plated	5	−55 to 175 °C	Excellent	5	1	Yes	Natural (silver)
Brass		4	−55 to 175 °C	Excellent	6	1	Yes	Brass
Marine bronze		5	−65 to 225 °C	Excellent	6	1	Yes	Bronze
Titanium		6	−55 to 175 °C	Excellent	3	1	Yes	Natural (silver)
Zytel® (PA66)		1	−55 to 175 °C	Good	1	3	Yes	Various
PEI		2	−55 to 175 °C	Very good	1	3	Yes	Natural (tan), black
PEEK	Unplated	3	−55 to 175 °C	Excellent	1	3	Yes	Natural (tan), black
	Ni-plated	3	−55 to 175 °C	Excellent	1	3	Yes	Nickel
PPS		2	−55 to 175 °C	Very good	1	3	Yes	Natural (tan), black
Estaloc™ RETPU		1	−55 to 175 °C	Good	1	3	Yes	Various

PTFE: Polytetrafluoroethylene; PEI: Polyethylene imine; PEEK: Polyether ether ketone; PPS: Polyphenylene sulfide

Restriction of Hazardous Substances (RoHS) [42] compliance is identified, and colors are noted for their availability in the material and finish listed.

12.19 Fiber-Optic Interconnect Requirements by Market

Fiber-optic interconnects have requirements to connect two terminated ends of a fiber and/or electrical cable repeatedly and reliably; isolate the electrical and fiber connections from the mechanical, thermal, electromagnetic, and chemical conditions found in harsh environments; work with a variety of cable configurations and sizes; and seal against water, dust, and electromagnetic emissions. Fiber-optic interconnect requirements for testing, performance, and material types are described for Amphenol products in Table 12.6.

12.20 High-Power Subsea Connector Selection

In high-power subsea connectors, various stress conditions exist at the same time, such as mechanical loading, temperature, pressure, and high voltage. Managing just one of these stress conditions can be difficult, but managing multiple stresses is even more challenging. Thoughtful selection of connector materials along with a robust reliability program is necessary for effective subsea equipment design.

Selecting the best materials for mechanical, electrical, thermal, chemical, or other performance measures is an important part of the design process. Systematic selection of the best performance material for a planned application starts with knowledge of the component requirements and the primary materials properties that affect those performance factors. When multiple criteria are considered, ranking candidates become a complex process. For example, high-power connector pin insulation needs to be stiff to maintain its shape in a high-pressure subsea environment while also possessing minimum dielectric strength degradation. A simple example relevant to subsea connector insulation is illustrated as a scatter plot that displays two or more properties of different materials (Ashby plot) in Figure 12.4, which shows the Young's modulus (stiffness) of various material classes plotted as a function of their respective strengths [42].

Another tool that is necessary today for the design of high-power connectors is accelerated life testing (ALT). At the component level, such as high-power ampacity pins, data can be generated by testing the multiple stress use and extrapolating the results back to field conditions, and then the time-to-failure under field conditions can be estimated. The reliability of the connector system eventually contributes to the reliability of the overall operating system. Interconnects, along with all subsea equipment, must meet the life cycle expectancies of the overall deployed system. The concept of ALT is testing under stress levels in excess of expected operating conditions for a shorter test time. The experimental data is used to estimate performance under actual use conditions. The factors that may affect the dielectric performance of a component include choice of thermoplastic material, geometry/contours of the pin, surface finish and plating of the conductor, percentage of glass content, molding process, heat treatment process, connector pin shell/fixture assembly, types of seals and elastomeric materials grades, and external loads and fixation points.

Table 12.6 Typical Amphenol fiber optic interconnect requirements by market [41].

Environmental Attributes Requirements	Shipboard	Tactical	Aero	Geophysical	Industrial
Operating temperature	−40 to 85 °C	−55 to 85 °C	−50 to 85 °C	−55 to 120 °C	−50 to 85 °C
Corrosion resistance	500-h salt spray test	500 hours salt spray test		>1000-h salt spray test	Application-specific
Conductivity	Required	Application- specific	Required	Application- specific	N/A
Weight	Medium priority	Medium priority	High priority	Medium priority	Medium priority
Color	Olive drab	Olive drab & black	N/A	Black	Black
EMI shielding effectiveness	80 dB	80 dB	80 dB	N/A	Application- specific
Typical materials used in each market	Cadmium-plated aluminum, 316 CRES, nickel-plated 316 CRES, hard anodized PTFE-plated aluminum	Zn-Ni-plated aluminum, Zn-Co-plated aluminum, hard anodized PTFE-plated aluminum, 303 CRES	Cadmium-plated aluminum, nickel-plated composite, Durmalon-plated composite	Hard anodized PTFE plated aluminum, 303 CRES, 316 CRES, brass, marine bronze	Anodized aluminum, hard anodized PTFE plated aluminum, 303 CRES

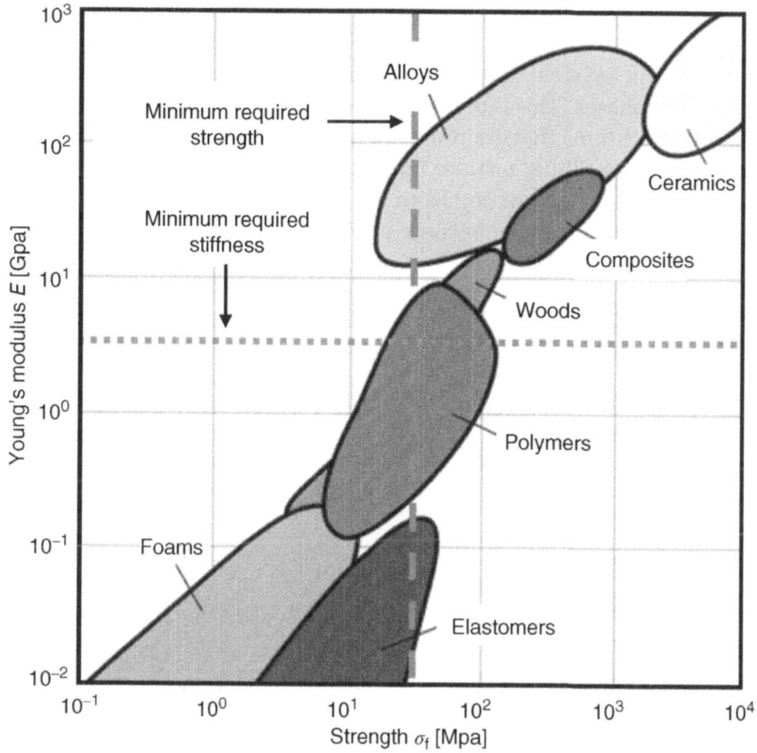

Figure 12.4 Materials selection based on an Ashby plot [43].

Optimizing component performance is achieved not only by careful material selection but also in conjunction with other design parameters. Such parameters are meant to address specific conditions and/or stresses that are dependent on the application and its environment. The combination of choices of certain parameters can also affect the outcome. The connector pin design can make use of a sufficient insulation level and appropriate heat treatment process, but if the insulation geometry and thicknesses are not carefully designed, this may lead to undesirable ramifications in the long term.

12.20.1 Undersea Connector Reliability

The earth's continents are connected with a web of undersea fiber-optic cables that join the world's population centers. The main advantage of underwater optical fiber communication is that the reliability of underwater cables is high due to extended repeater spacing, especially when multiple paths are available in the event of a cable break. For the manufacturer, reliability is defined during the time period after the connector is in place and gives a good indication of how long it will remain functional. The reason for limiting the area of reliability is based on the inability of the manufacturer to understand if handling, testing, and integration activities are performed by the customer in accordance with the manufacturer recommendations.

Reliability takes on a new definition when defined by an operator and where reliability statistics are not readily available. In this case, reliability would be defined as the number of connectors used in a system that remain operational through the integration, deployment, and operation phases. This is the area where reliability becomes a factor of exposure to untraceable conditions that are not controlled by the manufacturer. These conditions generally involve handling, testing, human error, and out-of-specification applications. Without the proper information from the manufacturer and proper training for its use, an undersea connector can become unreliable and susceptible to failure for a number of reasons as follows [44]:

- Excessive voltage during testing
- Conductive material or water in the interface
- O-ring surface damage
- Termination integration failures
- Cross-contamination in wet connectors
- Interchange in nonmatched configurations
- Live power demates
- Exposure to caustic fluids
- Impact loads in unrestrained shipping
- Excessive mates/demates during integration
- UV exposure
- Extended storage times
- Startup currents
- Misalignment during mating
- Mating with the wrong connector half
- Probing with test instrument
- Mating with damaged test connectors
- Excessive tension on connection
- Forklift intervention
- Demate via vacuum cleaner unplugging method
- Single-point bend over sheave
- Wet mate of dry style connector
- Fracture of exposed electrical pins before connection
- Excessive differential pressure during testing
- Excessive debris or sand during underwater connecting

The previous events lend themselves to further expansion by applying failure modes and effects analysis (FMEA) or failure modes, effects, and criticality analysis (FMECA) techniques to identify the possible failure modes and causes and to define the potential for each event to occur within a specific integration.

12.21 Screening Tests

Screening tests are intended to remove nonconforming parts (parts with random defects that are likely to result in early failures, known as "infant mortality") from an otherwise acceptable lot and thus increase confidence in the reliability of the parts selected for use. Product screening consists of electrical testing of the devices followed by a series of electrical and mechanical stress tests followed by further electrical tests. The main purpose of screening is to ensure the product is reliable by eliminating the early failures because early failures occur at a high rate (e.g. typically, for early failures with Weibull beta = 0.2, there are 27 failures before one year and only 6 failures from 1 to 10 years) [45]. Ideally, the screening method should be determined after analyzing the failures. This analysis should ascertain what types of failures occur during the early failure period and what stresses can cause these failures.

The following connector tests are performed to validate the connector according to its design specification for both structure and assembly and to determine whether the connector is manufactured and assembled properly:

- Axial concentricity
- Maintenance aging
- Contact engaging/separation force
- Contact insertion and removal force
- Cable pullout
- Contact tensile strength
- Insert retention
- Mating/unmating force
- Durability
- Contact retention
- Resistance to probe damage
- Crimp tensile strength
- Coupling torque
- External bending moment
- Coupling pin strength
- Gauge location
- Gauge retention
- Accessory thread strength
- Pin contact stability
- Solderability

A connector's electrical performance must also be measured to ensure that impurities, contaminants, or process issues have not affected the life of the product. The following connector tests for both connectors and terminals are performed for electrical performance:

- Contact resistance
- Low-level contact resistance
- Insulation resistance
- Dielectric withstanding voltage
- Shell-to-shell conductivity
- Discontinuity monitoring

The application of the connector can have different effects on different products and systems. Some of the connector tests to evaluate the service life of the product, connectors, and terminals are as follows:

- Drop test
- Vibration durability while monitoring discontinuity
- Water immersion
- Power wash testing
- Air leakage
- Magnetic permeability
- Highly accelerated life test (HALT)/highly accelerated stress screening (HASS)
- Solar radiation
- Salt fog
- Blowing dust
- Temperature/humidity/vibration/life cycle testing
- Dry circuit testing – current cyclic testing
- Micro-fretting
- Corrosion testing

EIA-364 establishes a uniform set of industry standards for evaluating the environmental performance of electrical connectors.

- EIA-364-1000, "Environmental Test Methodology for Assessing the Performance of Electrical Connectors and Sockets Used in Controlled Environment Applications"
- EIA-364-105A, "Altitude – Low Temperature Test"
- EIA/ECA-364-10E, "Fluid Immersion Test"
- EIA/ECA-364-110, "Thermal Cycling Test"
- EIA-364-11C, "Resistance to Solvents Test"
- EIA-364-17C, "Temperature Life with or without Electrical Load Test"
- EIA-364-26C, "Salt Spray Corrosion Test"
- EIA-364-27C, "Mechanical Shock (Specified Pulse) Test"
- EIA-364-28F, "Vibration Test"
- EIA-364-31C, "Humidity Test"
- EIA-364-32F, "Thermal Shock (Temperature Cycling) Test"
- EIA-364-49, "Ultraviolet Radiation Test"
- EIA-364-50B, "Dust (Fine Sand) Test"
- EIA-364-51A, "Ice Resistance Test"
- EIA/ECA-364-59A, "Low Temperature Test"

NASA defines three levels of screening: level 1 for highest reliability, level 2 for high reliability, and level 3 for standard reliability. Level 3 equates to standard lot acceptance inspection. Levels 1 and 2 call for additional testing. Spacecraft designers generally avoid the use of ferromagnetic materials, which can become magnetized and can interfere with sensitive instruments.

Visual and mechanical inspection performed in screening is critical for connectors, and defect criteria for each type of connector are listed in EEE-INST-002 (Table 12.7). Visual inspection will verify that connectors and contacts are properly marked, free of

Table 12.7 NASA EEE-INST-002 screening requirements for circular connectors.

Inspection/Test	NASA Level 1	NASA Level 2	NASA Level 3
Visual inspection	100%	100%	100%
Mechanical (dimensions)	2 pcs.	2 pcs.	N/A
Dielectric withstanding voltage (DWV)	2 pcs.	2 pcs.	N/A
Insulation resistance (room temperature)	2 pcs.	2 pcs.	N/A
Contact engagement and separation force (socket contacts)	2 pcs.	N/A	N/A
Coupling force	2 pcs.	N/A	N/A
Air leakage (hermetic connectors only)	100%	100%	N/A
Solderability/resistance to soldering heat	2 pcs.	N/A	N/A
Processing for outgassing (when contamination control is required)	100%	100%	100%

Table 12.8 NASA EEE-INST-002 qualification requirements for circular connectors.

Inspection/Test	NASA Level 1	NASA Level 2	NASA Level 3
Visual inspection	2 mated pairs A&B or C&D	2 mated pairs A&B or C&D	N/A
Mechanical (dimensions)	2 mated pairs A&B or C&D	1 A or C	N/A
Dielectric withstanding voltage (sea level)	2 mated pairs A&B or C&D	2 mated pairs A&B or C&D	N/A
Insulation resistance (room temperature)	2 mated pairs A&B or C&D	2 mated pairs A&B or C&D	N/A
Temperature cycling (except hermetics)	2 mated pairs A&B	1 mated pair A	N/A
Vibration	1 mated pair B or D	N/A	N/A

defects, and fabricated with good workmanship. Mechanical inspection will verify that connectors or contacts satisfy design, construction, and dimensional requirements and were manufactured with the specified materials and finish (see Table 12.8).

Depending on the application, outgassing, atomic oxygen, ultraviolet radiation degradation, and residual magnetism may need to be considered in selecting connectors and contacts for space applications.

NASA qualification requirements for D-subminiature connectors are tabulated in Table 12.9.

NASA connector workmanship requirements depending on the connector types are tabulated in Table 12.10.

Table 12.9 NASA EEE-INST-002 qualification requirements for D-subminiature connectors.

Inspection/Test	NASA Level 1	NASA Level 2	NASA Level 3
Visual inspection	3 pcs.	2 pcs.	N/A
Mechanical (dimensions)	3 pcs.	2 pcs.	N/A
Dielectric withstanding voltage (sea level)	3 pcs.	2 pcs.	N/A
Insulation resistance (room temperature)	3 pcs.	2 pcs.	N/A
Temperature cycling (except hermetics)	3 pcs.	2 pcs.	N/A
Vibration	3 pcs.		N/A
Shock	3 pcs.	2 pcs.	N/A
Residual magnetism (non-magnetic connectors only)	3 pcs.	2 pcs.	N/A
Mating and unmating force	3 pcs.	2 pcs.	N/A

12.22 Low-Voltage Automotive Single- and Multiple-Pole Connector Validation

General Motors' GMW3191 standard [46] defines the requirements, functional tests, and durability tests for low-voltage automotive single and multiple-pole electrical connectors, terminals, and related components. Procedures included within this specification are intended to cover performance testing at all phases of development, production, and field analysis of electrical terminals, connectors, and components that constitute the electrical connection systems. Table 12.11 describes the type of tests and the requirements depending on the terminal type.

12.23 Failure Modes, Mechanisms, and Effects Analysis for Connectors

FMEA or FMECA is used to identify the possible failure modes and causes of devices. FMEA or FMECA methodologies outline procedures to recognize and evaluate the potential failure of a product and its effects and to identify actions that could eliminate or reduce the likelihood of the potential failure occurring. Failure modes, mechanisms, and effects analysis (FMMEA) is a tool to support physics-of-failure-based design for reliability [47]. It can identify potential failure mechanisms for all potential failure modes and prioritize the failure mechanisms. FMMEA can aid connector manufacturers in the development of reliable designs, planning tests, and screens to validate nominal design and manufacturing specifications and determine the limits on the level of defects introduced by the variability in manufacturing and materials. FMMEA enhances the value of traditional FMEA methodologies by identifying the high-priority

Defect	Circular & Umbilical	DSUB-miniature	Micro-miniature	PC	RF	Plug-In Sockets & Strips	Twinax1553 Databus	Nano-miniature
Insert/insulator body								
Insert to shell positioning and orientation	√	√	√					√
Cracks, chips, busters, pinholes	√	√	√	√	√	√	√	√
Marking	√	√	√	√	√	√	√	√
Hermetic sealed connectors								
Negative meniscus (glass to contact & glass to shell)	√	√			√			
Solder cup misalignment, rear of connector	√	√						
Contact positioning (molded inserts with solder tails or solder cup contacts)								
Consistent centering between contacts	√	√	√	√	√	√		√
Solder cup misalignment, rear of connector	√	√	√			√		√
Grommet (as applicable)								
Nicks, gouges, tears, folds, discoloration	√		√		√	√		
Marking (as applicable)	√							
Shell/body								
Cracks, dents, burrs, sharp edges	√	√	√		√		√	√
Finish (peeled or blistered plating, scratches/exposed base metal, corrosion, discoloration)	√	√	√		√		√	√
Marking completeness, legibility	√	√	√		√		√	√
Threads (as applicable)								
Coupling (nicks, dents, voids, burrs)	√				√		√	
Attached hardware		√	√	√				√
Adhesives/molding material								
Excess bonding material (overflow)	√							
Voids		√	√	√	√	√	√	√
Leads (as applicable)								
Bent, nicked, cracked/broken leads, burrs	√	√	√	√	√	√	√	√
Finish (peeling, corrosion, exposed base metal)	√	√	√	√	√	√	√	√
Flexible circuit terminations (as applicable)								
Cracked insulation, separations, exposed conductors, delaminations				√				√

Notes
1. For additional details and illustrations of defects, the criteria in NASA-MSFC 85M03639, "Workmanship Limits for Connectors," may be consulted.
2. Perform visual inspection at 3X magnification, minimum. Progressively higher magnification may be used as necessary to examine anomalies.

Table 12.11 Connector validation test matrix.

Name of Test	Non-GMW3191 Approved Terminal System	GMW 3191 Approved Terminal System	Addition of New Connector in Existing Approved Series	Addition of a Second Source Terminal System	New Coding Added in Approved Connector tool	New Crimp Size Added in Approved Terminal Tool
Terminal-Mechanical Tests						
Crack corrosion	√			√		
Crimp/weld integrity	√			√		√
Terminal-to-terminal engagement force	√			√		√
Terminal cavity polarization	√			√		√
Terminal bend resistance	√			√		
Terminal push-out force (male terminals only)	√		√	√		
Terminal-Electrical Tests						
Maximum current rating	√			√		√
Voltage drop	√			√		√
1008-hour current cycling	√			√		√
Heavy-duty test	√			√		√
Connector-System Mechanical Tests						
Terminal-to-connector engagement force	√	√		√		√
Terminal-from-connector extraction force	√	√		√		√
Connector-to-connector engagement force	√	√	√		√	
Terminal position assurance (TPA)	√	√	√	√	√	√
Level and slide "open" position retention	√	√	√			
Mechanical assist integrity	√	√	√			
Connector mounting feature mechanical strength	√	√				
Connector audible feedback	√	√	√	√		
Connector lock mechanical overstress	√	√	√	√		

Connector position assurance (CPA)	√	√	√		
Connector seal retention-mated connector	√	√	√		
Connector seal retention-unmated connector	√	√	√		
Locked connector disengagement force	√	√	√		√
Unlocked connector disengagement force	√	√	√		√
Connector polarization (coding) feature effectiveness	√	√	√		√
Flammability	√	√	√		
Mechanical shock	√	√	√	√	√
Vibration with thermal cycling	√	√	√	√	√
Sealed Connector Environmental Tests					
Fluid resistance	√	√	√	√	√
Corrosion	√	√	√	√	√
Thermal aging	√	√	√	√	√
Thermal shock	√	√	√	√	√
Humid heat cyclic (HHC)	√	√	√	√	√
Humid heat constant (HHCO)	√	√	√	√	√
Water submersion	√	√	√	√	√
Pressure/vacuum leak	√	√	√	√	√
High-pressure spray	√	√	√	√	√
Unsealed Connector Environmental Tests					
Thermal aging	√	√	√	√	√
Thermal shock	√	√	√	√	√
Humid heat cyclic (HHC)	√	√	√	√	√
Humid heat constant (HHCO)	√	√	√	√	√
Corrosion	√	√	√	√	√

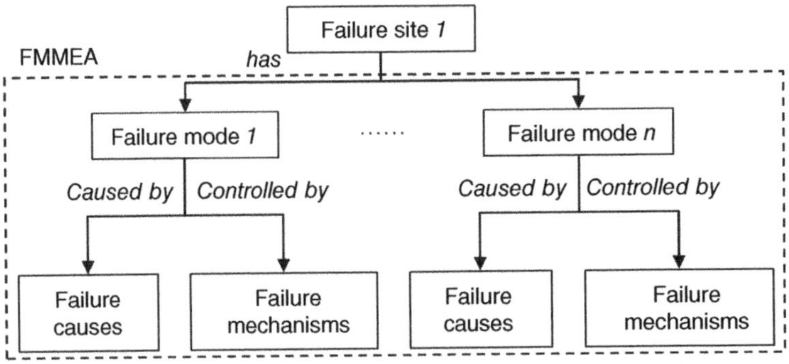

Figure 12.5 Failure site-mode-cause-mechanism structure for adverse event investigation [47].

failure mechanisms to create an action plan to mitigate their effects. FMMEA uses the LCP of a product along with the design information to identify the critical failure mechanisms affecting a product. An LCP is a forecast of the events and the associated environmental and usage conditions that a product may experience from manufacture to end of life. The device is divided into its lower-level subassemblies for investigation. A failure mode is the manner in which a failure is observed by methods such as visual inspection, electrical measurement, or other tests and measurements. For each failure mode, the potential failure causes are analyzed. A failure cause is the specific process, design, and/or environmental condition that initiates a failure and whose removal will eliminate the failure. Possible failure causes are investigated in the entire life cycle of the device, including design, manufacture, operation, and maintenance. Next, potential failure mechanisms are identified. Failure mechanisms are the processes by which a specific combination of physical, electrical, chemical, biological, and mechanical stresses induces failures.

During the life cycle of a product, several failure mechanisms may be activated by different environmental and operational parameters acting at various stress levels, although in general only a few operational and environmental parameters and failure mechanisms are responsible for the majority of failures. In the process of conducting FMMEA, the combinations of occurrence and severity of each failure mechanism are assessed, and the probability of occurrence is taken into consideration from the distributions of the loads and the geometric/material features, while the severity is obtained from the seriousness of the effect of the failure caused by a mechanism. The use of FMMEA will enable manufacturers to create an internal evaluation report organized in the failure site-mode-cause-mechanism structure, as shown in (Figure 12.5) [47]. Another benefit of conducting FMMEA is that it would help manufacturers monitor and improve the reliability of their products and provide them with useful information to investigate and correct the adverse events.

Table 12.12 shows an example of FMMEA inputs for a connector study.

Table 12.12 Sample FMMEA inputs for connectors.

Potential Failure Sites	Potential Failure Modes	Potential Failure Causes	Potential Effects	Potential Failure Mechanisms
• Housing • Contact spring • Contact finish • Body molding • Soldering	• Missing or incorrect hardware • Incorrect wiring • Other OEM workmanship • Assembly interference • Damaged or failed connector • Loose connector • Not seated/mis-mated • Broken wires/connections • Electrical shorts • Corrosion/contamination • Electrical opens • Misaligned connectors • Bent pins/connectors • Open circuit on a pin • Short circuit between adjacent pins	• Plating material • Workmanship • Insufficient inspection • Insufficient testing • Insufficient procedures for connector selection • Electrical requirements out of specifications • Mechanical requirements out of specifications • Connector is not selected for environmental requirements • Insufficient qualification • High/low temperature	• Electrically shorted, bent, and/or recessed contact pins • Contact pin/socket contamination leading to open or intermittencies • Connector plating corrosion or corrosion of connector components • Low or inadequate contact pin retention forces • Contact crimp failures • Unmated connectors and mis-wiring during installation or maintenance • Loose connectors due to manufacturing defects such as wavy washer and worn bayonet retention • Damaged connector elastomeric seals • Cryogenic connector failure • Less insulation resistance • Intermittent failure	• Corrosion • Contamination • Overstress cracks • Fatigue • Misapplication • Inadequate mounting • Wear-out

12.24 Connector Experiments

The reliability of connectors was tested using a salt spray chamber in one study [48] to compare the reliability of electrical connectors from two different manufacturers. The difference between the connectors was in the coating materials of the connector pins: one of them (A) used tin for the outermost coating layer, nickel for the second layer, and brass for the pin material; the other one (B) also used tin for the outermost coating and brass for the pin material, but used copper for the second coating layer. The test was conducted according to the ISO 9227 standard. The connectors were placed in a salt spray test chamber for 2000 hour. The test consisted of salt spray operation modes of 300 and 1100 hours and a period of 600 hour when the chamber was halted between the salt spray operation modes. During testing the connectors were measured using continuous real-time measurements. A 10% increase in the measured voltage was used as the failure criterion. Significant differences between the manufacturers could be found related to materials during testing. Cross-sections were analyzed by means of an optical microscope and a scanning electron microscope (SEM).

The pin itself is made of brass, which contains copper and zinc. In the elemental analysis of the untested connector it was observed that the brass contained about 65 wt% of copper and 35 wt% of zinc. The elementary analysis of the damaged connector revealed that the damaged area contained 95 wt% of copper and 5 wt% of zinc. The disappearance of zinc in the damaged connectors can be explained by dezincification. Connectors from manufacturer A started degrading substantially earlier than the connectors from manufacturer B. This is a result of the fractures of the pin coatings with the connectors from manufacturer A (Figure 12.6).

Connector terminals, particularly in automotive applications, can be prone to failure due to small-amplitude movements; vibrations have been recognized as one of the main causes of these degradations. The choice of connector design and of materials is often guided by the results of "imposed displacement" sine vibration tests [49]. A new experiment has been designed to measure the electrical properties of low-level connectors by Noël and Brézard-Oudot. A four-point measurement setup is used with a dedicated acquisition system allowing high-speed measurement of the voltage drop with the possibility to trigger ultra-high-speed acquisition when a threshold value is reached.

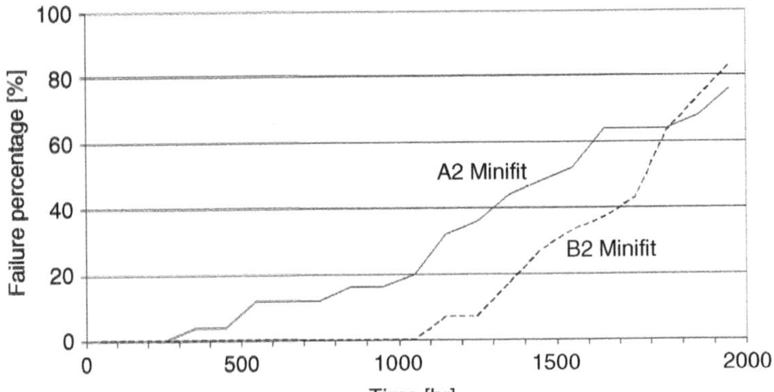

Figure 12.6 Failure percentage of A and B Minifit connectors as a function of time [48].

Fretting is known to be a cause of degradation and failure of electrical contacts. It is often due to vibrations and thermal expansion/contraction connector degradation, either by wear or a small-amplitude fretting mechanism. Fretting can be alleviated by using a proper lubricant film [49].

An experimental bench for the measurement of voltage drops in contacts during vibration runs has been set up. Low-noise recording of the voltage is achieved despite the amplification of the signal (×1000) necessary to record both milliohm values for undamaged samples and ohm values after the onset of degradation. No significant degradation of the gold/gold contacts could be recorded, even after 300 hours heat treatment at 125 °C. Dissymmetric Au/Sn contacts had to be used to measure resistance evolution. In this case, increase of the voltage values during vibration was recorded. The frequency of the peaks in voltage matched the vibration frequency. Evolution of the voltage within the vibration cycles can be visualized. Close examination of the signals showed that additional voltage peaks occurred due to the interface modification. The contact variations were several milliseconds long, and no intermittency could be measured [49].

In one study [50], the thermal stress deformation in high-density circuit configuration PCB connectors, which causes microscopic slide of the contact points of the connector contacts and results in contact failures due to fretting, has been evaluated by a holographic interferometry measuring system (HIMS). The compared test samples had 14 contacts with 2 mm of contact spring size (sample A) or 22 contacts with 1 mm of contact spring (sample B). The contact resistance of the connector contacts connected to the thermally deformed PCB was measured. The contact failure occurred in the higher-density connector (sample B). This result stemmed from its having less durability under thermal stress and suggests that the downsizing of the connector sacrifices the reliability. The microscopically deformed PCB connector surface was clearly observed with respect to the heat generated at the current flowing contacts (Figure 12.7).

In another study [51], the effect of corner/edge bonding epoxy on the board-level solder joint reliability of ball grid array (BGA) connectors (Figure 12.8) was investigated, including mechanical drop and 4-point bending tests, and the pre-conditioning effect of thermal aging (1000 hours at 125 °C) and accelerated temperature cycling (ATC,

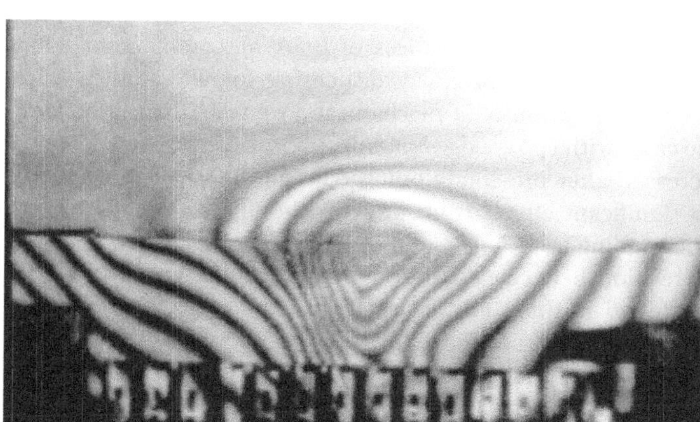

Figure 12.7 Reconstructed interferograms of the PCB connector surface for sample B due to current flow (after 60 days of thermal stress) [50].

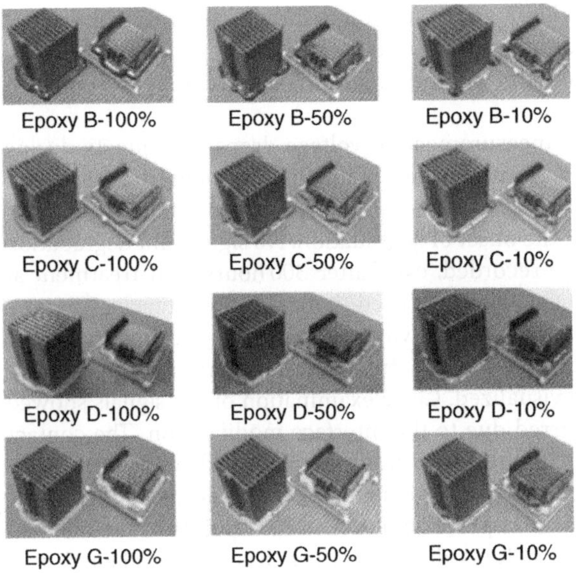

Figure 12.8 Test samples with different lengths of epoxy dispensing (Epoxy B, C, D, and G) [51].

−40 ∼ 125 °C, 500 cycles) on the reliability of connectors with corner/edge bonding epoxy is discussed. In an ATC test, devices are subjected to a number of cycles of alternate high and low temperature extremes, and this cyclic stress produced in temperature cycling is related to thermal expansion and contraction undergone in the material. The failure modes and fracture strengths of the button shear and board-level tests are cross-referenced for comparison and assessed for correlation.

The effects of length of epoxy dispensing, pre-conditioning ATC, and thermal aging on the mechanical reliability tests of the connectors were discussed as well. In addition, the button shear test and board-level reliability tests were cross-referenced for comparison and assessed for correlation. The major findings of this study can be summarized as follows:

- The mechanical drop and bending performances of BGA mezzanine connectors improved with the increasing length of epoxy dispensing for all samples. Such improvements appeared more obviously in mechanical drop tests than in bending tests, especially for the cases with epoxy edge bonding.
- The drop lifetime decreased after pre-conditioning ATC and thermal aging for all epoxies. However, no significant effects from pre-conditioning ATC and thermal aging were found in the bending test.
- A certain level of correlation between board-level drop/bending and epoxy button shear tests can be found in this study, especially for the correlation between drop/bending and button shear tests on connector LCPs.
- The cracks at the interface of epoxy and connectors can be found after drop/bending tests, which indicates that the bonding between epoxies and connector LCP during drop/bending tests is weaker than that between epoxies and PWBs.

Table 12.13 Specifications of the testing system [52].

Item	Operation range	Item	Operation range
Amplitude of plugging	0 ~ 16 mm	Channels of sample testing	4
Scope of insertion force	0 ~ 50 N	Sampling frequency	<500 K·s^{-1}
Scope of extraction force	0 ~ 50 N	Resolution of force	0.01 N
Plugging velocity waveform	Sine wave	Range of force	−50 ~ 50 N
Frequency of plugging	0 ~ 15 Hz		

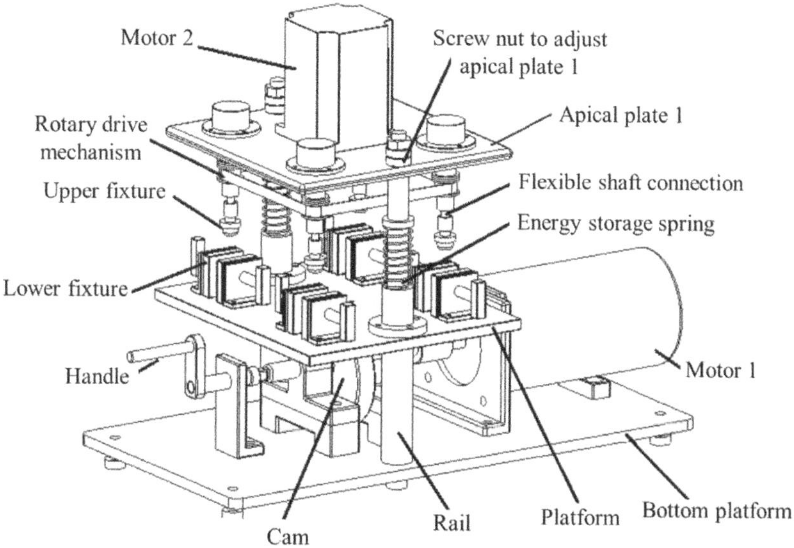

Figure 12.9 Assembly drawing of plug device [52].

Lu et al. [52] designed the connector life test system to study the connector service life in normal operating conditions. The life testing system was manufactured and has been used in connector life testing in the laboratory. The system consists of a mechanical unit, the human–computer interaction unit with the software console, a motion control unit, and insertion/removal force test units. Table 12.13 and Figure 12.9 show specifications and mechanical structural design of the testing system, respectively.

Figure 12.10 shows the relationship between the contact resistance and cycles. Contact resistance is measured using a four-wire method before the test, after the test, and in damp heat (six days, mode: mated, RH 90–100%, 25–55 °C). Their study [52] mentions that the mechanical and control devices of the life test systems are stable and the data acquisition is accurate and in real time. Their life test system was tested to verify the system performance to the design specifications.

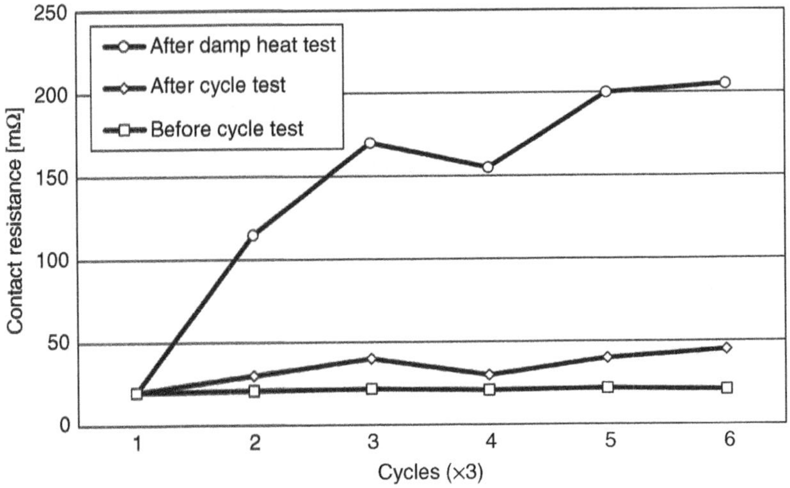

Figure 12.10 Diagram of the relationship between contact resistance and cycles [52].

12.25 Summary

The candidate parts selection process puts a part into the assessment flow. This step helps eliminate possible unpromising parts and balances between the need for functionality and the need to obtain costs and other practicality requirements for parts. It merges the input of functional groups besides electrical engineering into the parts selection during the initial decision-making process and reduces the cycle time of information flow. Ordering, receiving, and paying occur only after the part is assessed and a positive decision is made regarding its use.

References

1 Roos, G. and Gold, M. (2002). Software tools manage selection of components. *EE Times*: 91–92.
2 Spiegel, R. (2002). Parts database accurate? *Electronic News* 48 (29): 1–4.
3 Erickson, J. R., Hybrids make the connection. https://www.hydraulicspneumatics.com/technologies/other-components/article/21883993/hybrids-make-the-connection (accessed June 23, 2020)
4 Martens, R. Li, S., Ding, C., and Norris, N. (2012). Reliability study of low normal force LGA sockets. *Proceedings of the 58th IEEE Holm Conference on Electrical Contacts*, Portland, OR (September 2012), IEEE.
5 Amphenol Aerospace, Low mating force rectangular connectors. www.amphenol-aerospace.com/pdf/catalogs/12-035.pdf (accessed Nov. 12, 2019).
6 Amphenol, (2019) High performance line replaceable module (LRM) interconnects. www.amphenol-aerospace.com/pdf/catalogs/12-037.pdf (accessed Nov. 11, 2019).

7 Amphenol, (2019) High Density HDB3 and HSB3 Connector Series. www.amphenol.co.jp/military/catalog/SL-402.pdf (accessed Nov. 12, 2019).

8 Horn, J. and Egenolf, B. (1992). Shape optimization of connector contacts for reduced wear and reduced insertion force. *AMP Journal of Technology* 2: 42–46.

9 Electronics+Radio, Types of Connector- an overview, information or tutorial about the key facts about electronics connectors and some of the common connector types in use today. www.electronics-radio.com/articles/electronic_components/connectors/connector-types.php (accessed Nov. 12, 2019).

10 Cirrus Systems, Connector life cycles. www.cirris.com/learning-center/product-articles/other-products/227-connector-life-cycles (accessed Nov. 14, 2019).

11 Jensen, H., and Eakin, D. (2002). Demystify current ratings for connector selection. *Power Electronics Technology*. www.powerelectronics.com/content/demystify-current-ratings-connector-selection (accessed Nov. 12, 2019).

12 AirBorn Product Technical Bulletin #61, Operating Voltage. www.airborn.com/docs/OpenAccessProvider/product-documentation/ptb61.pdf?sfvrsn=10 (accessed Nov. 12, 2019).

13 BS EN 60512-5-2:2002 (2002). *Connectors for electronic equipment — Part 5-2: Current-carrying capacity tests — Test 5b: Current-temperature derating*. BSi.

14 AirBorn Interconnect, Inc. Product Technical Bulletin #54. Current carrying capacity of AirBorn's N-series connectors. www.airborn.com/docs/OpenAccessProvider/product-documentation/ptb54.pdf?sfvrsn=7 (accessed Nov. 12, 2019).

15 AirBorn Product Technical Bulletin #48. Current-Carrying Capacity of R-Series Connectors. www.airborn.com/docs/product-documentation/ptb48.pdf?sfvrsn=39 (accessed Nov. 12, 2019).

16 Mroczkowski, R.S. (1998). *Electronic Connector Handbook*. New York: McGraw Hill.

17 Zhang, X., Li, K., and Xu, J. (2011). The research of insulation failure on signals transmission in electrical connector. *Proceedings of the 2nd International Conference on Mechanic Automation and Control Engineering*, Hohhot, China (July 2011). IEEE.

18 IEC 60529:1989/AMD2:2013/COR1:2019. Corrigendum 1 – Amendment 2 – Degrees of protection provided by enclosures (IP Code). https://webstore.iec.ch/publication/64427 (accessed Jan. 26, 2020).

19 Cianciolo, D., 10 things to consider when selecting a connector, Fischer Connectors, Inc. www.ept.ca/features/10-things-to-consider-when-selecting-the-a-connector (accessed Nov. 12, 2019).

20 Ashby, M.F. (2011). *Materials Selection in Mechanical Design*, 4e, 57–96. Elsevier.

21 Orstad, R. (May 4, 2015). How to specify electrical insulating material. www.connectorsupplier.com/how-to-specify-electrical-insulating-material/ (accessed Jan. 26, 2020).

22 *High Speed and High Power Connector Design* [film]. Computer Simulation Technology, 2014.

23 McReynolds, D., Optimize your RF/MW coaxial connections. www.rfcoaxconnectors.com/Technical_wp-OptimizeConnections.htm (accessed Nov. 12, 2019).

24 Mahajan, S. and Kimerling, L.C. (1992). *Concise Encyclopedia of Semiconducting Materials and Related Technologies*, 51–57. Pergamon Press.

25 NASA-STD 8739.4 with Change 6, Crimping, interconnecting cables, harnesses, and wiring. http://snebulos.mit.edu/projects/reference/NASA-Generic/NASA-STD-8739-4.pdf (accessed Nov. 14, 2019).

26 Mroczkowski, R. S. (2004). A perspective on connector reliability. *Proceedings of the 50th IEEE Holm Conference on Electrical Contacts and the 22nd International Conference on Electrical Contacts,* Seattle, WA (September 2004). IEEE.

27 Cianciolo, D. Connector selection: Ten things to getting the connector right, Fischer Connectors. www.fischerconnectors.com/sites/default/files/whitepapers/white_paper_connector_selection_rev_1.0.pdf (accessed Nov. 14, 2019).

28 Crosby, P.B. (1996). *Quality Is Still Free: Making Quality Certain in Uncertain Times.* New York: McGraw-Hill.

29 Gaskell, T., Crosby's quality management maturity grid. www.qualityandproducts.com/2012/07/09/crosbys-quality-management-maturity-grid (accessed Nov. 14, 2019).

30 Reynolds, E., "Who owns connector selection responsibility?" Contributed article on March 31, 2014. www.connectorsupplier.com/owns-connector-selection-responsibility (accessed Nov. 15, 2019).

31 Murray, S., Boru, M., and Pecht, M. (2002). Tracking semiconductor part changes through the part supply chain. *IEEE Transactions on Components and Packaging Technologies* 25 (2): 230–238.

32 Syrus, T., Pecht, M., and Humphrey, D. (2001). Part assessment guidelines and criteria for parts selection and management. *IEEE Transactions on Electronics Packaging Manufacturing* 24 (4): 339–350.

33 Das, D., Pendse, N., Pecht, M. et al. (2000). Deciphering the deluge of data – understanding electronic part data sheets for part selection and management. *IEEE Circuits and Devices Magazine* 16 (5): 26–34.

34 Syrus, T., Ramgopal, U., and Pecht, M. (2001). Manufacturer assessment procedure and criteria for parts selection and management. *IEEE Transactions on Electronics Packaging Manufacturing* 24 (4): 351–358.

35 Jackson, M., Mathur, A., Pecht, M., and Kendall, R. (1999). Part manufacturer assessment process. *Quality and Reliability Engineering International* 15: 457–468.

36 Defossez, M. (2005). Selecting connectors for multi-gigabit transceiver designs. *Xcell Journal.* http://suddendocs.samtec.com/articles/pdfs/xc_v4connectors52%5B1%5D.pdf (accessed Jan. 26, 2020).

37 Digital Systems Connectors I CMPE 650, UMBC. http://ece-research.unm.edu/jimp/650/slides/connectors1.pdf (accessed Jan. 26, 2020).

38 NASA (2008). EEE-INST-002: Instructions for EEE Parts Selection, Screening, Qualification, and Derating. https://nepp.nasa.gov/docuploads/FFB52B88-36AE-4378-A05B2C084B5EE2CC/EEE-INST-002_add1.pdf (accessed Jan. 15, 2016).

39 Iannello, C., Davis, M., Kichak, R., and Slenski, G. (2009). Spacecraft electrical connector selection and application process. *Proceedings of the 2009 Aerospace Electrical Interconnect System Conference,* Portland, OR (October 2009). SAE.

40 "What is meant by a 'harsh environment' in electronics or photonics?" Frequently Asked Questions. www.twi-global.com/technical-knowledge/faqs/faq-what-is-meant-by-a-harsh-environment-in-electronics-or-photonics (accessed Jan. 26, 2020).

41 Amphenol Fiber Systems International, (2020) "Harsh environment connector material selection guide." www.fibersystems.com/pdf/whitepapers/Harsh-Environment-Connector-Material-Guide.pdf (accessed Jan. 26, 2020).

42 European Commission, (2020) "Restriction of hazardous substances in electrical and electronic equipment," Directive 2011/65/EU of the European Parliament and of the Council of 8 June 2011. http://ec.europa.eu/environment/waste/rohs_eee/legis_en.htm (accessed Jan. 26, 2020).

43 Spahi, S., and Jazowski, R. J. (2012). Reliability and qualification process for high power interconnects. *Proceedings of OCEANS 2012 MTS/IEEE Conference*, Hampton Roads, VA (Oct. 14–19, 2012). IEEE.

44 Toth, J. R., and Thumbeck, S. S. (2005). Reliability and economic considerations for connector applications in undersea system designs. *Proceedings of OCEANS, 2005 MTS/IEEE Conference*, Washington, DC (Sept. 17–23, 2005). IEEE.

45 Wilkins, D. J., (2020) The Bathtub Curve and Product Failure Behavior Part One - The Bathtub Curve, Infant Mortality and Burn-in. www.weibull.com/hotwire/issue21/hottopics21.htm (accesed Jan. 26, 2020).

46 General Motors, *Connector Test and Validation Specification* (GMW3191), June 1, 2012.

47 Cheng, S., Das, D., and Pecht, M. (2011). Using failure modes, mechanisms, and effects analysis in medical device adverse event investigations. *Proceedings of the International Conference on Biomedical Ontology*, Buffalo, NY (July 2011). CALCE.

48 Parviainen, A., Perälä, J., Frisk, L., and Kuusiluoma, S. (2009). Connector reliability testing using salt spray," *IEEE European Microelectronics and Packaging Conference*, EMPC 2009, Rimini, Italy (June 15–18, 2009). IEEE.

49 Noël, S., and Brézard-Oudot, A. (2014). Electrical behavior of golden automotive connectors under vibration tests. *27th International Conference on Electrical Contacts*, Dresden, Germany (June 22–26, 2014). ICEC.

50 Kubota, H., Taniguchi, M., Suziki, S., and Takagi, T. (2008). Application of holography to evaluation of printed circuit board connector due to thermal stress. *SICE Annual Conference 2008*, The University Electro-Communications, Japan (August 20–22, 2008). SICE.

51 Song, F., Yang, C., Wu, H. L. H., Lo, C. C. J., Lee, S.W. R., and Newman, K. (2010). Effects of corner and edge bond epoxy adhesives on board level solder joint reliability of BGA mezzanine connectors. *Proceedings of the IEEE Electronic Components and Technology Conference*, Las Vegas, NV (June 2010). IEEE.

52 Lu, N., Jiang, X. Y., Xu, L. J., and Li. Y. S. (2012). The connector life test system design. *Proceedings of the 26th International Conference on Electrical Contacts, ICEC-ICREPEC 2012*, Beijing, China, May 2012). ICEC.

13

Signal Connector Selection

Michael G. Pecht

Center for Advanced Life Cycle Engineering (CALCE), University of Maryland, College Park, MD, USA

As newer technologies have enabled transmission rates to increase to multigigabits per second, the increased transmission rates have led to increased challenges in controlling signal integrity. High-speed connectors have data rates greater than 100 Mbps, are physically small, and may use differential signaling. On the other hand, high-power connectors are designed to carry high current loads. These differences as well as reliability requirements and cost lead to different design, test, and analysis requirements. For high-speed connectors to work well, simulation and characterization analysis need to be performed to ensure that they adhere to uniform standards.

The performance of signal connectors is addressed in two broad topics: signal transmission quality (STQ) and electromagnetic compatibility (EMC). STQ refers to transmission of high-speed signals through connectors and interconnects without loss of quality of the signal. The losses may be due to cross talk, transmission delay, or characteristic impedance variation on signal propagation and reflections. The transmission and preservation of signal waveforms forms the crux of STQ. EMC refers to the ability of the system to prevent the degradation of the digital signal from extraneous noise and electromagnetic energy. It generally involves shielding, filtering, and grounding practices to control electromagnetic interference (EMI).

13.1 Issues Involving High-Speed Connectors

The main functional requirement of a signal connector is to preserve signal integrity. This involves preserving the shape, amplitude, and frequency of the voltage pulse being transmitted. The maximum pulse frequency is determined by the rise time of the voltage signal. When the rise time becomes shorter and shorter (e.g. in the nanosecond range), the connector needs to be considered as a transmission line because the wavelengths associated with the frequencies in such fast rise times are comparable to the dimensions of the connector. At such high rates of transmission, the designer must consider reflections at impedance changes, noise induced due to cross talk, and high-frequency attenuation [1]. Other serious problems that affect the signal integrity are ringing, ground bounce, and power supply noise.

Electrical Connectors: Design, Manufacture, Test, and Selection, First Edition.
Edited by San Kyeong and Michael G. Pecht.

The characteristic impedance in a connector revolves around the distribution of grounds. In typical open-pin field connectors, impedance and cross talk are controlled by the appropriate allocation of pins to ground. The signal-to-ground ratio is a reflection of such assignments, with the frequency capability of the connector increasing as the signal-to-ground ratio decreases.

In analog applications, information is intrinsically embedded in the waveform shape. The external noise is usually of very high frequency. Therefore, the noise can be filtered out in the receiver end, and as long as the shape is not affected or uniformly affected, there is no problem. In digital signal applications, the bits carry the information. There are two options – a 1 or a 0. If the noise affects the amplitude of a pulse and a 1 becomes a 0 or vice versa, it becomes difficult to retrieve the information. Hence, in digital signals the frequency of the signal has a dominant effect on the connector design requirements.

13.2 Signal Transmission Quality Considerations

Signal integrity must be maintained to meet the STQ requirement. This includes the rise time, voltage, and duration of the pulse. Several degradation mechanisms affect the signal integrity, which becomes more pronounced at high frequencies. These include interconnect delays and signal distortion from characteristic impedance integrity and cross talk. Each of these issues is discussed in this section.

13.2.1 Interconnect Delays

Interconnect delays are the sum of propagation delays within the interconnect (connector) plus the transit time through the interconnect (connector). Propagation delays are defined as the time taken by the signal to reach the receiver from the sender and are a function of the transmission path length and dielectric material properties. The signal should be of high quality and noise-free in a high-performance system. When the rise time of the signal decreases, the signal speed increases. Interconnect delays constitute a significant portion of the overall signal delays. For maximum speed to be realized at the system level, the response of the interconnection system must be optimized.

13.2.2 Signal Distortion

Signal distortion can result from cross talk and reflections due to the characteristic impedance mismatches. Both of these factors are affected by materials and geometry but not necessarily to the same degree or in the same direction. An analysis is generally required to determine which of the two factors is more crucial.

In cables, cross talk depends on the length of the connections and the ability to control the dielectric constant and geometry to tight tolerances. In connectors, impedance control is complicated by the variations in geometry and materials along the length of a connector, and cross-talk effects are often minimized by the relatively short lengths of connectors in the direction of propagation. When cross talk is critical, the value of the characteristic impedance is not the main concern; rather, the control of characteristic impedance within a component and across component interfaces is more important.

Control of the characteristic impedance is necessary for the quality of signal transmission along the line and to minimize signal reflections. The signal reflections at interfaces have a significant effect on the signal waveform, resulting in both attenuation and distortion.

The impedance control in cables is pretty easy as they are consistent in both geometry as well as materials. However, the impedance control in connectors is not so straightforward. In connectors, the intrinsic curvatures make consistency in geometry highly challenging. There are, however, two techniques to circumvent this problem: the addition of a dedicated ground pin and the introduction of ground planes within the connector. The introduction of ground planes in the connector is one way of reducing the requirement of ground pins in the connector and increasing the number of pins available for input/output (I/O).

The signal/ground ratio can determine signal integrity. As the signal/ground ratio decreases, the integrity of the signal increases. But if the number of ground pins increases, the pins available for the I/O requirement decrease. Minimizing the distance between a signal pin and a ground pin improves performance. This arrangement also reduces cross talk.

Cross talk occurs when there is EMI in the field due to one conductor on another conductor. The level of EMI depends on the distance between the conductors and the cross section geometries of the source and victim conductors. Cross talk can be minimized by reducing the spacing between the conductors or making the conductor thinner and shorter. Making the conductor thinner and shorter helps reduce the cross talk introduced by the connector. Another method of reducing cross talk is to shield the cable and contain internal grounds.

13.3 Electromagnetic Compatibility

EMC is the ability of an electronic device or system to operate in an environment without being a source of EMI or being affected by EMI generated from other sources. There are several techniques to help achieve EMC, including grounding, shielding, balanced lines, and filtering. All have their place as well as their limitations, depending on the applications and the severity of the EMI problem [1].

Good grounding can prevent EMI. Generally, an ideal ground is a surface whose potential does not change, irrespective of the amount of current supplied or drawn from it. Proper grounding is also necessary for the correct implementation of other techniques such as shielding and filtering.

Shielding protects a device from impinging electromagnetic radiation either by reflecting the wave or by absorbing the EMI. It is used with both enclosures and cabling. The effectiveness of the shield depends on the frequency of the electromagnetic wave. It also depends on the characteristics of the shielding material and the shape and number of discontinuities in the shield.

The need for a flexible cable and connectors to allow the cable to be connected and disconnected precludes this approach. The purpose of the connector is to provide a low impedance path from shield to ground. The connector should also present as little discontinuity as possible.

Table 13.1 Characteristics of Cat5e, Cat6, and Cat7 cables.

Category According to Telecommunications Industry Association	Maximum Specified Frequency (MHz)	Typical Application
Cat5	100	100 Mbps Replaced by Cat5e
Cat5e	100	100 Mbps Gigabit Ethernet 100 MBit Ethernet
Cat6	250	Gigabit Ethernet Most popular for new installs
Cat7	600	Unspecified

The shielding effectiveness of a connector is specified by the effectiveness of the performance of the cable to which it is connected. When attached to a cable, if a connector degrades the cable's performance, then the connector is not effective in shielding. Shielded connectors are often required for high-speed signal paths. The benefit from shielding the connector is in reduction of the noise that can be injected into the connector in the form of EMI from external sources.

Filtering is used for combating conducted EMI. A low-pass filter separates the high-frequency noise from the low-frequency signals and can be a simple capacitor as well as a variety of complicated circuits. Numerous filter capacitors are available, which are used for filtering the noise from the cables.

When choosing an Ethernet cable or connectors, one of the factors to consider is the bandwidth requirement. There are industrial standards such as Cat5, Cat6, or Cat7, and it is necessary that while installing or upgrading a network, all the network components are in compliance with the category level. Table 13.1 details the maximum frequencies and typical application that each category is used for. The performance of the network degrades if the connectors of Cat5, Cat6, or Cat7 are mixed on the same network. Connectors are matched with the environment where they are used. To protect against contaminants entering the connector and degrading performance, a connector with an ingress protection rating of IP65 or IP67 is preferred in most industrial applications. The IP code in IEC standard 60529 classifies and rates the degree of protection provided against intrusion (body parts such as hands and fingers), dust, accidental contact, and water by mechanical casings and electrical enclosures [2].

Fiber optics are used when the transmission distance is longer. Fiber-optic cabling for industrial applications is preferred to copper-based solutions because it has several advantages. For example, it has high transmission rates, can handle gigabit Ethernet loads, and does not require additional shielding.

13.4 Virtual Prototyping

Creating a connector starts with a layout design based on electrical and mechanical requirements. This design is then analyzed iteratively to ensure that it meets the required electrical and mechanical standards and specifications. The simulation analysis (virtual prototype analysis) is conducted in three stages: pre-processing, solver choice, and post-processing. Next, a physical prototype is created and run through real measurement tests. Once the prototype clears these tests, it can be sent for production.

In the pre-processing stage, general simulation preparations are involved such as importing figures from 3D computer-aided design (CAD), identifying the materials of the connector, and assigning material properties. There is currently computer simulation technology (CST) layout software that enables one to import data and conduct the analysis, but the material parameters must be correctly defined.

There are generally two types of solver choices: T-solver (time domain) and F-solver (frequency domain). For high-frequency connectors, either can be used. The advantages of T-solver are that it supports high geometric complexity, is memory-efficient, attains broadband results from a single excitation, and has true transient co-simulation. Comparatively, the advantages of F-solver are that it has many ports since it is a direct solver and makes the simulation efficient. F-solver has an electrically small structure, a narrowband frequency range, and conformational mesh abilities, and may have automatic energy-based mesh adaptations, which lead to high-accuracy results.

The ever-increasing clock and data rates in computer, communications, video, and network systems have made signal integrity a concern. Maintaining signal integrity means preservation of the signal's characteristics such as rise time, pulse width, timing, and jitter from source to receiver. The transmission impedance, including that of the connector, affects the signal's integrity. Therefore, impedance in the transmission environment needs to be controlled. Mismatches and variations in impedance can cause reflections that decrease signal quality as a whole. Impedance tolerances are part of the electrical specifications for many of today's digital system components and will require that the modeled connectors are verified with hardware measurements to ensure they adhere to the design specification.

13.4.1 TDR Impedance Measurements

A time-domain reflectometer (TDR) is an electronic instrument that is used to locate the discontinuities in a connector. The TDR allows the user to analyze the impedance profile inside the connector correctly. It measures the reflections that result from a signal traveling through a transmission environment. It sends a pulse through the medium and compares the reflections from the transmission environment to those produced from a standard impedance. Figure 13.1 shows a simplified TDR measurement block diagram [3]. The source is a step input. The oscilloscope displays a voltage waveform that is a superimposition of the incident waveform and its multiple reflections propagated back due to the impedance variation in the transmission line.

13.4.1.1 Reflection Coefficient

The reflection coefficient, ρ, is the ratio of the reflected pulse amplitude to the incident pulse amplitude. Most TDR instruments will perform the calculations of the reflection coefficient internally and display a numerical result. For a fixed terminal load, ρ can also be expressed in terms of the transmission line characteristic impedance, Z_c, and the load

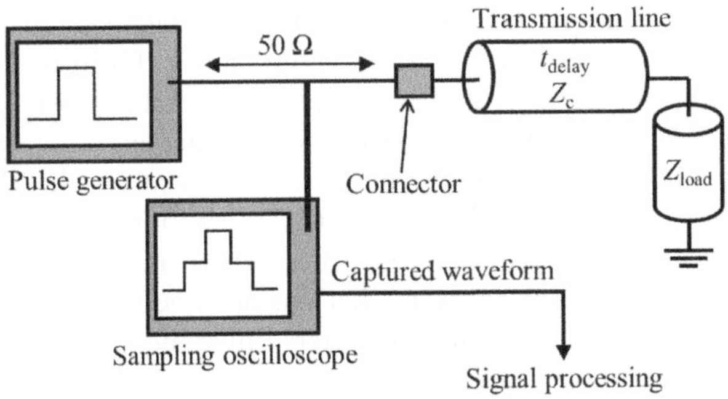

Figure 13.1 Block diagram of a TDR circuit [3]. Graphically renewed.

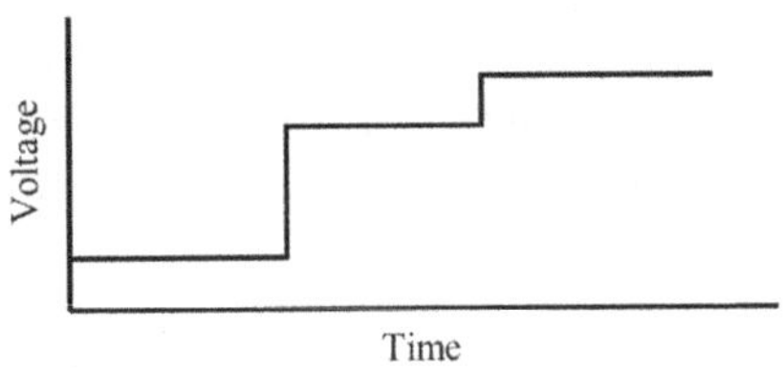

Figure 13.2 Simplified diagram of a TDR waveform.

impedance, Z_l.

$$\rho = \frac{V_{reflected}}{V_{incident}} = \frac{Z_l - Z_c}{Z_l + Z_c} \tag{13.1}$$

TDR measurements can be displayed in units of voltage, resistance, or ρ on the vertical magnitude scale. The horizontal axis represents units of time, as shown in Figure 13.2.

Figure 13.3 illustrates the TDR results for various kinds of loads including short- and open-circuit termination, matched and mismatched load terminations, only capacitive or inductive load terminations, or a combination of both. The discontinuities caused by connectors in a transmission line can be modeled as a combination of capacitors, inductors, and transmission lines.

13.4.1.2 TDR Resolution Factors

TDR provides useful insights into circuit impedance and signal integrity. The resolution plays a part in distinguishing between closely spaced discontinuities. If a TDR system has insufficient resolution, small or closely spaced discontinuities could be smoothed into a single aberration in the waveform. This may lead to incorrect impedance readings. Rise time, settling time, and pulse aberrations can also affect a TDR system's resolution.

The rise time is equal to the sum of the rise time of the incident step and the reflected waveform from the discontinuity. The distance between two discontinuities can be predicted from the distance between the two rise times in the TDR waveform. When the distance between the discontinuities is less than half the rise time, the two discontinuities become inseparable.

Aberrations are distortion in the TDR waveforms. They can occur due to the reflections from discontinuities in the transmission path before the main incident step arrives

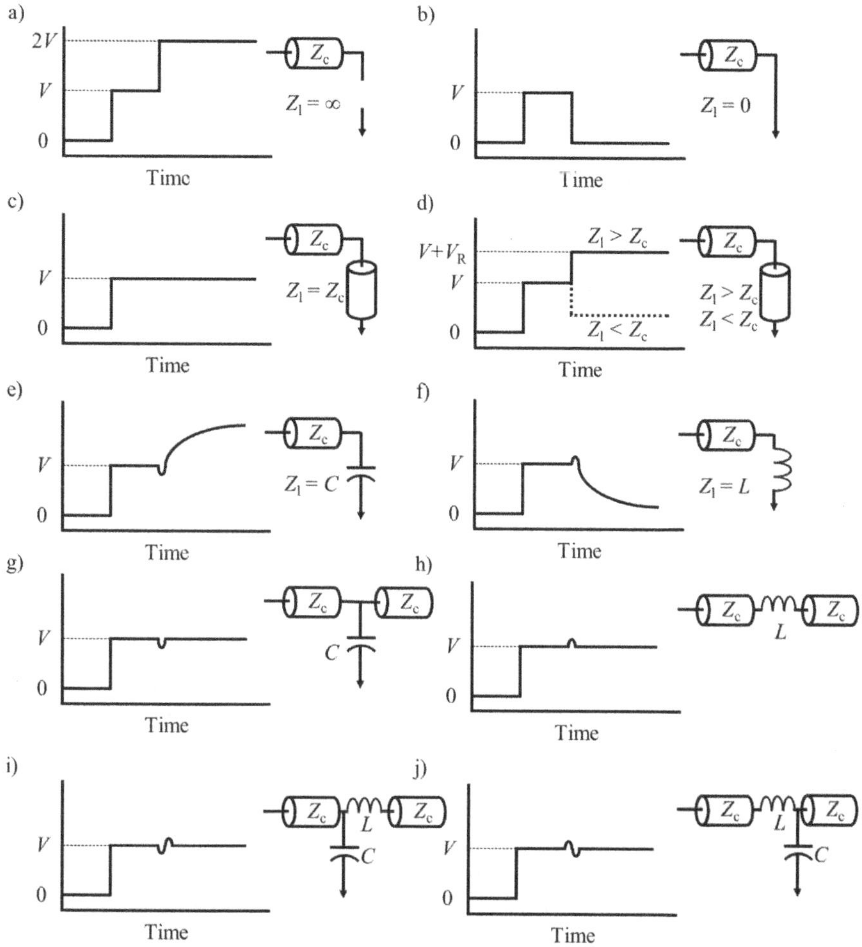

Figure 13.3 TDR results for (a) open- and (b) short-circuit terminations; (c) matched and (d) mismatched load terminations; (e) capacitive and (f) inductive load discontinuities; (g) shunt capacitance and (h) series inductance discontinuities; (i) shunt capacitance-series inductance; and (j) series inductance-shunt capacitance discontinuities.

at the output. These early reflections reduce resolution by obscuring closely spaced discontinuities.

Aberrations such as ringing that occurs after the incident step reaches the discontinuity will cause corresponding aberrations in the reflections. The aberrations will be difficult to distinguish from the reflections from discontinuity in the device under test (DUT). The aberrations in the TDR instrument's step generator and sampler have virtually the same effect. The aberrations of the TDR pulse rise are shown in Figure 13.4 and can be seen as the reflection causes a preshoot to occur [4].

13.4.1.3 TDR Accuracy Factors

Many factors contribute to the accuracy of a TDR measurement. These include the TDR system's step response, interconnect reflections and DUT losses, step amplitude

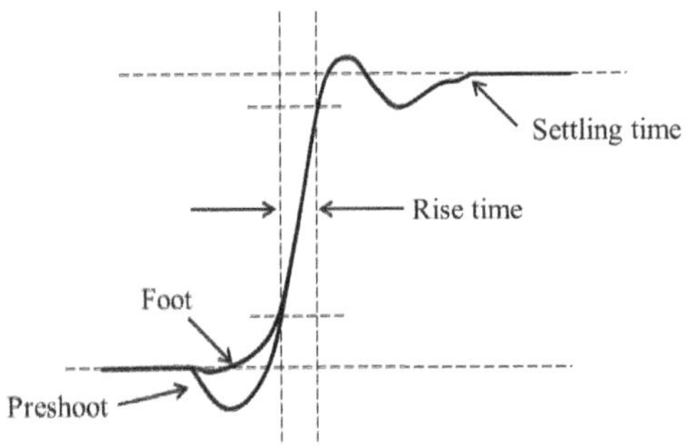

Figure 13.4 Aberration on TDR pulse [4]. Graphically renewed.

accuracy, baseline correction, and the accuracy of the reference impedance (Z_0) used in the measurements. All reference impedance TDR measurements are relative; they are made by comparing reflected amplitudes to an incident amplitude.

In modern TDR instruments, all the calculations are done internally by comparing the incident and reflected amplitudes. However, the TDR measurement still depends on the accuracy of the reference impedance (Z_0).

Modern TDR instruments automatically calibrate the incident step amplitude. They do this with known step amplitude and baseline levels.

The problem with the incident step aberration is that the magnitude of the reflected step cannot be measured accurately if the pulse does not settle in a time. This error is more important in a system where the DUT impedance is significantly different than 50 Ω. In these cases, the accuracy of the reflected wave amplitude is critical. In cases where the DUT amplitude is closer to 50 Ω, accuracy is more dependent on the reference impedance since the reflections are small. A second type of aberration is caused by the pre-shoot that precedes the incident step. If there is an open circuit at the end of the line there will be reflections from here before the incident step actually starts. This will cause an error at the end of the line being measured.

Noise can significantly affect the measurements of small impedance variations. However, modern instruments can perform signal averaging to remove the noise, which generally tends to be of high frequency. Noise can slow down the processing speed of the instruments.

The probe cable impedance can affect the accuracy of the measurement. The reference level will be shifted by the rho of the cable (ρ_{cable}), and the step amplitude incident on the DUT will be $(1 - |\rho_{cable}|)$. For maximum accuracy, these parameters should be calculated and correspondingly compensated. Reflections from interconnect components and the probe-to-DUT interface can also cause problems. The probe interface can cause an inductive reflection, which must settle down before a measurement can be made. It is important to keep the probe tip and ground leads short to minimize these problems.

While very fast rise times are desirable to prevent overlap with reflected signals, they could give misleading results in some cases on a TDR measurement. The problem is that a high rise time signal could compromise the environment, and the high rise time may affect the actual signal when it is passed through it. It is, therefore, preferable to see the transmission line's TDR response to rise times similar to the actual circuit operation. Some TDR instruments provide a means for increasing the apparent rise time of the incident step. The filtering is performed inside the instruments using math equations to show the filtered waveform as it will be in real time. This technique provides fully "live" filtered waveforms, quickly responds to changes, and requires no additional calibration steps.

13.5 Vector Network Analyzer

A vector network analyzer (VNA) is used to determine the magnitude and phase characteristics of connectors. It compares the incident signal leaving the receiver with the signal reflected from the test device. The system hardware is composed of an analyzer, precision components required for calibration and performance verification, optional use of synthesizers as a second source, and optional use of power meters for test-port leveling and calibration. The three basic components of a VNA are the source, the test set, and an analyzer.

The source functions as the input signal generator for the DUT. There are two types of signal sweeps possible: the analog sweep mode and the step-sweep mode. In the analog sweep mode, the signal is not locked to a stable reference. Hence, the measurement, especially of the phase, becomes difficult. To overcome this, the VNA should operate in step-sweep mode with phase-locked conditions.

The test module directs the signal from the source to the DUT. It also samples the transmitted and reflected signals. An auto-reversing feature helps to apply the input signal in both forward and reverse directions. This helps in avoiding manually reversing of the DUT to measure all four S-parameters. Auto-reversing also helps increase accuracy of VNA readings by reducing connector repeatability errors.

The third part of a VNA, the analyzer module, compares the input and reflected signal and calculates the magnitude and phase data. This result is displayed on a high-resolution display screen. This display can show all four S-parameters simultaneously, as well as a variety of other forms of displayed information such as group delay, time and distance, and complex impedance information.

13.6 Simulation Program with Integrated Circuit Emphasis (SPICE)

Simulation Program with Integrated Circuit Emphasis (SPICE) is simulation software used to model electronic circuits. It is generally used to predict the behavior of low- to mid-frequency (direct current to around 100 MHz) circuits.

The circuit diagram schematic is given as input and is automatically converted into netlists. There is always a trade-off between the complexity of the model and the simulation speed. With the ever-increasing speed of computers, it is now possible to simulate complex circuits. The broad purpose of a simulation is to validate the hand calculations and predict the circuit behavior.

The SPICE model of high-speed connectors is formed by cascading large numbers of unit cell circuits together, as illustrated in Figure 13.5. It is called a multisegment multiple transmission line (MMTL) model for simplification. Hundreds and even thousands of segments are needed in an MMTL model because high-speed connectors are large, and the model requires high precision. The parameters of the unit circuits are calculated from measured impedance profiles [5].

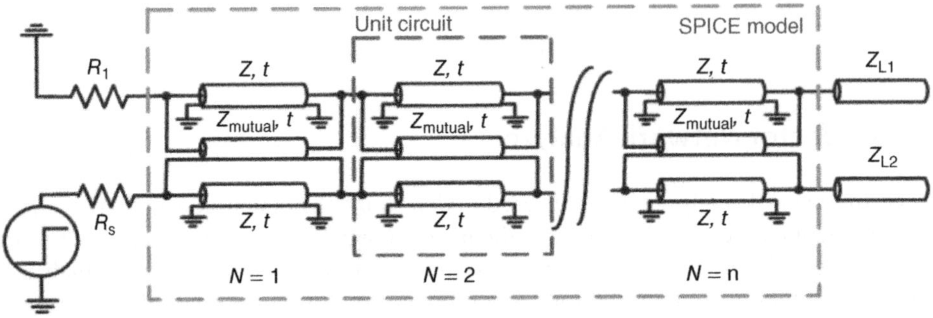

Figure 13.5 MMTL model of high-speed and high-density connectors [5].

References

1 Mroczkowski, R.S. (1998). *Electronic Connector Handbook*. New York: McGraw-Hill.
2 IEC 60529 (1989). *Degrees of protection provided by enclosures (IP Code)*, International Electrotechnical Commission.
3 Trebbels, D., Kern, A., Fellhauer, F. et al. (2013). Miniaturized FPGA-based high-resolution time-domain reflectometer. *IEEE Transactions on Instrumentation and Measurement* 62 (7): 2101–2113.
4 Lu, F. (2017). Calibration for time domain reflectometry modular in digital communication analyzer. *Proceedings of the 13th IEEE International Conference on Electronic Measurement and Instruments*, Yangzhou, China (October 2017). IEEE.
5 Zhang, M.S., Li, Y.S., Li, L.P., and Jia, C. (2008). Modeling and analyzing high-speed and high-density connectors by using multisegment multiple transmission lines model. *IEEE Transactions on Advanced Packaging* 31 (1): 203–210.

14

Advanced Technology Attachment Connectors

Neda Shafiei, Kyle LoGiudice and Michael G. Pecht

Center for Advanced Life Cycle Engineering (CALCE), University of Maryland, College Park, MD, USA

Advanced technology attachment (ATA), the standardized version of integrated drive electronics (IDE) [1], and serial ATA (SATA) are the mass storage device interfaces that connect hard disk drives (HDDs), optical disks, and solid-state drives (SSDs) to a computer motherboard. ATA connectors have important applications in data transmission for personal computer HDDs and optical drives alike. They are the intermediate link between a computer's motherboard and the disk controller on the HDD, allowing the HDD to read and write data. This chapter covers the ATA as well as the newer and faster SATA connectors.

14.1 ATA Connector and SATA Connector Overview

ATA connectors and their data transmission speed have undergone many different iterations to accommodate increased processor speeds, file sizes, and the demands of complex computer operating systems. When the data transmission speeds of computer processors and HDDs started to overwhelm the physical hardware of the ATA connector, alternative technologies using serial data transmission were developed. When SATA connectors were released in 2003, the name of ATA connectors was changed to parallel ATAs (PATAs); however, many sources still use ATA to refer to PATAs. In this chapter, ATA and SATA refer to parallel ATA and serial ATA, respectively.

There are two types of ATA/SATA interfaces, one for power supplement and the other for data transfer. Figure 14.1 shows the ATA power interface and data interface. The SATA power and data interface and their schematics are shown in Figure 14.2a and b, respectively.

The ATA data interface has two configurations. In the first configuration, the interface has two connectors embedded at the two ends of the cable. This type of ATA interface can be connected to one drive. In the second configuration, the interface has two connectors at the two ends of the cable and one connector in between, as shown in Figure 14.3. This type of ATA interface can be connected to up to two drives. The two drives are known as "master" and "slave."

The SATA data interface has only one configuration where two connectors are embedded at the two ends of a cable. Each SATA interface can connect to only one drive.

Electrical Connectors: Design, Manufacture, Test, and Selection, First Edition.
Edited by San Kyeong and Michael G. Pecht.

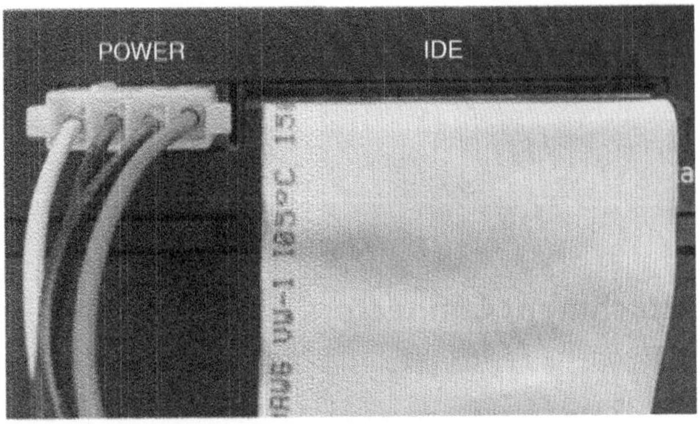

Figure 14.1 ATA power interface and data interface.

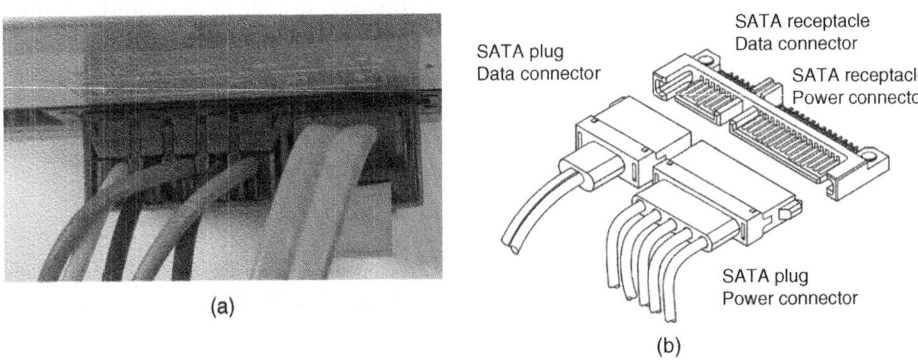

Figure 14.2 (a) SATA power interface and data interface, and (b) schematic of SATA power interface and data interface.

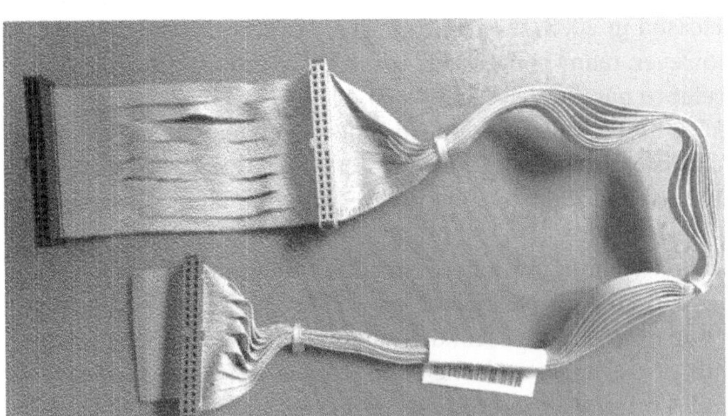

Figure 14.3 An ATA interface with three connectors.

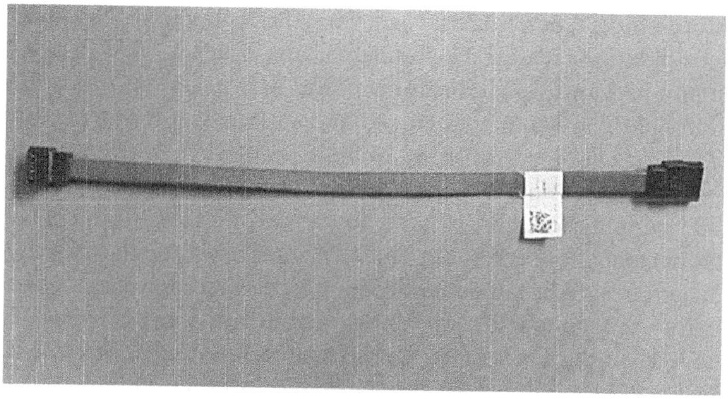

Figure 14.4 A SATA interface with two connectors.

Figure 14.4 shows the SATA data interface. SATA connectors are the successor that made ATA connectors obsolete in the twenty-first century.

SATA transfers data with two serial connections, one for transmission and the other for receiving. The transfer rates for the original SATA specifications begin at 187.5 MB/s. The encoding used by the host controller is 8 bits/10 bits, which takes an 8-bit data stream and converts it to a 10-bit data stream, which allows for no more than five 0 or 1 values to be in a row, helping the signal input become less distorted before going into the cable.

The commands used in SATA by the host controller in the motherboard are the same as the commands used in ATA with a packet interface standardization. SATA has several advantages over the traditional ATA, such as easy installation and configuration with true plug-and-play connectivity [2] (i.e. no need for user adjustment or reconfiguration when the device is first used or connected). SATA has a smaller number of conductors than ATA and is connected to a narrower ribbon cable, which facilitates the airflow inside the chassis, allows a smaller chassis design, and reduces the production costs. SATA uses hot swapping, which means that the interface can be replaced without shutting down the system. SATA has a faster and more efficient transfer rate than ATA through an input/output (I/O) queuing protocol [3]. SATA, however, has a few disadvantages over ATA. SATA can support only one storage device, whereas ATA supported two devices (known as the master drive and the slave drive). SATA HDDs sometimes need a special device driver (a computer program that operates the device) to recognize the drive. However, the SATA HDDs can work as an ATA HDD, thereby eliminating the need for a specific device driver [4].

14.2 History of ATA and SATA

Prior to ATA's standardization in the early 1970s, ST-506/412 was used as an interface between the HDD and its controller. In the 1980s, personal computer HDDs used Shugart Associates system interface (SASI) and enhanced small disk interface (ESDI) as a disk controller interface. The SASI connection was eventually rebranded into what is known today as the small computer system interface (SCSI). The SCSI connector was standardized in 1986 under document number X3.131-1986 by Technical Committee

T10, a division of the International Committee for Information Technology Standards (INCITS), which is responsible for creating IT standards. The original SCSI designs were compatible with many previous SASI controllers. The first generation of SCSI controllers could allow for eight drive connections and had a cable length of 6 m with a 5 MB/s transfer rate. SCSI was adopted for many popular computers at the time of its debut in 1986, but when ATA connectors were being introduced in 1994, many companies decided to add support for ATA to their lower-end PCs. SCSI and ATA became separated as ATA became more common. The ATA interface's popularity grew in personal computers, whereas SCSI became predominant in data storage centers and media companies [5, 6]. The ATA connector was standardized with several revisions. The initial standard, ATA-1, was approved by the American National Standards Institute (ANSI) on May 12, 1994, and further iterations were backward compatible, meaning new ATA drives could be used with older ATA connectors. The early ATA interface is often referred to as IDE, and ATA attachments may be referred to as IDE or PATA connections interchangeably.

The first version of the SATA connector was co-developed by APT Technologies, Dell, IBM, Quantum, Intel, Maxtor, and Seagate in 2000 and was officially released in 2001. The original developers of the SATA standard were later known as the serial ATA II working group. Systems using SATA interfaces were first released in 2002 and used discrete peripheral component interconnect (PCI) interface boards and chips. The extensions of SATA 1.0, released in 2002, made the SATA connector suitable for network storage. SATA was integrated into the motherboard chipset and was adopted in desktop systems in 2003. The SATA working group formed the SATA international organization (SATA-IO) in 2004 to provide storage device companies with guidance and support for implementing the SATA specifications [7]. Beginning in 2005, the SATA interface was adopted in laptops [8].

14.3 Physical Description of ATA Connectors, ATA Alternative Connectors, and SATA Connectors

HDDs are connected by an intermediary disk controller to the motherboard, allowing the motherboard to communicate through the CPU to the HDD or floppy drives. Modern drives incorporate the features of a disk controller within the drive housing or on the motherboard, and the ATA/SATA connector was designed to use those configurations.

The ATA connector, as shown in Figure 14.5, consists of a 2 × 20 matrix of female pin connections. Pin 20 is removed, and a cable key is added on the side to ensure the ATA connector is inserted in the correct orientation. The older versions of the ATA 40-pin connectors (ATA-1 to ATA-4, which will be discussed in Section 14.4) are connected to a 40-wire cable and are normally black or some other neutral color, while the newer versions of the connectors (ATA-5 and higher versions, which will be discussed in Section 14.4) are connected to a 80-wire cable to increase the transfer rate and are blue, black, or gray. The blue, black, and gray connectors should be connected to the motherboard, master drive, and slave drive, respectively. Although the color of ATA connectors might be different, their physical shapes are the same. The newer ATA connectors are downward compatible [9].

Figure 14.5 AT attachment 40-pin connector.

```
+----------------------------------------1-E-----+
| o o o o o o o o o o o o o o o o o o o o o o K C A |
| o o o o o o o o o o o o o K o o o o o o o o o K D B |
+44---------------------20----------------2-F-----+
```

Figure 14.6 Pin arrangement of the 44-pin connector [10].

Fifty-pin SCSI connectors look similar to ATA connectors, but they are wider, have more pins, and have no blocked pins. The reset pin 1 on an ATA connector is usually marked with a different colored cable. Newer ATA devices used in some notebooks use a connector with 50 pins and 44 wires, which includes the original 40-pin configuration along with additional pins used to combine the ATA connector with the power connector. The ATA connector is a more compact spacing of the original pin configuration because the pin spacing is 2 mm, whereas the original design spacing is 2.54 mm.

The 44-pin connector developed by the Small Form Factor (SFF) committee is an alternative for ATA connectors and is used for 2.5-in. and smaller drives. The additional four pins are for power. The distance between the pins in the 44-pin connector is less than the distance between the pins in the 40-pin connectors, and the 44-pin connector is physically smaller than the 40-pin connector. Figure 14.6 shows the schematic of the pin arrangement of the 44-pin connector. The first four pins of the connector plug are for the manufacturer's use and thus are not connected to the host. Pins A, B, C, and D may be utilized by some drives for option selection via physical jumpers. Pins E, F, and 20 are keys and are removed [11].

SFF has two unitized connectors defined by SFF8057 and SFF8058 for 3.5-in. and larger drives. Each of these connectors has a signal portion, a power portion, and a jumper connector portion that comply with the 40-pin ATA connector. Figures 14.7 and 14.8 show the schematics of the connectors.

The SATA interface has a cable, two connectors at the ends of the cable, and two PCBs. The connectors are soldered onto the pads in the PCB. The PCBs connect the connectors to the cable. The width of the connector is 13.87 mm. The connectors are embedded only at the ends of the cable and are interchangeable (e.g. the ends of the cable are identical) [13]. The connectors in SATA data and SATA power plugs are notched to ensure they are oriented properly. The SATA data connector has seven pins that are spaced apart by 1.27 mm. Figure 14.9 shows a SATA keyed connector with pins. The end of the connector may be straight, right-angled, or left-angled, which allows the cable to connect the drive from different directions. Figure 14.10 shows an angled- and a straight-SATA connector.

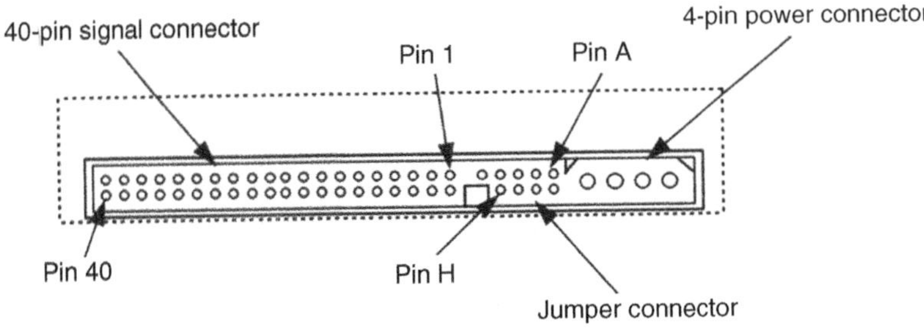

Figure 14.7 The SFF8057 connector, an alternative for ATA 40-pin connector [12].

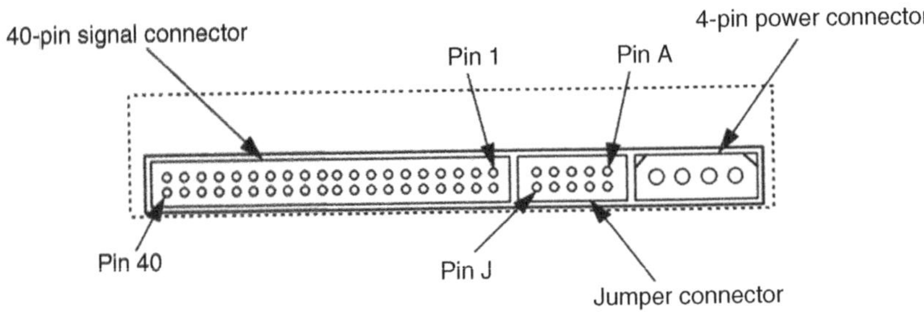

Figure 14.8 The SFF8058 connector, an alternative for ATA 40-pin connector [12].

Figure 14.9 SATA connector magnified.

The SATA data connector and the SATA power connector are almost the same width, but the data connector is thinner than the power connector. The connectors in both SATA data and power plugs are notched to ensure they are oriented properly. An adapter can be used for systems with no SATA power connector to convert a standard power connector to a SATA power connector.

The connection is point-to-point, meaning that only a single drive may be connected to the motherboard using a connector, without an intermediary connection for additional drives along the cable. The point-to-point serial link, however, allows as many SATA devices as SATA connectors to be connected to the motherboard at the same time via several interfaces. Figure 14.11 shows two SATA cables that are connected to a motherboard at the same time.

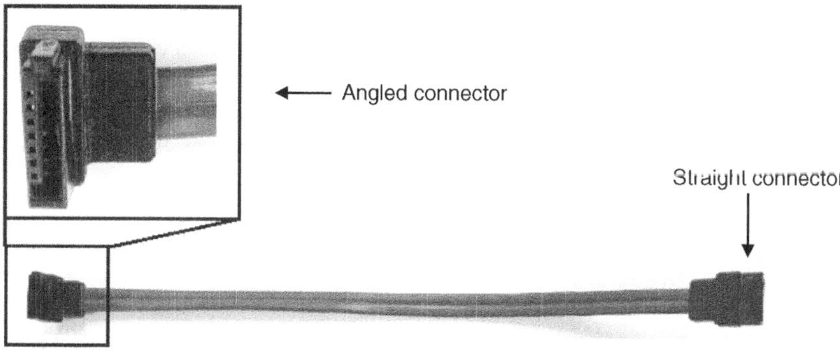

Figure 14.10 A SATA interface with one L-shaped connector and one flat connector.

Figure 14.11 Connection of two SATA interfaces to a motherboard.

The SATA connector materials are required to be selected based on the UL94V-0 flammability standard [14]. Some materials used for SATA receptacle/plug connector housing include high-temperature thermoplastics [15], glass filled [16], and heat-resistant polyimide [13]. The receptacle connector contact and the plug connector contact are made from different materials. As an example, brass with a layer of gold on the mating area and a layer of tin on the solder area, over nickel on the entire area, was used for plug connector contacts, whereas the receptacle connector was made from phosphor bronze with a layer of gold on the mating area and a layer of tin on the solder area, over nickel on the entire area [17].

Revisions to the ATA interface protocol throughout the years have allowed ATA to incorporate a higher bit rate along with more supported types of devices, such as the packet interface expansion for compact disc read-only memory (CD-ROM) and tape. Eventually the transfer rate required by modern computing became too challenging for the ATA design, and it was replaced with SATA. ATA supporting drives and connectors continued to be supported and improved up to ATA-7 in 2005 [3, 4].

14.4 ATA Standardization and Revisions

The ATA standards are the responsibility of the INCITS and filed through ANSI. ANSI oversees the Information Technology Industry Council (ITIC) division; the INCITS, formerly known as Accredited Standards Committee X3 and National Committee for Information Technology Standards, which appear on some standardization documents. INCITS focuses on standardizing file formats, computer communication methods, and storage media. Within INCITS' subcommittees, Technical Committee T13 was specifically created for the ATA standards and maintains ATA standards that have yet to be withdrawn. The first ATA standardization was under ANSI X3.221-1994, usually referred to as ATA-1 or IDE, while later versions of the attachment have different suffixes associated with their standardization number [3, 4, 18].

New generations of ATA occurred with several upgrades and modifications of the ATA standard. Technical Committee T13 is responsible for all standards relating to the ATA interfaces. The ATA standard requires HDDs that use the later versions of ATA to be able to work with the systems produced by the earlier version of the standard. All versions of the ATA standard define the electrical requirements to transfer the data between the HDD and controller across a 40-pin connector and cable.

Officially published in January 1994, ATA-1 was the first revision of the ATA standard developed by Control Data Corporation, Western Digital, and Compaq. ATA-1 defined the original ATA interface. ATA-1 interfaces had two connectors and could be connected to two HDDs named as master (drive 0) and slave (drive 1). Through ATA-1 the designation of master or slave drive was made in the following three ways: (i) a switch on the drive, (ii) a jumper plug on the drive, and (iii) use of a cable select (CSEL) pin. ATA-1 interface had 40/44-pin connectors and cabling. The 44-pin connectors were used for devices that did not have independent power interfaces, and the power was supplied through the extra four pins. ATA-1 defined two methods of data transfer supported by all versions of the ATA standard including programmed input/output (PIO) and direct memory access (DMA). In the PIO method, the central processing unit (CPU) managed the information transformation, whereas in DMA the drive sent the information directly to memory [17]. For each of the transfer methods, ATA-1 defined three levels of transfer rate that ranged between 800 KB/s and 8.33 MB/s and were known as modes 0, 1, and 2, where higher modes of PIO indicate faster read and write times. Although the ATA-1 interface was designed to support HDDs having up to 136.9 GB capacity, the basic input/output system (BIOS) could only support HDDs of up to 528 MB capacity. The ATA-1 standard was withdrawn in 1999 [3, 10].

ATA-2, known as enhanced IDE (EIDE) by Western Digital, and fast ATA/fast ATA-2 by Seagate and Quantum, was drafted in 1993 and was officially published in 1996 as ANSI X3.279-1996 ATA interface extensions [16]. ATA-2 was updated to define an interface between the host system and different kinds of storage devices not limited to HDD. ATA-2 could support faster PIO and DMA transfer mode, power management, removable devices, and PC cards known as personal computer memory card international association (PCMCIA) cards.

In addition to supporting the oldest modes defined in ATA-1, ATA-2 added modes 3 and 4 to increase the maximum transfer rate from 8.33 to 16.6 MB/s. ATA-2 could not support a data transfer rate of above 16.6 MB/s because of low-performance cabling and poorly arranged signal processing that lead to signal degradation during transmission and poor signal quality. ATA-2 included methods to enable BIOS support up to 8.4 GB and improvements in the identify drive command.

Subsequently, the ATA-3 standard was approved by ANSI in 1997 under document number X3.298-1997. ATA-3 did not increase transmission speed from past designs but added security features to provide password security option for device access, more power management, and a new self-monitoring analysis and reporting technology (SMART) feature. SMART was designed to warn of a drive failure in advance from hard drives going out of specified tolerances. ATA-3 made logical block addressing (LBA) mode support mandatory and recommended some solutions for the noise issue at higher transfer speeds. ATA-3 document ANSI X3.298-1997 has been withdrawn since 2002 [3, 4, 6].

The ATA-4 standard with packet interface extension (ATAPI-4) was approved by ANSI in 1998 under document NCITS 317-1998. ATA-4 includes ATAPI (packet interface) commands, enabling an ATA interface to use SCSI-style commands such as ejecting mainly for optical and tape devices. It introduces transfer mode UDMA/33, also known as ultra-DMA/33 or ultra-ATA/33, which supports data transfer rates of up to 33 MB/s. The speed of data transfer doubled from regular DMA to ultra-DMA as the frequency of data operations doubled. Using a method called "double transition clocking," the controller clock began performing data transfer operations during the rise and fall of the controller's digital clock frequency, while DMA operated only when the clock frequency increased. The redesign of DMA into UDMA allowed for much better drive performance needed at the time, and by reducing cycling time between data transfers, higher modes of UDMA benefitted twice as much. With the introduction of ATA-4 also came changes to the cables, with the addition of an 80-wire connector. The additional 40 wires connected to ground to eliminate cross talk between the original 40-data wires and became necessary as the speed required of ATA was increased multiple times from the original design. After UDMA/33 transfer speeds, 80-wire connectors were mandatory for all future iterations of ATA connectors (5–7). The host protected area (HPA) was first introduced in ATA-4. HPA allows a portion of the device to be reserved for information that cannot be easily accessed or modified by the user, BIOS, or operating system. HPA contains information such as HDD utilities, diagnostic tools, and boot sector code [19]. ATA-4 maintained compatibility with the previous specifications and, while providing additional functions, no past devices required changes as it still used the 40-pin connector [3, 4, 7].

ATA-5 standard was approved by ANSI in February 2000 under document NCITS 340-2000. ATA-5 added support for ultra-DMA/66, also known as ultra-ATA/66, with data transfer rates of up to 66 MB/s, and could detect between 40- or 80-wire cables to ensure 80-wire cables were connected for high-speed UDMA/66. Using a 40-pin 80-connector cable, which is a CSEL cable, became mandatory in ATA-5 and later versions of the standard. A 40-pin, 80-connector cable increased the data transfer accuracy and reduced the noise and electromagnetic interferences. ATA-5 standard maintained a high degree of compatibility with ATA-4 with packet interface extension (ATA/ATAPI-4) and previous iterations [3, 6].

The ATA-6 standard with packet interface extension (ATAPI-6) was approved in 2002 under document ANSI NCITS 361-2002. ATA-6 added support for ultra-DMA/100 (mode 5), also known as ultra-ATA/100, with data transfer rate of up to 100 MB/s. This is when ATA began to be referred to as PATA. In April 2005, the ATAPI-7 standard was approved under the document INCITS 397 – 2005 [20] (Vol. 3), which added support for UDMA/133 and SATA/150. Only Maxtor created a hard drive capable of using UDMA/133 because all other manufacturers had focused on SATA [21]. In ATA-6, sector per command increased from 8 to 16 bits, and LBA was extended from 2^{28} sectors to 2^{48} sectors to support devices up to 144.12 PB. ATA-6 made cylinder-head-sector (CHS) addressing absolute. According to ATA-6, drives must use 28-bit or 48-bit LBA addressing.

The ATA-7 standard, also known as ATA133, was officially introduced in 2004 by the T13 Technical Committee. ATA-7 had three volumes. Volume 1 defined the register delivered command sets. Volume 2 defined the connectors and cables, the electrical and logical characteristics of interconnecting signals, and the protocols for transporting commands, data, and status over the interface for parallel interface. Volume 3 defined the same features as volume 2 but for a serial interface. ATA-7 added support for UDMA/133 (mode 6), which enabled the transfer rate of 133 MB/s, and for long physical/logical sectors. The ATA-7 long physical sector feature set increased media format efficiency. Through ATA-7, the platters were formatted such that each long physical sector contained an error correction code (ECC) and multiple logical sectors. This increased the format efficiency with fewer ECC bytes stored. The ATA-7 long logical sector feature allowed additional words per sector for server applications. Sectors with 520 or 528 bytes were typical instead of 512 bytes, which was previously used. Devices that implemented the long logical sector feature set were not backward compatible with applications that used 256-word logical sectors, e.g. desktop and laptop systems. SATA revision 1.x was incorporated into ATA-7. The SATA I interface, also known as SATA 1.5 Gb/s, was the first generation of SATA interfaces with a bandwidth throughput of up to 150 MB/s. ATA-7 was divided into volumes 1–3, which covered the command set and logical register (for both ATA and SATA), parallel transport protocols and interconnects (for ATA), and serial transport protocols and interconnects (for SATA), respectively.

ATA-8 has had several updates since 2004 when the initial part of the standard was published. SATA revisions 2.x and 3.x were incorporated into ATA-8. ATA-8 replaced the read long/write long function and improved HPA management.

14.5 SATA Standardization and Revisions

The SATA standard has had three revisions, and all of the revision have used the same cable and connector. The SATA versions differ in their transfer rate performance [22]. The first version of the standard was Serial ATA 1.0, also known as SATA 1.5 Gbit/s, SATA-150, and SATA I, and was officially released in August 2001 [8, 22]. The extension of SATA 1.0 made SATA suitable for network storage.

In April 2004, SATA revision 2.0, also known as SATA 3 Gbit/s, SATA-300, and SATA II, was released. It had a maximum performance of 300 MB/s and was backward compatible with SATA 1.0. The eSATA standardization was also created in 2004 and was mechanically different from the typical SATA connector design.

In 2005, SATA revision 2.5 was introduced by adding native command queuing (NCQ) support, staggered spin-up, hot plug, port-multiplier, and eSATA support. NCQ allows the hard disk drive to optimize the path the read/write head takes when circling the disk in order to reduce the number of additional rotations around the disk and thereby improving read/write speeds [23]. The staggered spin-up is a method used in computers with multiple HDDs to reduce the power consumption during spinning. Spinning consumes a higher power than the other HDD's mechanical activities, which might lead to power shortage. The staggered spin-up provides a mechanism that allows the SATA host bus adapter (HBA) to sequence HDD initialization and spin-up [24]. Hot plug allows an additional component to be added to a running system without shutting down the system. The port-multiplier is a device that allows a single SATA port to be connected to multiple drives [25]. If an external hard drive uses the SATA bus interface, then it is referred to as eSATA. eSATA connectors are more robust and have a different physical design for greater usage and to conform to external computer cable safety specifications. eSATA connectors use higher voltage to support a longer cable, yet they have the same pin connections as a typical SATA connector. The cable has more shielding and recessed pins to prevent against electrostatic discharge, and the way both connectors are notched prevents either one from being plugged into the wrong port. The SATA and eSATA connectors are not interchangeable due to regulations on external computer wire components. The eSATA connector is designed to survive 5000 insertion and removal cycles, while the original SATA cable is rated at 50 cycles [14, 26].

SATA revision 2.6 was released in 2007 with the cooperation of Dell, Hewlett Packard, Hitachi Global Storage Technology (HGST), Intel, Seagate Technology, and Vitesse Semiconductor corporations. SATA revision 2.6 added an internal slimline cable and connector, an internal micro SATA connector for 1.8-in. HDDs, a mini SATA (mSATA) internal/external multilane cable and connector, modifications to NCQ such as NCQ priority and NCQ unload, enhancement to the built-in self-test (BIST) activate frame information structure (FIS), and enhancement for FIS. The slimline connector was designed to enable connection of a "slimline" form factor drive to the SATA interface. The micro SATA connector was used for 1.8-in. form factor HDDs [25], and the mini SATA was designed for SSDs used in laptops and tablets. In 2009, SATA revision 3.0, also known as SATA 6 Gbit/s or SATA-600, was released, and the maximum performance rose to 600 MB/s. SATA 3.0 supported backward compatibility with its former version 2.5/2.6. SATA revision 3.0 had many enhancements for improved functionality. Some of them included introduction of a small low insertion force (LIF) connector for more compact 1.8-in. hard drives and a connector to accommodate 7-mm optical disk drives for thinner notebooks [6, 11, 15, 27].

Subsequently, in 2011, SATA revision 3.1 continued to extend SATA technology to a wider variety of devices and electronics. The enhancements of revision 3.1 ranged from new power management methods to improved drive efficiency and reduced power demand on SATA devices. Key features included the new mSATA connector for notebook storage, zero power optical disk drive support for reduced power in idle SATA optical drives, and queued trim command for SSDs to maintain fast write times while erasing data [28].

In January 2013, the SATA revision 3.2 specification was introduced with a maximum performance of 750 MB/s. Key updates featured in the revision 3.2 included the new SATA Express (SATAe) connector and enhancements for emerging solid-state hybrid drives. The SATA Express specification enables SATA and peripheral component interconnect express (PCIe) standards to coexist. They do not work together but give users the option of choosing which interface to use for their storage drives. PCIe technology enables increased transfer speeds up to 2 GB/s (2 lanes of PCIe 3.0), compared with 2013's SATA technology at 750 MB/s. SATA Express is not to be confused with eSATA. Additional key features also included a universal storage module (USM), which enabled removable/expandable storage for electronics, DevSleep power reduction for computer sleep modes, along with hybrid information and rebuild assistance, which enabled greater drive performance for hybrid drives using combined solid-state and hard disk storage technology, along with enterprise storage methods of multiple drives with redundant information [29, 30].

SATA revision 3.3 was released in February 2016. The shingled magnetic recording (SMR) support released in revision 3.3 allows for a 25% capacity increase over previous hard drive recording methods. It overlaps the data tracks on a hard disk drive like shingles so that more tracks can be written within the same space, eliminating the unwritten gaps between each track [31].

14.6 SATA in the Future

In the future, SATA technology may be contested by lower-cost storage methods for smaller computer devices such as low-end laptops and smartphones. Embedded multimedia controller (eMMC) storage is a method of delivering the flash storage and controller together in a small package, which eliminates the issues of compatible connections. This technology currently has a big foundation in smartphone internal storage and two-in-one laptops. Newer hybrid drives incorporate both solid-state and hard disk technology into the same package for faster boot times and larger overall storage, while still using predominantly SATA. Newer SSDs are much smaller in comparison to previous iterations, making them ideal for permanent internal storage methods, and different kinds can be compatible within the same motherboard connector [32].

The M.2 solid-state card has phased out mSATA and SATA Express for SSDs as it allows for SATA and faster PCIe modes up to 4 GB/s. This internal connector allows an SSD or similar card to be directly plugged into the motherboard and can use SATA serial transfer up to 600 MB/s. However, the four lanes of PCIe are more preferable for large files. SATA and PCIe have been added together in different connector styles through SATAe and M.2, but since SATA was originally designed for the speeds of a hard disk drive, it is unable to keep up with newer SSD drives. Nonvolatile memory express (NVMe) is an SSD protocol specification that improves upon methods of accessing storage in SSDs and uses PCIe buses exclusively. Higher-performance SSDs may choose to be compatible with PCIe buses for this reason. Ultimately, SATA has retained its popularity because it is compatible in almost any kind of device, and the added speed of PCIe buses and the newer NVMe standard is not as necessary to most computing needs other than enterprise storage solutions with high volumes of data storage. SATA will continue to be the computer bus standard of choice for desktop and netbook PCs alike for years to come while coexisting with PCIe technologies through M.2 slots and SATA Express technologies [33–35].

References

1 Casey, J. (2015). *Computer Hardware: Hardware Components and Internal PC Connections,* Guide for undergraduate students. Dublin Institute of Technology.

2 Seagate (2008). Product manual: Barracuda 7200.11 Serial ATA. www.seagate.com/staticfiles/support/disc/manuals/desktop/Barracuda%207200.11/100507013c.pdf (accessed July 1, 2020).

3 Wang, S.P. and Ledley, R.S. (2012). *Computer Architecture and Security: Fundamentals of Designing Secure Computer Systems*. Hoboken, NJ: Wiley.

4 Computer Hope (Jan. 24, 2018). Advantages of SATA over PATA. www.computerhope.com/issues/ch001325.htm (accessed Sept. 29, 2019).

5 Beal, V. (2017). Difference between SAS, SATA, SCSI and ATA. www.webopedia.com/DidYouKnow/Computer_Science/sas_sata.asp (accessed Oct. 29, 2019).

6 Fachat, A. (2013). SCSI interface information. www.6502.org/users/andre/hwinfo/scsi.html (accessed Oct. 28, 2019).

7 SATA-IO Corporation. About SATA-IO. sata-io.org/about-sata-io (accessed Oct. 31, 2019).

8 Mueller, S., Soper, M.E., and Sosinsky, B. (2006). *Upgrading and Repairing Servers*. Pearson Education.

9 Quentin, D., Dulaney, E., and Skandier, T. (2012). *CompTIA a+ Complete Study Guide: Exams 220-801 and 220-802*. Indianapolis, IN: Wiley.

10 American National Standard of Accredited Standards Committee-X3 (1994). Information technology -AT Attachment Interface for Disk Drives, Revision 4c.

11 American National Standard of Accredited Standards Committee-X3 (1996). Information Technology -AT Attachment Interface with Extensions (ATA-2).

12 American National Standard of Accredited Standards Committee-X3 (1997). Information Technology - AT Attachment with Packet Interface Extension (ATA/ATAPI-4), Revision 14.

13 Kim, M. (2014). Electrical characterization on signal transmission of SATA cable assembly. *International Journal of Control and Automation* 7 (8): 173–180.

14 SATA International Organization (2007). Serial ATA revision 2.6. read.pudn.com/downloads157/doc/project/697017/SerialATA_Revision_2_6_Gold.pdf (accessed July 1, 2020).

15 Goh, C. Qualification Test Report, TE Connectivity, 2018.

16 Mueller, S. (2015). *Upgrading and Repairing PCs*. Que Publishing.

17 Tyson, J., How IDE controllers work. computer.howstuffworks.com/ide.htm/printable (accessed July 1, 2020).

18 Technical Committee T13 (2017). Technical Committee T13 AT Attachment. www.t13.org/Standards/Default.aspx?DocumentType=3&DocumentStage=2 (accessed Oct. 29, 2019).

19 Gupta, M.R., Hoeschele, M.D., and Rogers, M.K. (2006). Hidden disk areas: HPA and DCO. *International Journal of Digital Evidence* 5 (1): 1–8.
20 Technical Committee T13 AT Attachment, T 13 Standards. www.t13.org/Standards/Default.aspx?DocumentType=3 (accessed Oct. 29, 2019).
21 MiniTool Solution Ltd. (2015). www.powerdatarecovery.com/data-recovery/ata.html (accessed Oct. 29, 2019).
22 Deming, D.A. (2014). *The Essential Guide to Serial ATA and SATA Express*. Auerbach Publications.
23 SATA International Organization (SATA-IO). Native command queuing. sata-io.org/developers/sata-ecosystem/native-command-queuing (accessed Oct. 28, 2019).
24 Intel. (2004). Serial ATA Staggered Spin-Up. https://sata-io.org/system/files/member-downloads/Implementing%20Staggered%20Spin%20Up%20v1.pdf (accessed July 1, 2020).
25 SATA International Organization. Port multipliers. sata-io.org/developers/sata-ecosystem/port-multipliers (accessed Oct. 28, 2019).
26 Silicon Image (2004). External serial ATA. www.sata-io.org/sites/default/files/documents/External_SATA_WP_11-09.pdf (accessed Oct. 29, 2019).
27 SATA International Organization (2009). FAQ about SATA 6Gb/s and the SATA Revision 3.0 Specification. www.sata-io.org/sites/default/files/documents/SATA-Revision-3.0-FAQ-FINAL.pdf (accessed Oct. 29, 2019).
28 SATA International Organization (2011). SATA-IO releases revision 3.1 specification. sata-io.org/system/files/member-downloads/SATA-IORevision31_PRfinal_0.pdf (accessed Oct. 29, 2019).
29 Elliott, R. (2003). Serial attached SCSI general overview. www.also.com/pub/pdf/hp_serial_attached_scsi.pdf (accessed Oct. 29, 2019).
30 SATA International Organization (2013). SATA-IO unveils revision 3.2 specification. sata-io.org/sites/default/files/documents/SATA_v3%202_PR__Final_BusinessWire_8.20.13.pdf (accessed 29 Oct. 2019).
31 SATA International Organization (2016). SATA-IO expands supported features in revision 3.3 specification. www.sata-io.org/sites/default/files/images/SATA%20Spec%203%203%20Press%20Release_FINAL.pdf (accessed Oct. 29, 2019).
32 Datalight Inc. (2017). What is eMMC?" www.datalight.com/solutions/technologies/emmc/what-is-emmc (accessed Oct. 29, 2019).
33 Kyrnin, M. (2017). How the M.2 SSD is going to make your PC even faster. www.lifewire.com/what-is-m2-833453 (accessed Oct. 28, 2019).
34 Lee, J. (2016). PCIe vs. SATA: Which type of SSD is best for you? www.makeuseof.com/tag/pcie-vs-sata-type-ssd-best (accessed Oct. 29, 2019).
35 Landsman, D. (2013). AHCI and NVMe as interface for SATA Express devices. sata-io.org/sites/default/files/documents/NVMe%20and%20AHCI%20as%20SATA%20Express%20Interface%20Options%20-%20Whitepaper_.pdf (accessed Oct. 29, 2019).

15

Power Connectors

Michael G. Pecht and San Kyeong

Center for Advanced Life Cycle Engineering (CALCE), University of Maryland, College Park, MD, USA

A power cable consists of an assembly of conductors usually held together within an overall sheath [1]. The assembly is used for transmitting electric power from a source to a destination. These cables can be found within a building [2], buried under the ground [3], overhead [4], or even under the sea [5]. The connector requirements for each of the previously mentioned applications are different. When power connectors are used in a submarine environment, for example, several elements should be considered, such as the water-blocking sheath, insulation, additional armoring materials against corrosion, and thermal reliability. The connectors for underwater usage will be discussed in detail in Chapter 16.

The current rating of power connectors is higher than that of signal connectors. Power connectors are normally operated at high voltage, whereas signal connectors work at low voltage and are shielded to prevent data loss. Temperature rating is another distinguishing characteristic of power connectors. Compared to signal connectors, the temperature rise is much higher in power connectors because of the Joule heating due to higher currents. Temperature rise increases the energy of the electrons. Since the electrons of high energy collide more frequently, the electrical resistivity increases. To control the Joule heating, the bulk resistance should be minimized, and the stability of the interface resistances must be ensured. These requirements are reflected in the selection of higher-conductivity spring materials and larger cross sections to minimize bulk resistance, and higher normal forces and/or redundant contact beams to minimize and stabilize separable interface resistance.

15.1 Requirements for Power Connectors

Selecting a power connector requires electrical, thermal, and mechanical considerations. The electrical considerations involve proper material selection and dimensions to withstand voltage breakdown effects. That is, the insulation material and the spacing between the connector contacts become design criteria for voltage ratings. The thermal considerations depend on the resistance of the connector. The mechanical properties of power connectors also should be considered because most power connectors are used in outdoor environments where physical collisions can occur.

Electrical Connectors: Design, Manufacture, Test, and Selection, First Edition.
Edited by San Kyeong and Michael G. Pecht.

Several factors affect the ability of power connectors to transport high electrical energies without energy loss. The resistance of the connector must be minimized to reduce thermal effects as well as voltage drop. The resistance includes both the bulk resistance and the interface resistance. To minimize the bulk resistance, low-resistivity conductors should be used. However, other properties should also be considered to ensure the power connectors perform effectively in the environments they encounter.

15.2 Power Connector Materials

Copper and its alloys are good candidates for base materials of power connectors because they have high conductivity [6]. Due to their malleability and ductility, it is easy to manipulate and produce products with various shapes using copper-based materials. The electrical resistivity of pure copper is $1.68 \times 10^{-8}\ \Omega\,\text{m}$, which is the highest value of any nonprecious metal. Annealed copper has a resistivity of $1.72 \times 10^{-8}\ \Omega\,\text{m}$ and is referred to in terms of the International Annealed Copper Standard (IACS). Electrolytic tough pitch copper, Cu-ETP, is the most common high-conductivity copper and has a minimum conductivity of 100% IACS. Cu-ETP is produced by refining with electrolytic deposition, melting, and oxidization. However, Cu-ETP should not be used in a hydrogen environment with a temperature of more than 370 °C. When Cu-ETP is heated in hydrogen, it becomes vulnerable to embrittlement. For high-temperature applications, oxygen-free copper, Cu-OF, is a solution. Cu-OF can be produced under controlled conditions with minimal contaminants. Although the addition of phosphorus is also effective to deoxidize copper, the prepared copper alloys have low conductivity, which is not appropriate for some electrical applications. Copper-silver alloys are preferred in applications of high creep resistance at high temperature. Furthermore, they offer high softening temperature without losing electrical conductivity. These alloys maintain the conductivity of 100% IACS. Copper-beryllium alloys are used for springs and connectors owing to their fatigue resistance. They also do not have magnetic properties that could induce electromagnetic heating. However, copper-beryllium has a low electrical conductivity of 20–70% IACS.

Aluminum and its alloys are also used as power connector materials [7, 8]. They have a remarkably low density of $2.70\ \text{g/cm}^3$, which is more than three times lighter than copper, and a resistivity of $2.64 \times 10^{-8}\ \Omega\,\text{m}$. Moreover, aluminum-based metals are resistant to corrosion because they form a passivation layer. However, there are some drawbacks when using aluminum as an alternative to copper. To supply the equivalent amount of energy, a greater volume of aluminum wire is required than for copper wire. Aluminum also has lower current-carrying capacity, also described as "ampacity," compared to copper. Furthermore, aluminum has low mechanical strength, which makes it difficult to manipulate into diverse shapes for power connector applications. The mechanical vulnerability of aluminum can be improved by adding magnesium, silicon, iron, or cobalt to form alloys.

Silver is also used as a material for power connectors. However, its utilization has been mainly focused in electroplating and coating rather than bulk material, not only because it is expensive but also because it has a low melting point and mechanical strength. Mechanical and thermal loads to the silver bulk could cause several failures such as wear formation, silver ion migration, and whiskers [9]. Silver sulfides also could be formed as a product of silver corrosion in a sulfuric atmosphere. These failures, which are explained in Chapter 8, can result in an increase of electrical contact resistance, showing fluctuation or intermittence of energy transfer [10]. Nevertheless, silver is a good candidate for the power connector applications due to its electrical conductivity of $1.59 \times 10^{-8}\ \Omega\,m$, which is the lowest value of all metals. Its mechanical strength and chemical inertness can be improved by forming alloys with metals such as palladium [11], copper [12], and cadmium [13].

In applications where the operating currents are low, platinum and palladium are good options because of their hardness and durability against wear formation [14, 15]. However, as explained in Chapter 8, frictional polymerization can occur under organic vapor-surrounded conditions.

15.3 Types of Power Connectors

There are many types of power connectors. Each type of connector is designed to meet the mechanical, electrical, and thermal requirements in diverse operating fields. The proper choice of connector design depends on the operating environment and affects the performance and reliability of delivering energy. Power connectors are categorized as three types according to their electrical requirements. Light-duty connectors are used in the operating conditions at current below 5 A and voltage below 250 V. Medium-duty connectors are used in the operating conditions of current above 5 A and voltage above 1 kV. Heavy-duty connectors are used in extreme conditions such as high current of more than 10 kA and voltage of hundreds of kV. Cost efficiency is also considered in the design of power connectors [16]. Figure 15.1 shows general designs of power connectors.

Plug-and-socket connectors, which are widely used in electrical appliances, are designed to allow quick and easy electrical coupling and decoupling according to user needs. These connectors are composed of a pin-type male plug and a receptacle-type female socket. Figure 15.2 shows a socket and a plug connector used for charging electric vehicles [17]. The electrical and thermal requirements might be relatively low because these connectors would serve as light-duty connectors; however, the materials

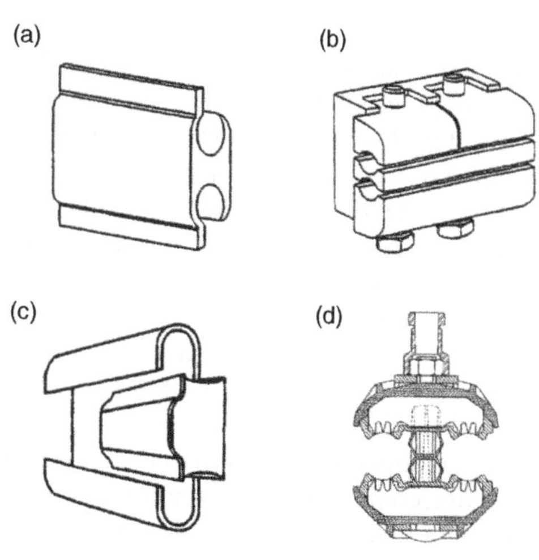

Figure 15.1 Power connector designs: (a) compression connector, (b) bolted connector, (c) wedge connector, and (d) insulation-piercing connector [16].

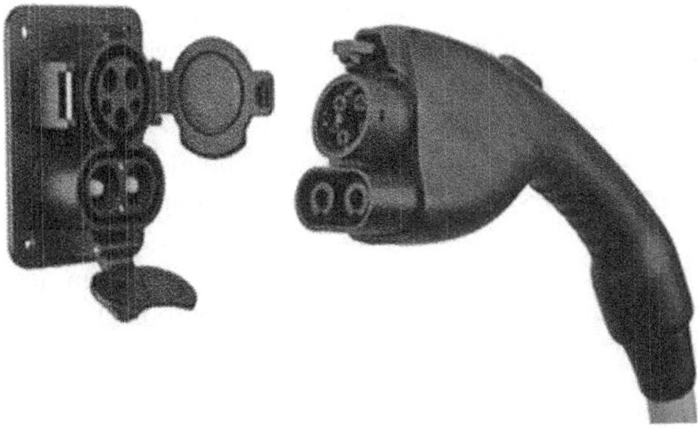

Figure 15.2 SAE J1772 connector for electric vehicles [17].

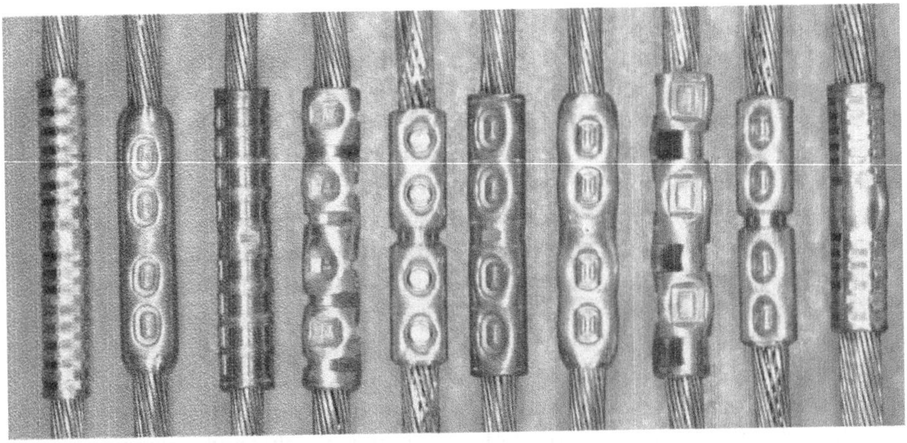

Figure 15.3 Various types of compression connectors [18].

should be carefully chosen to withstand wear or corrosion failures that can occur during frequent coupling/decoupling actions with a power source.

Compression connectors are used to transmit and distribute high-voltage power. Wire conductors are inserted into a compression connector and then immobilized by various coupling methods such as welding, soldering, or crimping. Decoupling of these connectors from conductors is normally irreversible. Figure 15.3 shows examples of compression connectors for stranded aluminum power conductors. The conductors are compressed with the connectors by mechanical or hydraulic force [18].

Wire conductors can also be immobilized by bolted connectors. Bolted connectors are easy to use and cost-effective compared to compression connectors. A wrench or a screwdriver is the only required tool for a bolted connector, whereas a complex machine that provides mechanical or hydraulic force is needed to assemble a compression connector. However, the utilized bolts should maintain their axial force during the lifetime of the connector. Figure 15.4 shows examples of bolted connectors, including a cross-sectioned image comparison with a compression connector [19].

Wedge connectors are considered reliable for use in harsh environment conditions including high temperature, frequent thermal cycling, high-velocity winds, and high

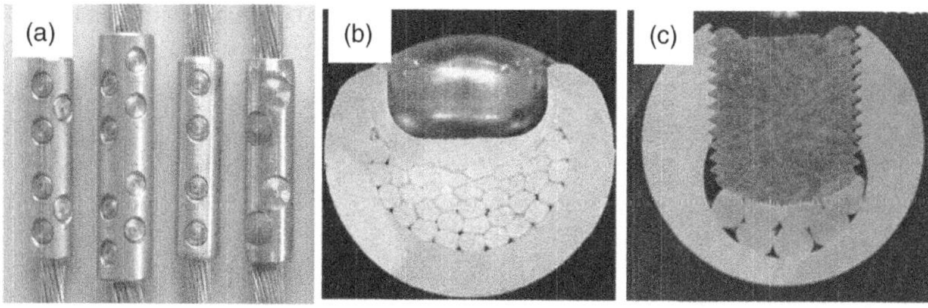

Figure 15.4 Various types of bolted connectors (a); cross-section images of compression connector (b) and bolted connector (c) [18, 19].

Figure 15.5 Wedge connector installation.

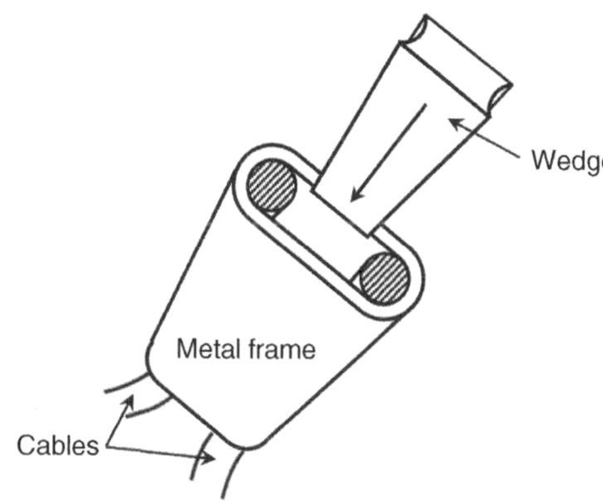

pollutants [20]. As shown in Figures 15.1c and 15.5, a wedge connector is made of a wedge placed between two cables located at opposite ends of a curve-shaped metal frame. When the wedge is inserted between the two cables, the insertion force causes plastic deformation of the curved metal frame since the frame tends to spread perpendicular to the insertion direction of the wedge. This plastic deformation mainly occurs at the flat area of the metal frame and generates an elastic restoring force. The restoring force makes the metal frame secure both cables tightly.

Insulation-piercing connectors are designed to withstand humid and corrosive conditions. The connector is composed of two compression blades, screw bolts, and their housing, as shown in Figure 15.1d. Sharp teeth are located at both ends of each blade and pierce the conductor wires to make an electrical contact between the wires. The piercing action is managed by the coupling torque from the screw bolts. With enough pressing force, the insulation-piercing connector can protect the pierced wires from moisture and pollutants [7]. Figure 15.6 shows a typical insulation-piercing connector with two wires [21].

Performance tests for the compression, bolted, wedge, and insulation-piercing connectors have been conducted by Braunovic [16]. The four connectors were subjected to current-cycling tests in a salted environment. The results showed that the insulation-piercing and wedge connectors maintained their low electrical resistance, whereas the compression and bolted connectors failed the tests, as shown in Figure 15.7 [16]. The unsatisfactory performance of the latter two connectors was caused by insufficient elastic restoring energy.

Figure 15.6 Insulation-piercing connector with wires [21].

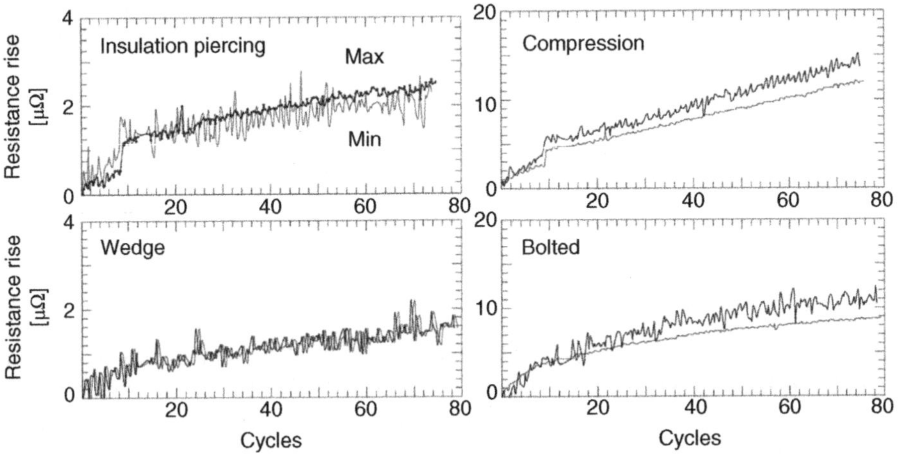

Figure 15.7 Effect of current cycling on the contact resistance and temperature of the four connectors [16]. "Min" and "Max" denote the resistance measured at the end of cooling and heating cycles, respectively.

15.4 Power Contact Resistance

In contacts intended for signal applications, the bulk resistance of the contacts is usually higher than the permanent and separable interface resistances by a significant amount. Therefore, the power loss due to the bulk resistances will be an important factor in determining the temperature rise. On the other hand, with power connectors, temperature rise and voltage drop must also be considered. In general, industries follow the 0.01 Ω change in resistance and 30 °C temperature rise criteria that are required by Underwriters Laboratories Specification UL 498, a well-known specification standard for power contacts [22]. The voltage drop and heating are closely related.

Since there are asperities at the surface of the practical connectors, the real conducting contact area is only a small portion of the expected contact area between the two connectors. The presence of insulating materials such as oxides on the contact surfaces can also cause a decrease of the conducting area of the contact region. These can result

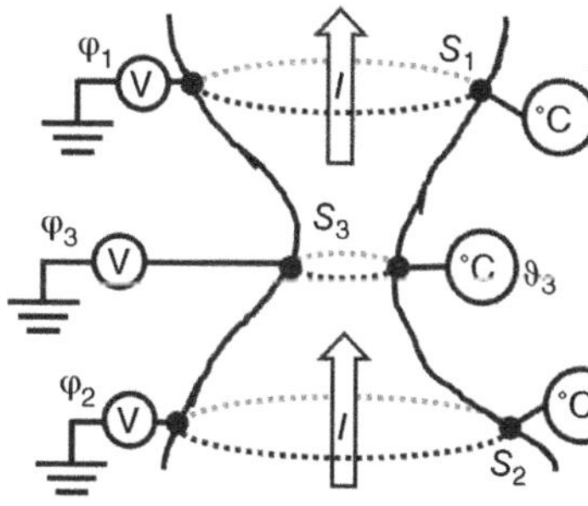

Figure 15.8 Arbitrarily shaped conductor, thermally and electrically insulated from the environment [26]. Graphically renewed.

in a constriction current at the contact point [23]. The constriction current provides the Joule heating at the interface of contacts, increasing the local temperature. The difference between the bulk temperature T_b and the elevated temperature T_c in the contact spot by constriction current is called "super temperature."

A relationship model between voltage and temperature was led in 1900 by Kohlrausch [24, 25]. In his work, an arbitrarily shaped conductor that is electrically and thermally insulated from the environment is considered, as shown in Figure 15.8 [26]. The current flows through the conductor, and the system is in the electric-thermal steady-state. It is also assumed that every equipotential surface is also an isothermal surface under the stated conditions [27]. Two governing processes can be given as

$$\mathrm{div}(\sigma \mathrm{grad}\varphi) = 0 \tag{15.1}$$

$$\sigma(\mathrm{grad}\varphi)^2 = -\mathrm{div}(\kappa \mathrm{grad}\vartheta) \tag{15.2}$$

where κ is the thermal conductivity, σ is the electrical conductivity, φ is the electric potential, and ϑ is the temperature, respectively. Eq. (15.1) describes the source-free flow of electric current, and Eq. (15.2) describes the electric power losses as the source of the thermal flow.

The relation between temperature and potential of the three equipotential and isothermal surfaces in the conductor of Figure 15.8 can be derived as [24]

$$\varphi_1 \int_{\vartheta_2}^{\vartheta_3} \frac{\kappa}{\sigma} d\vartheta + \varphi_1 \int_{\vartheta_3}^{\vartheta_1} \frac{\kappa}{\sigma} d\vartheta + \varphi_3 \int_{\vartheta_1}^{\vartheta_2} \frac{\kappa}{\sigma} d\vartheta = \frac{1}{2}(\varphi_1 - \varphi_2)(\varphi_2 - \varphi_3)(\varphi_3 - \varphi_1) \tag{15.3}$$

On the other hand, a theoretical approach can be derived from the thermal-electrical relationship, which is given by the Wiedemann–Franz law as

$$\frac{\kappa}{\sigma} = LT \tag{15.4}$$

where L is the proportionality constant, which is known as the Lorenz number, $2.44 \times 10^{-8}\ \mathrm{V^2/K^2}$.

If the contact system is thermally symmetric, both surfaces S_1 and S_2 have the same temperature $\vartheta_1 = \vartheta_2$, which can be altered as the bulk temperature T_b. The total voltage drop V is $\varphi_2 - \varphi_1$. Combining Eq. (15.3) with Eq. (15.4) and plotting ϑ_3 as a function of φ_3 yields a parabolic curve, as shown in Figure 15.9. At $\varphi_m = V/2$, this φ–ϑ curve has its maximum ϑ_m, which can be called the super temperature T_c.

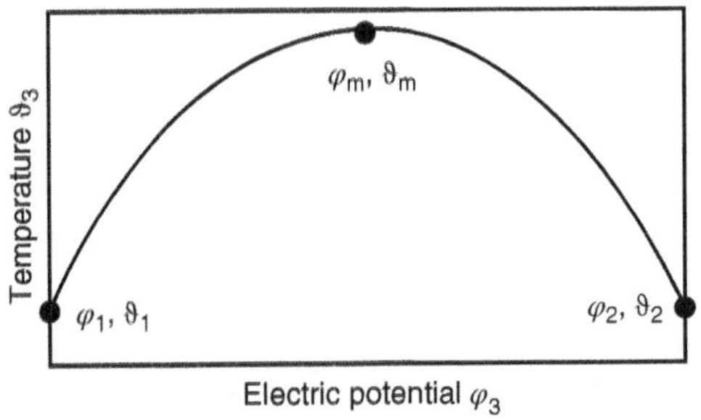

Figure 15.9 φ–ϑ curve of a thermally symmetric contact system [26].

Solving Eq. (15.3) by considering this contact system provides the well-known Kohlrausch relation, which is given as

$$V^2 = 4L(T_c^2 - T_b^2) \tag{15.5}$$

Since $V = IR_c$, where R_c is the constriction resistance, the voltage drop could be varied by the constriction resistance. In other words, as the constriction resistance increases due to contact age or any other reasons, the voltage drop and the contact temperature increase [24]. Hence, it is possible to evaluate the performance of power contacts by measuring the voltage drop instead of the standard techniques using thermocouples.

15.5 Continuous, Transient, and Overload Current Capacities

There are three current capacities for fixing the rating of the contacts – continuous, transient, and overload. The transient current capacities are of small durations (in milliseconds). The continuous current capacity is a magnitude the connector can withstand for a long duration. The continuous current capacity is commonly determined by the bulk temperature rise of 30 °C [22]. The transient current capacity should be determined by the super temperature because the super temperature could cause melting of the contact interface [28]. In addition, there is an overload current capacity for high-magnitude currents, but the overload current duration is slightly longer than the transient current. For determining the overload current capacity, the pulse shape, the amplitude of the current, and the duration must be considered. All three phases could be observed when an induction motor is started with a direct online starter [29]. As an example, Figure 15.10 shows a typical starting current of an induction motor.

15.5.1 Continuous Current Capacity

A current rating indicates the maximum current that the connector can carry continuously. The connector should have a current rating in excess of the continuous current. The temperature rise is a result of the heat balance between the Joule heating generated by the current and the heat dissipation to the external environment. The heat generated by direct current (DC) will be simply related to I^2R. For alternating current (AC) current, the temperature will depend on the current frequency, amplitude, and duration. The average of the Joule heating can be normally described as $I_{rms}^2 R$.

Heat can be dissipated by conduction, convection, and radiation. Radiation is the least significant among the three. Convection depends on the temperature and airflow

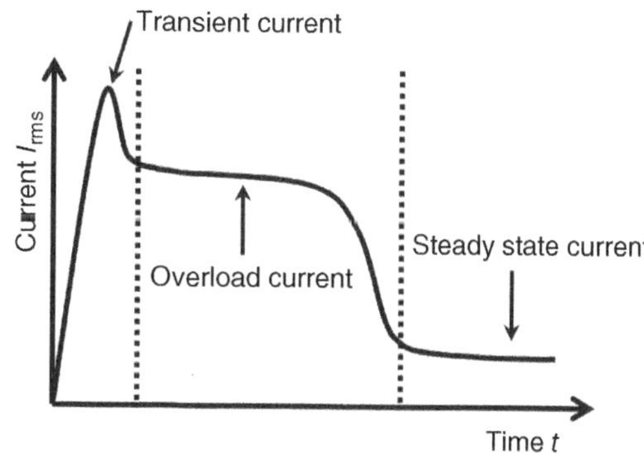

Figure 15.10 Starting current of an induction motor.

surrounding the connector. The basic relationship for heat transfer by convection is given as

$$\dot{Q} = hA(T_f - T) \tag{15.6}$$

where $\dot{Q}$ is the heat transferred per unit time, A is the area of the object, h is the heat transfer coefficient, T is the surface temperature of the connector, and T_f is the temperature of surrounding airflow. If the connector is sealed from the external environment by the connector housing, the heat dissipation by convective transfer will be limited. Hence, the presence of the connector housing should be considered when determining the current rating. Conduction of heat from the contact to an attached conductor depends on the ambient temperature, the thermal conductivity of the wire, and the cross-sectional area and length of the wire/trace. The conductive heat transfer is well established by Fourier's law, which is given as

$$\dot{Q} = \frac{\kappa}{s}A(T_f - T) \tag{15.7}$$

where s is the length of the wire.

15.5.2 Transient Current Capacity

Transient current can be observed when a typical induction motor starts. This type of current lasts normally several milliseconds, while its duration and magnitude depend on the application. The transient current barely contributes to the total Joule heating and temperature rise of the connector because the duration is short and the Joule heating caused by transient current occurs only in a localized area [27]. However, when the transient current phase is at the peak current, which may be more than 10 times the continuous current, the super temperature of the contact area can reach the melting temperature of the contact material. This could cause welding between the contacts, resulting in further failures including disabling separable connectors.

The result from calculation using Eq. (15.5) is consistent with the common sample. Bowden [23] described the super temperature of the interface between gold-to-gold contact while applying current. With a contact made of gold, which melts at 1063 °C, the voltage drop required for melting the interface is calculated as 0.41 V using Eq. (15.5). The interface melted where a voltage drop across the interface was about 0.4 V, while the super temperature of the contact was observed as about 1000 °C. The critical current to melt the contact interface was varied in accordance with the constriction resistance of the samples, retaining $I_{melt}R_c = 0.4$ V. Hence, power connectors having lower contact resistance are preferred for high currents.

15.5.3 Overload Current Capacity

The overload current differs from the transient current in the duration of the current. The overload current is of higher duration than the transient current, but the magnitude and duration depend on the applications. Some examples of overload current include the inrush current and motor startup current. The overload current is typically 5–10 times the continuous current of the application [30] and lasts several seconds. Since the peak point of the overload current does not exceed that of the transient current, it is rare that the interface of the contact is melted due to the super temperature of the overload current condition. However, some applications that involve long-duration overload currents require high current rating.

15.6 Current Rating Method

Specification standards of the current rating for power connectors have been established by related industry associates. Such standards include various testing methods, reliability analysis techniques, and safety instructions. The qualification for power connectors follows a standardized procedure, as shown in Figure 15.11. A general standard procedure has been proposed by Electronic Industries Association EIA CE 2.0 National Connector and Socket standard.

The qualification procedure starts with a shape examination of the power connector. The contact or housing design is observed to determine where temperature measurements should be made on the product. Subsequently, measurements for dry circuit resistance and temperature rise are made. The current measurements are then followed by several tests for reliability against repeated coupling/decoupling action, thermal cycling, and aging by temperature, humidity, flowing gases, and vibration.

For power connectors, the current rating should be based on the temperature rise. First, a dry circuit resistance of the connector is measured in order to set a baseline of the contact resistance. To measure the dry circuit resistance, the connectors should be tested at an open circuit voltage of about 20 mV and a short circuit current less than 100 mA. This is to avoid unnecessary heat generation, which may skew the results of the following temperature rise measurement [31].

After the measurement of the dry circuit resistance, the temperature rise is measured. The temperature rise is normally defined as the difference between the hottest point of

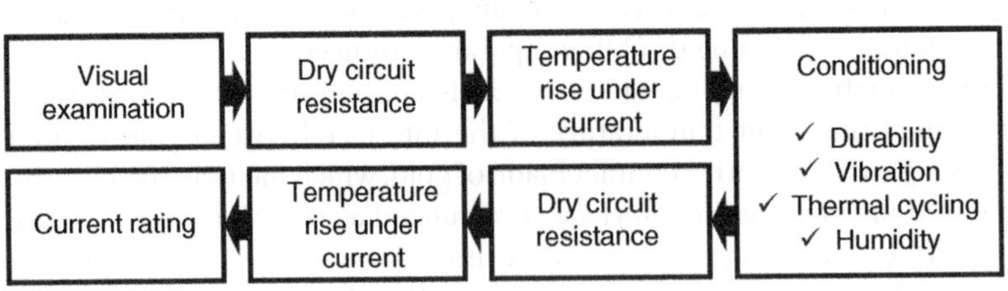

Figure 15.11 A general qualification procedure for current rating of power connectors.

the connector and the ambient temperature. A thermocouple or an infrared thermal kit can be used. Measuring the ambient temperature is simple, whereas measuring the hottest point of the connector requires consideration of the connector housing. If the hot point of the contact is inaccessible due to the connector housing, an accessible point that is the nearest to the hot point should be chosen and measured. In addition, a proper correlation should be addressed. The electrical current of the connector should be kept consistent to acquire thermal stability during the temperature measurement procedure. According to one of the common standards for current rating of connectors, EIA-364-70, thermal stability can be achieved when the temperature measured by each thermocouple is maintained within ±1 °C during a minimum of three consecutive readings taken at five-minute intervals maximum [30]. The current rating will be set where it causes a temperature rise of 30 °C.

Mechanical durability tests determine the effects of repeated coupling/decoupling on power connectors [32]. To measure durability, the specimen is mounted in a suitable coupling/decoupling test machine. The total number of coupling/decoupling cycles follows the referencing document. Generally, the cycle rate per hour is 500 ± 50 for automatic equipment and 300 at maximum for manual action. Electrical load can be applied to the specimen during the test in case there are other specifications in the referencing documents. After the load conditioning, the specimen is examined visually to determine whether there are any mechanical damages. The failure criteria of the damages should also be set according to the applications.

Thermal cycling tests can be conducted for power connectors if necessary. These tests evaluate the effects of extreme temperature conditions [33]. Prior to conducting the thermal cycling test, a chamber is prepared using materials that will not be degraded or affected by the testing temperature. The temperature loads should be provided using a convective air flow. Controlling the temperature via radiative heat suppliers is not recommended, unless there are other specifications in the referencing document. The fixtures should be lightweight and made of low thermal conductive material to avoid heat sink effects. A thermocouple is placed at the middle point of the specimen. The thermal cycle procedure is then performed in the following manner: ramp up or down to the specified temperature level, and dwell at the temperature level, respectively. The ramp rate should be an average of 5 °C/min. The general test conditions for thermal cycling are +15 °C/+85 °C, −15 °C/+85 °C, and 0 °C/+100 °C. The highest and lowest temperature should be controlled within ±3 °C, whereas the dwell time should be longer than 10 minutes. The number of cycles can differ from 10 to several thousand times depending on the applications.

To perform vibration tests for power connectors [34], the specimen is mounted on the test table with a defined axis. The test equipment that provides vibrational motion to the specimen should be capable of generating sinusoidal or random excitation. There are common frequency and peak-level conditions for sinusoidal vibration testing. For testing at low frequency, the frequency range should be 10–55 Hz. For testing at high frequency, the frequency range should be 10–2000 Hz. The peak level should be set to 10–20 g_n, where g_n is an SI unit for the standard acceleration due to the earth's gravity, of which the value is 9.8 m/s^2. Since the maximum amplitude can be doubled at a critical resonant frequency, the resonant frequencies should be monitored using resonance-detecting equipment. The measurement duration should be set at a range of several hours. If the equipment provides randomly excited vibration, the amplitude

Table 15.1 Test chamber temperatures and connector internal temperatures with electrical load.

Condition #	Connector exposure temperature (°C)	Connector internal temperature (°C)
1	55	65
2	85	102
3	125	150
4	200	238
5	350	400

of vibrational motion should be controlled in a Gaussian distribution. The maximum magnitude of electrical loads is 100 mA for any connectors unless there are other specifications in the referencing document.

To examine the reliability of power connectors under elevated temperatures, a temperature life test can be performed [35]. This test is divided according to whether electrical load is applied. The chamber used in this test is prepared similarly to the thermal cycling test, as explained earlier. However, if the test includes electrical load, the inner space of the chamber should be large enough to dissipate the Joule heating generated by the connector specimen. For tests without electrical load, the temperature of the test chamber should be set as specified. The common conditions for this test are 55–500 °C. If the connector specimen is subjected to a temperature condition below 125 °C, the temperature tolerance of the chamber should be set at 2 °C. If the temperature condition is between 125 and 200 °C, the tolerance should be set at 5 °C. For higher temperature conditions such as 350 or 500 °C, the tolerance should follow the referencing document. For tests with electrical load, the connector specimen is mounted in the test chamber. The chamber stabilizes at a temperature condition that is similar to the test without electrical load. After the specimen is mounted, a direct current is applied until the maximum connector internal temperature is achieved. Table 15.1 shows common combinations of chamber temperature and connector internal temperature. If the maximum of the connector internal temperature is not reached even at the connector's rated test current, the chamber temperature should be increased. For both tests with or without electrical load, the test durations are commonly set at 250, 500, and 2000 hours. When the test is completed, the specimen is examined for the following: dimensional changes, hardening/softening of dielectric materials, opening of seals, and cracking/delamination of components.

After the conditioning processes are completed, the dry circuit resistance and the temperature rise by current are measured again to verify the effects of the applied loads on the current capacity of the connector. If the temperature rise measurement shows acceptable results, the current rating of the connector is now certified.

References

1 Mroczkowski, R.S. (1998). *Electronic Connector Handbook*. New York: McGraw-Hill.

2 IEEE 837-2014 (2014). *Standard for Qualifying Permanent Connections Used in Substation Grounding*. IEEE Standards Association.

3 Sacks, N.M. (1972). Development of a separable connector system for underground power distribution. *IEEE Transactions on Power Apparatus and Systems* PAS-91 (3): 986–991.

4 Zhou, Y., Lan, F., and Kong Z. (2015). The effect of electrical joint compound on the life of aluminum power connector. *Proceedings of the 2015 Prognostics and System Health Management Conference*, Beijing, China (October 2015). PHM.

5 Worzyk, T. (2009). *Submarine Power Cables: Design, Installation, Repair, Environmental Aspects*. Springer Science & Business Media.

6 Chapman, D. (2016). *High Conductivity Copper for Electrical Engineering*. Copper Development Association.

7 Braunovic, M. (1981). Effect of current cycling on contact resistance, force, and temperature of bolted aluminum-to-aluminum connectors of high ampacity. *IEEE Transactions on Components, Hybrids, and Manufacturing Technology* CHMT-4 (1): 57–69.

8 Dang, C. and Braunovic, M. (1990). Metallurgic and contact resistance studies of sleeve connectors in aluminum cable splices. *IEEE Transactions on Components, Hybrids, and Manufacturing Technology* 13 (1): 74–80.

9 Chudnovsky, B. H. (2002). Degradation of power contacts in industrial atmosphere: silver corrosion and whiskers. *Proceedings of the 48th IEEE Holm Conference on Electrical Contacts*, Orlando, FL (October 2002). IEEE.

10 Fu, R., Choe, S. Y., Jackson, R. L., Flowers, G. T., Bozack, M. J., Zhong, L., and Kim, D. (2010). Experimental study of the vibration-induced fretting of silver-plated high-power automotive connectors. *Proceedings of the 56th IEEE Holm Conference on Electrical Contacts*, Charleston, SC (October 2010). IEEE.

11 Cohen, U., Walton, K.R., and Sard, R. (1984). Development of silver-palladium alloy plating for electrical contact applications. *Journal of the Electrochemical Society* 131 (11): 2489–2495.

12 Simon, D., Bardolle, J., and Bujor, M. (1980). Study of the reactivity of silver, copper, silver-copper, and silver-palladium alloys used in telephone relay contacts. *IEEE Transactions on Components, Hybrids, and Manufacturing Technology* CHMT-3 (1): 13–16.

13 Furtado, H.C. and Antunes da Silveira, V.L. (1988). Metallurgical study of Ag-Cd and Ag-CdO alloy electrical contacts. *IEEE Transactions on Components, Hybrids, and Manufacturing Technology* 11 (1): 68–73.

14 Coutu, R.A., Kladitis, P.E., Leedy, K.D., and Crane, R.L. (2004). Selecting metal alloy electric contact materials for MEMS switches. *Journal of Micromechanics and Microengineering* 14 (8): 1157–1164.

15 Antler, M. (1982). The application of palladium in electronic connectors – continuing studies result in growing use. *Platinum Material Review* 26 (3): 106–117.

16 Braunovic, M. (2004). Effect of connector design on the performance of service entrance power connectors. *IEEE Transactions on Components and Packaging Technologies* 27 (1): 72–78.

17 Yilmaz, M. and Krein, P.T. (2013). Review of battery charger topologies, charging power levels, and infrastructure for plug-in electric and hybrid vehicles. *IEEE Transactions on Power Electronics* 28 (5): 2151–2168.

18 Runde, M., Jensvold, H., and Jochim, M. (2004). Compression connectors for stranded aluminum power conductors. *IEEE Transactions on Power Delivery* 19 (3): 933–942.

19 Runde, M., Magnusson, N., and Lenes, A. (2008). Bolted connectors for stranded aluminum power conductors. *IEEE Transactions on Power Delivery* 23 (2): 523–530.

20 Schindler, J.J., Axon, R.T., and Timsit, R.S. (1996). Mechanical and electrical contact properties of wedge connectors. *IEEE Transactions on Components, Hybrids, and Manufacturing Technology* 19 (3): 287–294.

21 Gacek, Z., Maźniewski, K., and Sosiński, R. (2005). Comparative testing of chosen piercing connectors. *IEEE Transactions on Power Delivery* 20 (2): 1122–1128.

22 Malucci, R.D. (2006). Stability and contact resistance failure criteria. *IEEE Transactions on Components and Packaging Technologies* 29 (2): 326–332.

23 Bowden, F.P. and Williamson, J.B.P. (1958). Electrical conduction in solids I. Influence of the passage of current on the contact between solids. *Proceedings of the Royal Society A: Mathematical, Physical, and Engineering Sciences* 246: 1–12.

24 Kohlrausch, F. (1900). Ueber den stationären Temperaturzustand eines elektrisch geheizten Leiters. *Annalen der Physik* 306 (1): 132–158.

25 Holm, R. (1967). *Electric Contacts*, 60–63. Springer-Verlag.

26 Gatzsche, M., Lücke, N., and Großmann, S. (2015). Validity of the voltage-temperature relation for contact elements in high power applications. *Proceedings of the 61st IEEE Holm Conference on Electrical Contacts*, San Diego, CA (October 2015). IEEE.

27 Greenwood, J.A. and Williamson, J.B.P. (1958). Electrical conduction in solids II. Theory of temperature-dependent conductors. *Proceedings of the Royal Society A: Mathematical, Physical, and Engineering Sciences* 246: 13–31.

28 Timsit, R.S. (1983). On the evaluation of contact temperature from potential drop measurements. *IEEE Transactions on Components, Hybrids, and Manufacturing Technology* CHMT-6 (1): 115–121.

29 Silva, F.B.B., Baldan, C.A., Fardin, J.F., and Simonetti, D.S. (2015). Starting current limitation method using HTS for induction motors. *Journal of Superconductivity and Novel Magnetism* 28: 697–700.

30 EIA 364-70B (1997). *Temperature Rise Versus Current Test Procedure for Electrical Connectors and Sockets*. Electronic Industries Alliance Technical Standards.

31 Song, J. and Schmidt, H. (2009). Electrical resistance of connectors with different contact finishes under low level conditions. *Proceedings of the 55th IEEE Holm Conference on Electrical Contacts*, Vancouver, Canada (September 2009). IEEE.

32 EIA 364-09C (1999). *Durability Test Procedure for Electrical Connectors and Contacts*. Electronic Industries Alliance Technical Standards.

33 EIA 364-110 (2006). *Thermal Cycling Test Procedure for Electrical Connectors and Sockets*. Electronic Industries Alliance Technical Standards.

34 EIA 364-28E (2006). *Vibration Test Procedure for Electrical Connectors and Sockets*. Electronic Industries Alliance Technical Standards.

35 EIA 364-17B (1999). *Temperature Life With or Without Electrical Load Test Procedure for Electrical Connectors and Sockets*. Electronic Industries Alliance Technical Standards.

16

Electrical Connectors for Underwater Applications

Flore Remouit[1], *Jens Engström*[1] *and Pablo Ruiz-Minguela*[2]

[1] *Division for Electricity, Uppsala University, Uppsala, Sweden*
[2] *Division for Energy and Environment, TECNALIA, Derio, Spain*

In 1858, the first electrical communication was established between North America and Europe through a transatlantic telegraph cable [1]. More than 3200 km of cable, weighing nearly 550 kg/km, were deployed at a depth of often more than 3 km. One hundred and fifty years later, the techniques to deploy and protect undersea cables remain similar. Connectors are the weak point of most electrical cables, and significant effort has been devoted to improve their connectivity and water tightness, simplify their manufacturing process, and decrease their cost [2].

Despite the variety of connectors and the number of manufacturers, there is little information on the different connectors available [1] on the market and their respective properties. Moreover, there are few papers dealing with underwater electrical connectors. In addition, it appears that unsuccessful offshore and underwater projects are mostly due to connection failures. Those failures were generally caused by using connectors that were inadequate for the environment [3] or incorrectly designed [4]. In the case of Underwater Communications Cable Failures, according to the cable owners, most fiber-optic cable failures out to a depth of 1200 ft result from man-made incidents, i.e. ship anchor drags and fishing/trawling rigs snagging the cable.

Nowadays, numerous companies offer a large choice of underwater connectors and assemblies, and it can be complex to distinguish the different technologies employed for each of them. However, the use, deployment, maintenance, and lifetime of any subsea equipment, from a simple sonar to a wave energy converter, relies on its connectors. Hence, the design of an underwater electrical connector needs to be carefully evaluated, especially for tailor-made applications that have more specific requirements. To produce a good connector it is necessary to account for thermal, electrical, and mechanical properties, as well as to determine the best materials that should be used for the application. Finally, connector issues go hand in hand with the deployment and operation of

★This chapter was originally published as "Review of electrical connectors for underwater applications" in *IEEE Journal of Oceanic Engineering*, vol. 43, pp. 1037–1047, Oct. 2018. It has been edited to fit this book. The research leading to these results is part of the OceaNET project, which has received funding from the European Union's Seventh Framework Programme for research, technological development and demonstration under grant agreement no. 607656.

Electrical Connectors: Design, Manufacture, Test, and Selection, First Edition.
Edited by San Kyeong and Michael G. Pecht.

any electrical equipment, and it is of interest to review the different techniques for cable connection, as well as the challenges related to cable layout.

Those challenges can be of a different nature, but they should all be taken into account for any subsea connection.

This chapter provides an overview of the types of connectors that exist for underwater and subsea electrical connections. The chapter is divided into three parts; in the first part different types of commercial-off-the-shelf electrical connectors will be presented, and their performances and defaults will be listed, as well as typical failure modes and their known causes. The second part contains the theory on sealing and on connectors' thermal, electrical, and mechanical properties. The last part provides information on connection procedures and other details about connecting subsea cables with a focus on the connection of offshore renewable energy (ORE) farms.

16.1 Background and Terminology

Underwater electrical connections are mainly used in the following sectors [5–8]:

Remotely operated vehicle (ROV)/autonomous underwater vehicle (AUV): The ROV industry is growing, and the need for connectors is hence increasing. In particular, the volume and weight of the vehicles are decreasing, and consequently the connectors should meet those new requirements of small size and low weight for the same specifications [9].

Oil and gas: The oil and gas industry is the biggest consumer of underwater connectors. Many new challenges occurred with the increasing depth of production and exploration, and with them the issues to develop robust connectors that are resistant to high pressure, seismic risks, and harsh environments [10].

Offshore renewables: With the development of many ORE technologies, the need for high-performance connectors is also increasing, and suppliers are focused on the design of affordable technologies with robust connectivity [11]. ORE technologies have different demands than the oil and gas industry because they need higher power and extended bandwidth communications; for instance, offshore wind transmission is currently done at 33 kV in most countries [12]. This is why a new market for high-voltage connectors is emerging and developing.

Research: Underwater electrical connections have always been necessary for oceanographic, meteorological, and other research activities. The usual requests are for tailor-made connectors that suit particular and singular applications [2, 13, 14].

Military/defense: The military/defense sector is looking for high reliability and adaptability of its connectors, due to the variability of the operations, causing dynamic effects and pressure variations on the equipment and the dynamics involved. The connectors require hence robust performance.

Communications: The communications industry uses mostly optic fiber for its connections between countries and continents. Those connectors are not the main focus of this chapter, since optical communication has been widely studied in various papers [15–18]. In addition, Ref. [19] presents a map of global submarine communication links showing the multiplicity of communication routes across the oceans.

16.1.1 History

The history of underwater connectors can be found in several papers (e.g. [20–22]), but deserves to be shortly written here. There was no real industry for underwater connectors before World War II, and homemade methods were commonly used to create subsea electrical connections. One of these methods was to wrap rubber around the junction of a regular connector so it overlapped the connector's sleeve. This rubber tube would then be sealed with string [22].

It was only in the 1950s that the growing number of submarines used for military defense created the need for real underwater connectors. The first connectors were either rubber-molded or rigid shell and are still in use [21]. These connectors were "dry-mate," which means that they were to be mated above the surface before being deployed.

New technologies to connect and disconnect cables directly underwater began to be developed in the early 1960s, specifically, the "wet-mate" connectors. The first type of wet-mate connector was the pressure-balanced oil-filled (PBOF) connector, which isolates contacts from sea water by surrounding them with sealed oil chambers. Inductive coupling connectors were subsequently developed in the 1970s. These connectors contain a conductive coil, and each time a signal fluctuates in one of them the other transmits the signal with an induced current. Underwater optical connectors were developed in the 1980s, and hybrid wet connectors one decade later [21].

Today, thousands of different connectors are available off the shelf for underwater application, each supplier having its own technology.

16.1.2 Terminology

According to the ANSI/NEMA CC 1-2009 standard [23], an electric connector is defined as "A device that joins two or more conductors for the purpose of providing a continuous electrical path."

The plug is the male body and contains pins. The receptacle is the female body and contains sockets. All connectors are equipped with either dummy plugs or dust caps to keep electrical contacts clean when the connector is not mated, but the dust caps are used only for the surface zone (splash-proof) [24].

A connector contains basically two systems [25], as follows:

- The front wet-end system, which is the active part of the connector and whose role is to create the mating between the two halves
- The back-end and termination system, which is responsible for the sealing between the cable and the connector

The number of components in both systems can vary depending on the complexity and the technology of the connector.

There are basically two types of underwater connectors. Dry-mate or dry-mateable connectors need to be connected and disconnected outside the water; they are less expensive and simpler, with none or few moving parts. Wet-mate or wet-mateable connectors, which can be mated subsea, are much more complex because of the need to electrically isolate the pins/receptacles from saltwater. We can also distinguish wet-mate and underwater connectors, the first ones being used for shallow water connections, and the second ones for deep water operations.

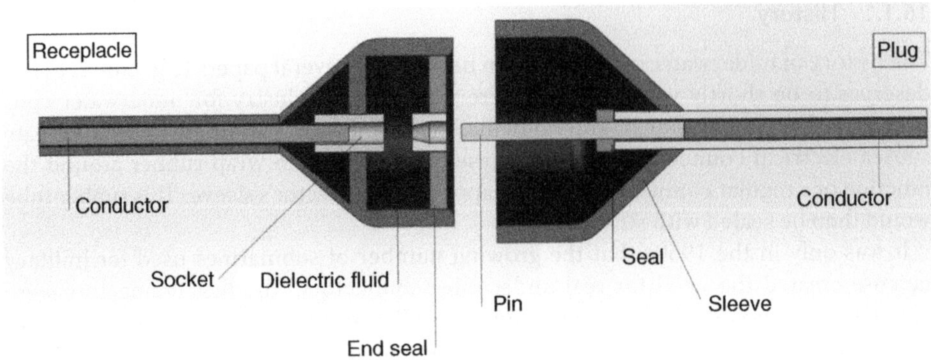

Figure 16.1 Sketch of a basic fluid-filled connector.

Figure 16.1 shows a basic schematic of a one-pin oil-filled connector.

16.2 Commercial Off-the-Shelf (COTS) Connectors

Here we present the main types of underwater connectors available on the market. Although each manufacturer has its own designs and technologies, underwater connectors can be grouped into five main categories: rubber-molded, rigid-shell or bulkhead assemblies, fluid-filled underwater mateable, inductive coupling, and assemblies (non-unmateable).

16.2.1 Rubber-Molded

Rubber-molded (RM) connectors are the most common and simple type of subsea connectors, and general descriptions of them can be found in [22, 24, 25]. They are basically cables with male and female terminations that are molded at the conductor's end. The molding material is typically neoprene or polyurethane. These connectors are generally rated up to 600 V [25], but high-voltage connectors are available up to 3 kV. Polarization is made through the contact of the pins inside the sockets, and the molded area provides a seal between the two halves, as well as a strain transition between the contact area and the cable [26].

RM connectors are unable to control their internal pressure owing to their soft internal and external structure. For this reason it is difficult to unmate them underwater at a big depth, the pressure between the pins and receptacles being too high [22]. Moreover, the seal can be breached because their material (rubber) is subject to differences of elasticity depending on the temperature. Rubber has also a high thermal resistance, which makes the connector unfit to withstand high temperatures in the cable. On the positive side, these connectors are comparatively cheap, lightweight, and easy to manufacture. The dielectric material (rubber) will not be affected by corrosion and saltwater. Figure 16.2 shows RM dry- and wet-mate connectors.

Plastic-molded (PM) connectors evolved from the RM connectors. They also have a simple structure, with epoxy-reinforced glass fiber molded directly on the conductor, which provides a higher resistance to mechanical stress than rubber. However, PM

Figure 16.2 Rubber-molded connectors, dry mate on the left, and wet mate on the right (where the pins have rubber that can wipe the water off the socket) [27].

connectors are more brittle than RM connectors. The material is also sensitive to temperature changes.

16.2.2 Rigid-Shell or Bulkhead Assemblies

Molding a rubber connector into a rigid body increases the strength and stability of the connector and improves coupling and locking. The connector is then able to withstand harsher environments. The keying is performed by screwing together the two half shells, and the sealing is provided by O-rings. The voltage rating usually goes up to 600 V for regular connectors, as for the RM and PM connectors, and up to 3 kV for more specifically designed connectors [5]. The rigid-shell assembly is small and low cost, but also heavier than the molded connectors. The bulkhead materials are typically metal, plastic, or an epoxy – glass fiber composite. Figure 16.3 shows a metal bulkhead connector.

The drawback with rigid-shell connectors is the pressure differential between the outside of the shell and its interior, which could lead to early breakage. The void in the assembly can cause condensation in the contact area. Moreover, a metallic bulkhead can be subject to corrosion. Special care is also needed for the molding between the shell and the cable at the back end of the connector. This interface is particularly

Figure 16.3 12-pin bulkhead connector with metallic housing [28].

susceptible to damage caused by cycling loads, especially where different materials are used together. If welding is used, the material must be adequately protected from the welding temperature [26]. More information on rigid-shell or bulkhead assemblies can be found in [8, 21, 22, 24, 25, 27, 28].

16.2.3 Fluid-Filled Underwater Mateable

Fluid-filled connectors use a chamber filled with a dielectric fluid, typically oil, to isolate the contact from seawater. Seawater is wiped off as the conductive pins go through a diaphragm in the face of the receptacle [26]. A principal advantage of these connectors is that they do not need to be keyed. These connectors are generally physically larger and more expensive than the standard metal shell electrical connectors, but they are inherently more robust [25] by design and experience. Their high price is mostly due to the complexity of the assembly with the different moving parts.

In this case, the connector is not exposed to pressure differentials because the oil inside the cavity and the seawater are balanced. As a consequence, the connector can be mated or unmated at any depth with no extraordinary force. Since the pins are immersed in oil, they are not subject to corrosion and can be electrically charged before mating.

Figure 16.4 is a schematic of one of the first PBOF connectors from Ref. [25], while Figure 16.5 is a much newer PBOF connector from 2003 from Seacon [27, 28].

One of the difficulties when designing oil-filled connectors is to provide seals that allow the pins to be inserted repeatedly inside the oil chambers without losing the oil or allowing water intrusion. One possibility, as described earlier, is to wipe the water off the pins with the help of a diaphragm installed ahead of the receptacle. Another widely used design uses dielectric pistons, where each socket is connected to a spring that is pushed inside when the pin is penetrating the receptacle.

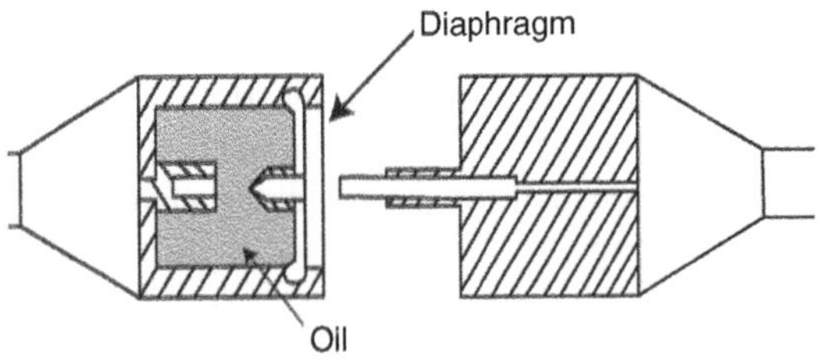

Figure 16.4 Scheme of a pressure-balanced oil-filled connector [25].

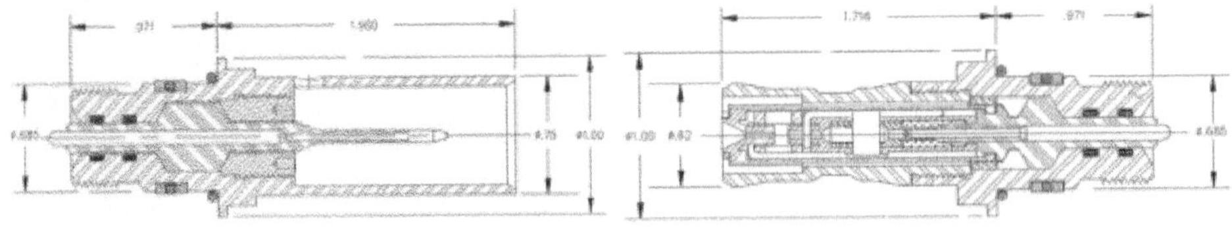

Figure 16.5 Scheme of a pressure-balanced oil-filled connector [27, 28].

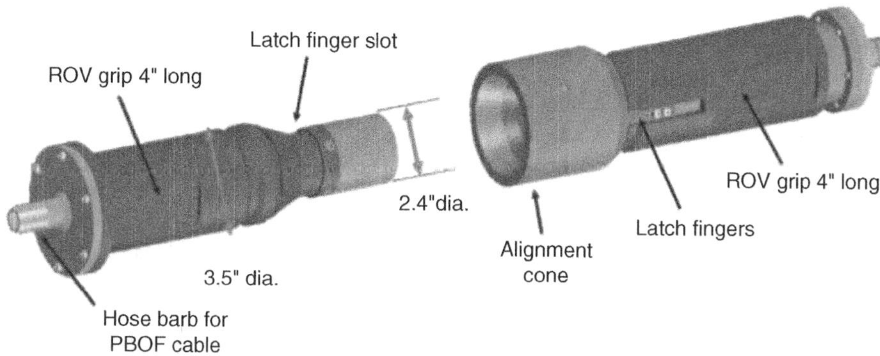

Figure 16.6 Non-contact wet-mateable connector [2].

PBOF connectors are still evolving, and some interesting designs are presented in references [2, 20, 21, 29].

16.2.4 Inductive Coupling

Each half of an inductive coupling connector contains a coil that allows electrical current to pass from the female to the male and vice versa. The advantage is that no conductive parts are exposed to seawater [21]. Inductive coupling connectors are also mechanically simple and easy to mate and demate since no pin insertion is required. However, they are not commonly used because they are usually rated for low power (<20 W) and they only allow AC current [22]. This is why they are mostly used for data transfer only.

References [2, 30] describe an innovative noncontact wet-mateable connector that is presented in Figure 16.6.

16.2.5 Assemblies (Non-unmateable)

The term "connectors" includes products that allow a connection between electrical cables, but without the possibility to mate or demate them. Among those products we

Figure 16.7 Cable termination [31] (top) and Y-splice [24] (bottom).

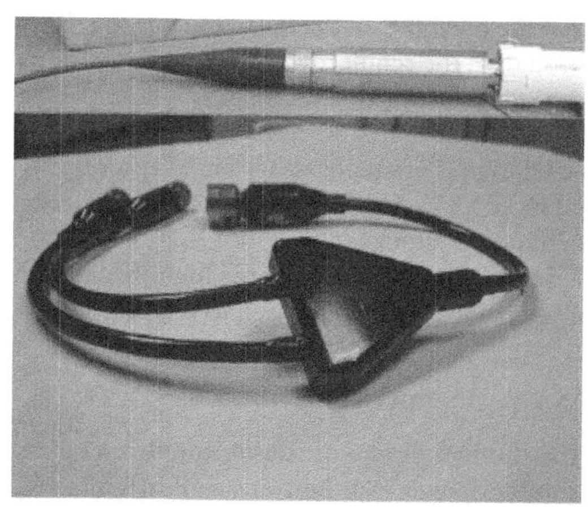

Table 16.1 Summary of COTS connector types and their characteristics.

Connector	Mating type	Voltage/power rating	Cost	Pros
Rubber-molded	Dry	Up to 1 kVAC/850 VDC	*	Cheap and available
Rigid shell/bulkhead	Dry	Up to 3 kVAC/2.55 kVDC	**	Robust, comparatively cheap, easily available
Oil-filled	Wet	Up to 3.3 kVAC	***	Reliable, easy to deploy and to maintain on-site
Inductive coupling	Wet	20 W	***	Easy to deploy and maintain on-site, no conductive part exposed to seawater
Assemblies	Dry/wet	Up to 600 V AC/510 VDC	**	Allows equipment to be deployed before connection

distinguish terminations or penetrators [31], which connect cables to a device, a piece of equipment, or a pressure housing, and joints that link two cables together [32]. Cable glands protect the cable from water insertion but are rated neither for deep submerged use nor for dynamic applications. T-splice and Y-splice connector group involve putting together two cables into one, respectively, in a T- or a Y-form. A field installable termination assembly (FITA) allows a connection on-site directly underwater, which is advantageous [33]. Figure 16.6 shows examples of a cable termination and a Y-splice.

Reference [34] gives a nonexhaustive list of dry- and wet-mate connector manufacturers, and Table 16.1 summarizes the main characteristics of the types of connectors.

16.3 Connector Design

Although many different types of connectors are commercially available, it is common to purchase tailor-made ones to fit exactly the project's needs. One reason for this is simply that the choice of connectors is not large enough to cover the demand. Another reason is that the reliability of the connectors is so critical for some underwater projects that they need to fit the requirements exactly. In many cases, once the connector is deployed, the time and cost needed for repairing or replacing it are simply too high to be considered; the connector must withstand its whole lifetime without any failure. This is why a special design is often necessary for AC power cables, optical communications, permanent joints as splices and terminations, or other connectors that must operate for a long time without maintenance.

The design of a connector is highly related to the design of the cables, and the interested reader is hence advised to read through references [35, 36].

16.3.1 Thermal Design

The thermal design of subsea connectors aims to define the right housing size so the required power can be transferred from one cable to another without exceeding the temperature limits of the conductor or its environment. The connector's size mainly depends on the outer diameter of the cable; therefore, it is necessary to consider the power cables' thermal properties when designing or choosing the appropriate connectors [36, 37].

The most basic thermal model for DC power cables corresponds to a single-core conductor. Heat is produced as current passes through the cable, and this heat is transferred from the wires to the external environment (here water) through the successive layers of the connector.

The ohmic losses in the conductor can be written as

$$P_L = RI^2 \tag{16.1}$$

where P_L is the power losses in W/m, R is the conductor's electric resistance in Ω/m, and I is the current passing through the conductor in A.

R depends on both the conductor's material (usually copper or aluminum) and its cross section. The thicker the wire, the smaller the resistance.

Under steady-state conditions, all the heat generated in the conductor must flow to the outside environment through the cylindrical dielectric insulation. The relationship between the temperature drop across the connector insulation, the power losses, and the thermal resistance of the insulation wall can be described as

$$\Delta\theta = P_L R_T \tag{16.2}$$

where $\Delta\theta$ is the temperature drop, P_L is the power losses through the conductor, and R_T is the thermal resistance of the insulator, with

$$R_T = \frac{\rho_T}{2\pi} \ln\frac{D_o}{D_i} \ln\frac{D_o}{D_i} \tag{16.3}$$

where ρ_T is the specific thermal resistivity of the insulation material, D_o is the connector's outside diameter, and D_i its inside diameter.

If the insulator consists of several layers, the temperature drop can be calculated as the sum of the different temperature drops between each layer:

$$\Delta\theta = \sum_n \Delta\theta_n = \sum_n P_L R_{Tn} = P_L \sum_n R_{Tn} \tag{16.4}$$

where R_{Tn} is the thermal resistance of the insulation wall's nth layer, as described in the previous equation.

Table 16.2 presents the thermal resistance values for different materials used in the isolation of electrical cables.

Temperature changes due to load cycling induce weaknesses in the material's properties, as strength and elasticity, which drastically reduces the connector's lifetime.

For AC cables, additional losses are due to electromagnetic factors, such as the skin effect or the corona effect. Although these factors impact thermal studies, they are discussed in Section 16.4.2.

16.3.2 Electrical Properties

The electromagnetic field distribution is the most influential factor in the failure of insulation systems, and these failures may be boosted by the presence of pollutants such as water or moisture [38].

Table 16.2 Thermal resistivity of isolating materials for electrical cables.

Material	Thermal Resistivity (Km/W)
Paper insulation, oil-filled cables	5.0
PPL (polypropylene layers, insulation material)	5.5
XLPE (cross-linked polyethylene, used for extra-high-voltage cables) or PE thermoplastic (polyethylene)	3.5
EPR (ethylene propylene rubber) for cable ratings above 3 kV	6.0
Outer serving	
PE (polyethylene)	3.5
PVC (polyvinyl chloride)	6.0

For AC cables, the alternating magnetic field causes the "skin effect" around the conductor, making the current density higher close to the conductor's surface and lower at its core. The larger the conductor's area, the more significant this skin effect. In addition, another effect called the "proximity effect" is created by the closeness between the wires when the cable contains several cores (e.g. three-phase cables). The consequence is an inhomogeneous current density in each wire, the current from one core trying to stay as far as possible from the others. These two effects increase the power losses in the cable and need to be taken into account in the calculation.

Although electrical design affects all connectors, the ones that need specific studies are mostly used for power transmission, at high voltage (>1 kV) or extra-high voltage (>275 kV), because at this voltage level, if the strength of the electrical field exceeds a certain value, an effect called the "corona effect" occurs [39]. This effect produces audible noise, radio interference, and power losses, and generates harmful chemical compounds such as ozone and nitrogen oxides [39]. Therefore, the corona effect is a focus of concern, and many studies are being conducted to investigate this phenomenon. Unfortunately, the corona discharge is a complex phenomenon that is difficult to model since it depends on several variables, including experimental arrangement geometry, surface roughness, operating voltage, and environmental conditions (e.g. pressure, temperature, humidity, salinity). To reduce the strength of the electric field, corona shields are used on electrical connectors. Thus, corona activity, including power loss, television and radio interference, and audible noise levels, is significantly reduced [11, 38].

For low-voltage connectors, the electrical design must take into account the choice of amperage and how the power transfer is done. The design will give the type of connector (electrical, optical, hybrid), its size and diameter, the number of necessary pins/fibers, and the type of contact. If the connection must be done with power on, then a wet-mate technology must be used; if not, the design possibilities are then broader (at least concerning the electrical side).

Electrical insulation is also critical for protecting current transfer from the conductive seawater. In that matter, the choice of material, as well as the mechanical seals, are critical factors.

16.3.3 Mechanical Properties

Underwater connectors must withstand the stresses of manufacturing, transportation, deployment, and lifetime subsea. Dynamic forces are applied to the connector during the operation and maintenance of the equipment. Hydrostatic pressure is present during the connector's lifetime, and the magnitude depends on the water depth. Too little precaution in the mechanical design could lead to damages that will force early maintenance, replacement, or decommissioning before the end of the connector's electrical lifetime [36].

Cycling from low to ambient system pressure is the major contributor to mechanical failure in an underwater electrical system. The more severe conditions occur with a higher magnitude of pressure [39]. These pressure variations can cause the void and pressure barriers to collapse, the conductors to become fatigued, and the bonds and sealed parts to break.

Figure 16.8 presents the typical endurance to pressure cycles' function of the ambient pressure for two cables with different cross sections. Endurance to pressure cycles can be defined as the capacity of a component to withstand fatigue loads created by cycles of pressure differences. The bigger the cable, the less resistance to the cyclic loads.

The connector's mating and unmating procedure is also a source of great stress that will increase quickly depending on if the connector is to be plugged once or a few times (power cables) or a large number of times (ROV connectors).

The critical mechanical aspects related to connectors are as follows:

The connection between the conductor and the pin/socket: The connecting of cable conductors to contact components is generally performed by crimping or soldering. A bad connection could lead to a short circuit or a break, and this operation is especially sensitive for miniaturized connectors with many pins.

The bonding between the connector and the cable: Usually, the connector's sleeve is molded on the cable, but depending on the type of connector one could also use bonding agents [40, 41] or welding if some metal parts are involved. It is absolutely essential that the bond prevents any water ingression between the connector and the cable.

The contact plug-receptacle: Depending on the procedure to connect the two halves, the mating force will typically vary from ca 100–200 N to 500–1000 N. For wet-mate connectors, the pressure at great depth makes the connection forces even larger. The

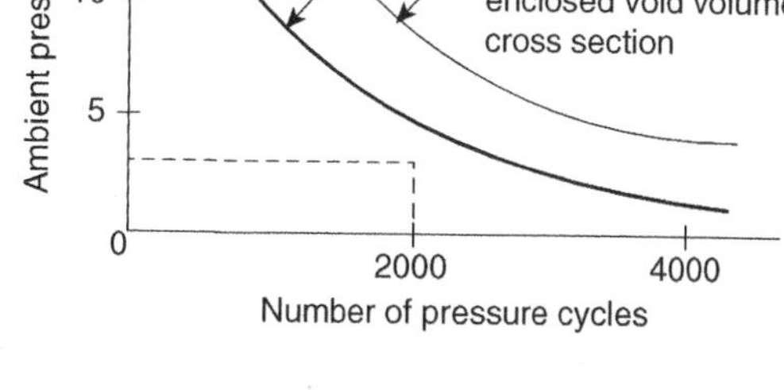

Figure 16.8 Characteristic curves of the endurance to pressure cycles depending on the ambient pressure [39].

mechanical design should include a study on the contact plug-receptacle in order to allow an insertion of the pins into the sockets within a certain range of forces. The required force should not be too high so that contact cannot be performed by a human or a machine, and not too low to be easily loosened.

The sealing between each connector half or between the joint and the device/junction box (see e.g.[42]): Sealing is commonly done using O-rings, and references [43, 44] give detailed information in this matter. However, some other methods can be used for permanent joints, such as pressure gasket-plates, internal plugs, or internal feedthrough [39]. Often one would increase the number of seals inside or around the connectors as backup protection.

Once the contact is made, the keying would typically use threaded locking sleeves, although sometimes the contacts lock with no additional tool, as the RM connectors whose rubber helps the bonding of the two halves. Some connectors are equipped with only a small plastic band that is slipped around each half when mated. This band is not as reliable as the thread locking sleeve, but it saves weight and size [45].

Weight and size: With the miniaturization of electronics and electrical components came the need for smaller connectors and thinner electrical cables. Offshore activities have been subjected to the same development, and the weight and size of subsea connectors are now becoming critical mechanical design factors. Innovative products have appeared on the market, and many suppliers offer dry- and wet-mate off-the-shelf connectors in a normal, mini, or even micro format with approximately the same electrical and thermal properties [27, 28]. Progress in the area is mainly due to innovations in materials science with the development of both light and strong materials, as discussed in Section 16.3.4.

16.3.4 Material Choices

Different connector types require different materials, but the same properties requirements apply: high degree of dimensional stability, high impact strength, low mold shrinkage, low water absorption, high-compressive strength, and nonflammability [26]. Commonly used materials are the following:

- Elastomers and their varieties are used in many types of connectors because they are elastic and have a high compressive strength and low water absorption coefficient. Such a rubber material could typically be polyurethane or neoprene. Those materials are commonly used to wrap the connector and cable together. However, it appears that rubber can be employed as a cement, as in reference [46], where the material 3140 RTV, a silicone-based rubber, is used to bond a splice to a cable.
- Plastics are known for their low water absorption coefficient. They can be utilized as a shell in the bulkhead assemblies. Polymers, on the other hand, are used as coatings, adhesives, or covering materials, and some types can be used underwater [47]. They include resins that allow a bonding between the connector and the cable, or between the shell and the plug. A few epoxy resins for subsea use are suggested in Table 16.3.
- Composites and, in particular, glass-fiber-reinforced epoxy can serve as housing for the connector. They are much stronger than rubber.
- Metals should be noncorrosive, or a sacrificial anode should be added to the system (see e.g. Ref. [48]).

Table 16.3 Epoxy resins selection for underwater use.

Manufacturer	Product	Comments/Use
Rescon	UV-L	Bonding aid/adhesive
Colebrand	CXL 600	Small cracks injection
	CXL 78T	Small splits, adhesive
	CXL 194	Adhesive
Sika Inertol	Sikadur 53LV	Low viscosity injection

Source: from Ref. [47].

- Dielectric fluids, or oil, are used to make the pins and sockets watertight and seal them from seawater. Moreover, they keep a pressure balance between the inside and outside of the connectors, usually with the help of a vent that adapts the amount of oil inside the cavity, or a piston that counteracts the space containing oil with the water pressure.

The challenges for those materials are their capacity to withstand high pressure, corrosion, and seawater absorption. References [49, 50] give good insights in the choice of materials used for electrical connections, as well as failure analysis concerning corrosion.

In fact, thermal properties are always related to either electrical or mechanical design. Many studies analyze the effects of one property on another, for example, by combining high temperature cycles and high power transfer [51, 52], by modeling thermal effects due to electrical loads [53, 54], or with thermomechanical studies [55]. Some other articles study the influence of electrical and thermal properties on a connector's lifetime and aging capacities [56, 57].

16.3.5 Specifications for Underwater Connectors

The design and development of underwater electrical connectors includes many steps: design and feasibility studies; prototyping; concept reviews; critical design reviews (CDRs); failure mode, effects, and criticality analysis (FMECA); final design reviews (FDRs); test readiness reviews (TRRs); detailed qualification programs; and pilot project deployments.

All those steps require specifications and qualifications, but so far only a few industry-standard specifications for underwater connectors exist. However, there are a variety of organization-specific requirement documents created by oil and gas operators and associated major equipment providers [58].

Two military specifications, MIL-C-24217 and MIL-C-24231, cover dry-mate style electrical connectors for underwater use [59, 60]. Both of these military specifications were developed in the 1960s and were last revised in the early 1980s [4]. MIL-C-24217 specifies the use of glass seals on the electrical conductors of the receptacle assemblies. MIL-C-24231 is a specification for molded assemblies that are contained in a metal housing.

API 16A and API 16D are two specifications for underwater drilling equipment, and API 17E specifies the use of subsea umbilicals. ISO 13628-6 was created by the oil and gas industries and concerns the design and operation of subsea production systems. Finally, STATOIL TR2390 is a specification for electrical/optical connectors and jumpers for subsea control systems [61].

16.4 Connector Deployment and Operation

There is no established specification for performing an underwater connection. However, industry working groups such as Subsea Fiber Optic Monitoring (SEAFOM) are working on creating industry-recommended practices for subsea interconnected systems [20, 23, 62].

16.4.1 Connection Procedure

The simplest connection procedure is the dry-mate connection that is done manually onboard. However, this procedure means that the length of cable to be lifted on each side of the connector will be approximately three times the water depth, which introduces associated problems such as heavy lift and stress on the cable.

Wet-mate connections can be performed with the help of either ROVs, divers, or stab plates. Divers can only operate in shallow water and for a limited amount of time (e.g. 40 minutes for a 25-m-deep operation), which reduces drastically the possibilities. They could operate up to 300 m deep, but then the procedure becomes expensive and hazardous as the divers would work as a team in a pressurized cabin, taking turns for up to one month. On the other hand, divers are fast and precise. However, any subsea activity that includes human work is subject to a certain risk that increases with the depth and distance of the working station, and also with the weather. ROVs can be used in deep and shallow water, but they are much slower to operate, which increases the time of the procedure. Their efficiency also depends on weather conditions, namely, winds, currents, and waves. However, using ROVs is obviously less risky because no human lives are at stake.

Finally, stab plates can be advantageous for grouped connections. The half connectors are attached to two different plates that contain many connectors. Several connectors can be mated at the same time, sparing time and space [63].

Wet-mate connections surpass dry-mate connections in many ways, and Table 16.4 lists these differences.

However, wet-mate connections are more expensive than dry-mate connections, especially with ROV installation. Wet-mate connections need a frame support to be grasped by the ROV, which increases the weight and complexity of the connector, as shown, for example, in the Nautilus products from Teledyne [64], as shown in Figure 16.9.

Papers [65] and [66] present two innovations to connect dry-mate connectors underwater: the first one uses air-filled pockets that create a dry environment surrounding the connector, the second one employs an oil-filled box to seal the two halves before removal of the caps and mating, as shown in Figures 16.10, 16.11, and 16.12.

Table 16.4 Advantages of wet-mate compared to dry-mate connectors concerning the connection procedure.

Dry-mate drawbacks	Wet-mate advantages
Manipulation of cable	Disconnection at the bottom (easiness and shortness of operation)
Additional cable length	Reduced lifting capabilities
Amount of cable to be lifted	Cable connection independent from device installation
Cable cannot be fixed to the seabed	

Source: from Ref. [65].

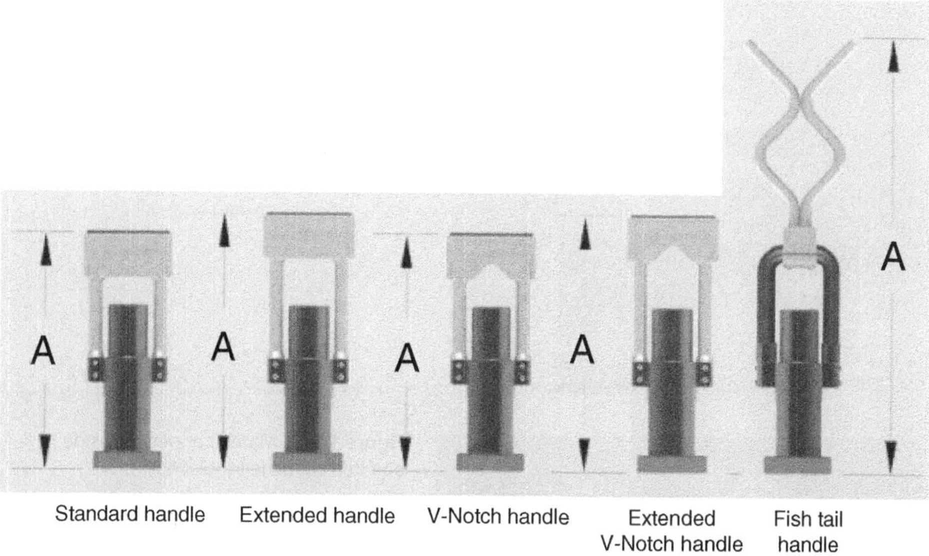

Figure 16.9 Nautilus connector, ROV mateable [64].

The use of a junction box is also common for oceanographic and observation projects, as seen in [29, 55]. A junction box serves as a hub between one or several devices and the shore. It enables both communication and power supply from and to the equipment and has the advantage of channeling all the connectors to one single unit.

16.4.2 Connection Layout

With the emergence of ORE, the question of connecting multiple cables in an optimal layout was raised. The vast majority of ORE farms require many units working in arrays to produce sufficient power, which is why multiple connections are necessary within the same area. Several strategies to optimize the cable layout for smooth subsea connections have been investigated. Other related challenges include maneuvering in a farm of

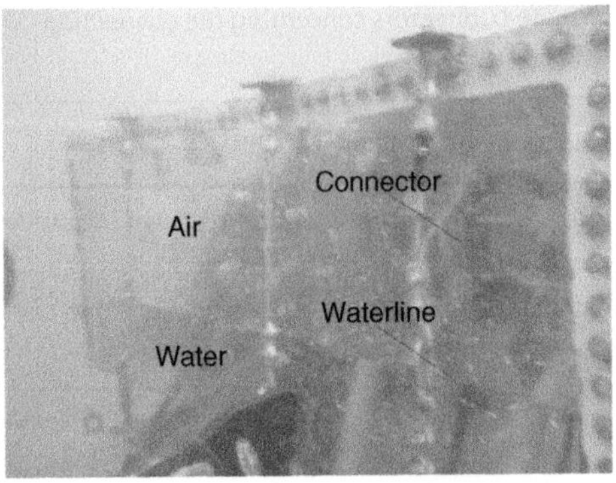

Figure 16.10 A diver filling the pocket with air to perform the dry connection [65].

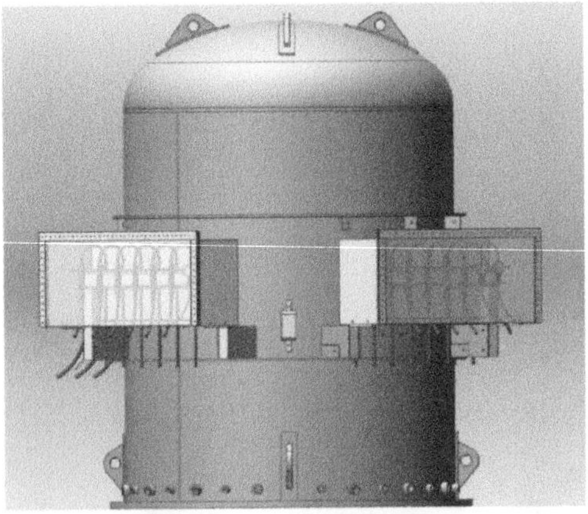

Figure 16.11 CAD file of the air pockets for underwater cable connections at the substation of Uppsala University, Lysekil project [65].

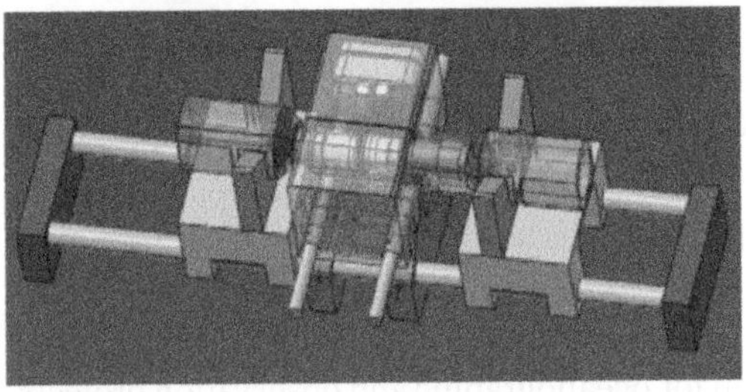

Figure 16.12 Mated connector inside oil-filled clamshell tool [66].

devices and identification and location of connectors and cables in a complex network of cables.

The main design criteria for a connection layout are fault tolerance, ease of assembly and management, and reasonable costs.

Fault tolerance can be translated as the need to limit the number of connectors since they are an important source of failure. Although connectors facilitate the assembly, they should be kept to a minimum. Ease of assembly and management is achieved by dividing the system into subsystems that are easy to deploy, test, and maintain without disconnecting the whole farm/grid [11, 67]. As a consequence, the cost of operation and maintenance will decrease, but those criteria must also meet other requirements that impact the project. These requirements can be environmental: for instance, two devices must be kept at a respectable distance from each other for fish and marine mammals to go through the array.

Offshore renewable energies are facing additional challenges related to optimization of energy absorption and farm layout. Depending on the incoming wind/waves/current parameters, a park could gain in efficiency by following a specific pattern (see e.g. Ref. [68]), and this model can conflict with the optimal connection layout.

Cabling layout optimization has been studied in [69–72], while [73] concerns power cable connections in general.

16.4.3 Reliability

Most underwater connection projects, especially within big offshore industries (offshore wind, oil, and gas, etc.), include a reliability study that analyzes the failure possibilities in the system and their consequences in terms of time, costs, and security. FMECA is the most popular method among the investigations of underwater connections. FMECA evaluates all possible failure modes for a connector/connection and studies both the probability of occurrence of these failures, from very unlikely to frequent, and the severity of their impacts on the system, from minor to catastrophic [61]. FMECA is typically conducted at an early stage of the project. The predictions are qualitative at first, owing to the lack of data, and evolve to quantitative measurements with more or fewer details. It is possible to survey the entire system (i.e. the connector manufacturing, deployment, and connection procedure) as well as to deepen the study up to the smallest part or component (i.e. pin, socket, or material) [74]. The resulting tables serve as feedback to improve a design or a procedure. They are meant to be used ahead of the project; however, the experience of failures will help enhance the process and prototypes.

One innovative method for reliability estimation is described in paper [75], using a model based on Weibull-covariate. Several other models that predict mean failure rates of different underwater systems are also presented in this article, such as the accelerated life testing (ALT) model, Monte Carlo models that are based on the failure history, and proportional hazard models.

Finally, paper [76] gives a detailed insight into the complexity of reliability modeling depending on the connection layout for a substation used in the oil and gas industry.

16.5 Discussion and Conclusion

Although underwater electrical connections have existed for more than 150 years, only a relatively small number of papers have been published on this topic. This scarcity of research is caused by the necessity for industries to patent any new product and hence retain complete confidentiality. Underwater electrical connectors have always been designed by industries, and research institutes have usually not taken part in the process. Hence, publications in this field are scarce but valuable.

It is interesting to mention that many of the articles reviewed here were written in the 1970s and are still up-to-date and relevant, which means that many of the challenges related to subsea connection have remained constant. The current literature shows that the same failure causes have persisted over the years: corrosion, leakage, and bond failures.

Many types of connectors are available today, especially optical fiber connectors and hybrid joints. This variety allows a multiplicity of underwater projects to emerge, which before were set aside for oil and gas, military, and telecommunications sectors. For example, the production of mini and micro connectors came together with the emergence of small ROVs that can be used for research projects and oceanographic studies with little funding. Those commercial off-the-shelf (COTS) connectors are presented in Section 16.3.

Despite the increase of available underwater electrical connectors, the need for tailor-made mating concepts remains. These concepts are so critical to system reliability that it is essential to purchase the best item to fit the requirements as well as possible. Particularly, the projects concerning power cables, telecommunications, deep water industry, or the military, must not fail, and the connectors are sometimes to be deployed once, with no maintenance for their whole lifetime, usually around 20–25 years. Consequently, they need a unique design that merges the right properties in the electrical, thermal, and mechanical areas. There are many critical factors in these different aspects, as explained in Section 16.4, and it illustrates the complexity of some connectors.

From this chapter review it must be emphasized that the emergence of new offshore and subsea markets requires enhanced development of COTS underwater electrical connectors. As said before, offshore renewable energies have specific requirements: high-voltage connections at a moderate depth compared to oil and gas (up to 200 m), with both dynamic and static cables. The market for offshore renewable energies is growing fast, and the lack of adapted connectors could hinder their full development. The same challenge exists for small ROV connectors: although their numbers are growing, the demand for small and light ROVs working at moderate depth does not meet the supply in terms of size, weight, cost, and reliability of the COTS connectors.

It is relevant to point out that the number of papers studying optimal procedures and layouts for connection architectures is increasing. The development of offshore renewable energies comes with the need to decrease costs and facilitate underwater operations, due to the multiplicity of devices in one farm (unlike in the oil and gas industry). It could be also said that after a first phase whose objective was to produce reliable underwater electrical connectors, it is now most important to decrease their costs and produce smaller products in larger quantities. The popularization of scuba diving, for example, the emergence of companies selling private ROVs, or the fast-growing subsea communications industry will surely accelerate the dissemination of those valuable underwater connectors.

References

1 Finn, B., "Underwater cables," Engineering and Technology History Wiki. http://ethw.org/Underwater_Cables (accessed Nov. 19, 2019).

2 Baer, C. M., Alten, M., Bixler, G., et al. (2009). Non-contact wet mateable connector. *Oceans '09 MTS/IEEE*, Biloxi, MS (Oct. 26–29, 2009). IEEE. pp. 1–6.

3 Walker, S.R.J. (2013). Barriers to the deployment of a 100 MW tidal energy array in the UK. *International Journal of Energy Engineering* 3 (3): 80–92.

4 Galford, T. C., and Purvinis, G. (2011). A plea for the standardization of underwater connectors. *Oceans '11 MTS/IEEE,* Kona, HI (Sept. 19–22, 2001). IEEE. pp. 1–6.
5 MacNeill A., and El-Hawary, M. (2014). Underwater power cable approximation models for offshore applications. *Oceans '14 MTS/IEEE,* St. John's, NL (Sept. 14–19, 2014). IEEE. pp. 1–7.
6 Virr, L.E. (Jun. 1987). Role of electricity in subsea intervention. *IEE Proceedings A: Physical Science, Measurement and Instrumentation, Management and Education, Reviews* 134 (6): 547–576.
7 Brown, G. E. (2003). Development, testing and track record of fiber-optic, wet-mate, connectors. Seacon. http://seaconworldwide.com/downloads/technical-papers/dev-test-track-fo-wet-mate-2003 (accessed Nov. 19, 2019).
8 Newell C., and Brown, G. (2005). Underwater optical and electrical connector systems, technical innovations. *Offshore Technology Conference,* Houston, TX (May 2–5, 2005). OTC.
9 Christ, R.D. and Wernli, R.L. Sr., (2014). The ROV business. In: *The ROV Manual,* 2e (eds. R.D. Christ and R.L. Wernli), 3–20. Oxford: Butterworth-Heinemann.
10 Shukla, A. and Karki, H. (Jan. 2016). Application of robotics in offshore oil and gas industry— a review, part II. *Robotics and Autonomous Systems* 75 (Part B): 508–524.
11 Hernández-Guiteras, J., Riba, J.R., and Romeral, L. (Apr. 2014). Improved design of an extra-high-voltage expansion substation connector through magnetic field analysis. *Simulation Modelling Practice and Theory* 43: 96–105.
12 Ng, C. and Ran, L. (eds.) (2016). *Offshore Wind Farms: Technologies, Design and Operation.* Elsevier.
13 Ruppe, C.V. and Barstad, J.F. (2013). *International Handbook of Underwater Archaeology.* Springer Science & Business Media.
14 Ageron, M. et al. (Nov. 2012). Search for neutrinos from transient sources with the ANTARES telescope and optical follow-up observations. *Nuclear Instruments and Methods in Physics Research Section A: Accelerators, Spectrometers, Detectors and Associated Equipment* 692: 184–187.
15 Snyder, J., Slater, M., and Chwaszczewski, R. (2011). Electro-optical-mechanical termination technology for low-cost field repairs of cabled systems. *OCEANS '11 MTS/IEEE,* Kona, HI (Sept. 19–22, 2001). IEEE. pp. 1–5.
16 Lamare, P. and Vernin, P. (2009). Underwater mateable electro-optical connectors: the feedback from ANTARES. *Nuclear Instruments and Methods in Physics Research Section A: Accelerators, Spectrometers, Detectors and Associated Equipment* 602 (1): 252–254.
17 Barlow, S. Advances in undersea connector technology improve efficiency. http://www.lightwaveonline.com/articles/print/volume-16/issue-11/special-report/advances-in-undersea-connector-technology-improve-efficiency-53489852.html (accessed Nov. 19, 2019).
18 del Río, J., Toma, D., O'Reilly, T.C. et al. (Jul. 2014). Standards-based plug and work for instruments in ocean observing systems. *IEEE Journal of Oceanic Engineering* 39 (3): 430–443.
19 Telegeography, Submarine cable map. http://www.submarinecablemap.com (accessed Nov. 19, 2019).

20 Jenkins, D., and Christiansen, M. (2012). Enabling technology: High capacity wet mateable optic connection. *OCEANS '12 MTS/IEEE*, Hampton Roads, VA (Oct. 14–19, 2012). IEEE. pp. 1–7.

21 Cairns, J., "Hybrid wet-mate connectors: Writing the next chapter," *Sea Technology*, July 1, 1997.

22 Cairns, J. (1983). Recent advances in underwater mateable electrical connectors. *OCEANS '83*, San Francisco, CA (1983). IEEE. pp. 507–511.

23 T&D World, NEMA Publishes ANSI/NEMA CC 1–2009 Electric Power Connection for Substations - NEMA. https://www.tdworld.com/test-monitor-amp-control/nema-publishes-ansinema-cc-1-2009-electric-power-connection-substations (accessed Nov. 19, 2019).

24 Christ, R.D. and Wernli, R.L. Sr., (eds.) (2014). Cables and connectors. In: *The ROV Manual*, 2e, 163–220. Oxford: Butterworth-Heinemann.

25 Brown, G. E. (2003). Operational considerations for underwater-mateable connectors," *Proceedings OCEANS 2003*. San Diego, CA (Sept. 22–26, 2003), IEEE. Vol. 4, pp. 1843–1848.

26 Defense Technical Information Center (1976). *Reference Manual on Interference Seals and Connectors for Undersea Electrical Applications*. https://apps.dtic.mil/dtic/tr/fulltext/u2/a165743.pdf (accessed Nov. 19, 2019).

27 Brown, G. E., Advancements in the reduction of size and weight of electrical and optical connectors for extreme environments, Seacon. http://seaconworldwide.com/downloads/technical-papers/advancements-reduction-size-weight-electrical-optical-extreme-environments (accessed Nov. 19, 2019).

28 Brown, G., Low-cost, miniature connectors for underwater use on homeland security applications, Seacon. http://seaconworldwide.com/downloads/technical-papers/low-cost-min-connectors-homeland-security-apps (accessed Nov. 19, 2019).

29 Connector ensures reliability of high-voltage operations on the seabed. *Sealing Technology* 2016 (7): 2.

30 Granger, R. P., Baer, C. M., Gabriel, N. H., Labosky, J. J., and Galford, T. C. (2013). Non-contact wet mateable connectors for power and data transmission. *OCEANS 2013 MTS/IEEE*, San Diego, CA (Sept. 23–27, 2013). IEEE. pp. 1–4.

31 Simpson, R., Heybrock, K., and Linden, J. P. (2005). High-reliability, modular ocean cable termination," in *OCEANS 2005 MTS/IEEE*, Washington, DC (Sept. 19–23). IEEE. Vol. 3, pp. 2701–2704.

32 Multon, B., Ed., *Marine Renewable Energy Handbook*, Wiley-ISTE, 2012. http://www.researchandmarkets.com/reports/2182846/marine_renewable_energy_handbook.pdf (accessed Dec. 8, 2016).

33 Kihara, M., Koyama, R., Abe, Y. et al. (Aug. 2013). Novel field installable optical fiber splice and connector for optical access networks. *Optical Fiber Technology* 19 (4): 269–278.

34 "Deliverable 3.1 - DTOcean." http://www.dtocean.eu/Deliverables-Documentation/Deliverables/Deliverable-3.1 (accessed Jan. 10, 2017).

35 Powers, W. F. (1994). The basics of power cable. *Proceedings of IEEE 36th Cement Industry Technical Conference*, Seattle, WA. IEEE. pp. 37–45.

36 Worzyk, T. (2009). *Submarine Power Cables: Design, Installation, Repair, Environmental Aspects*. Dordrecht: Springer-Verlag.

37 Huang, Z., Pilgrim, J.A., Lewin, P. et al. (2015). Thermal modelling and analysis for offshore submarine high-voltage direct current cable crossings. *IET Generation, Transmission & Distribution* 9 (16): 2717–2723.
38 Hernández-Guiteras, J., Riba, J.R., and Romeral, L. (Mar. 2014). Redesign process of a 765 kVRMS AC substation connector by means of 3D-FEM simulations. *Simulation Modelling Practice and Theory* 42: 1 11.
39 Berian, A. (1976). Compatability of underwater cables and connectors. *OCEANS '76.* IEEE. pp. 197–205.
40 Khandeparker, L. and Anil, A.C. (Mar. 2007). Underwater adhesion: the barnacle way. *International Journal of Adhesion and Adhesives* 27 (2): 165–172.
41 Waite, J.H. (Jan. 1987). Nature's underwater adhesive specialist. *International Journal of Adhesion and Adhesives* 7 (1): 9–14.
42 Felitney, R. (ed.) (2014). Chapter Six – Failure guide. In: *Seals and Sealing Handbook*, 6e, 437–474. Oxford: Butterworth-Heinemann.
43 Felitney, R. (ed.) (2014). Chapter two – static seals. In: *Seals and Sealing Handbook*, 6e, 7–103. Oxford: Butterworth-Heinemann.
44 Felitney, R. (ed.) (2014). Chapter three - rotary seals. In: *Seals and Sealing Handbook*, 6e, 105–288. Oxford: Butterworth-Heinemann.
45 Lamborn, O. (1977). Underwater electrical cable terminations. *OCEANS '77 Conference Record*. IEEE. pp. 389–391.
46 Edgerton, G. (1975). Underwater repair of electromechanical cables," in *OCEANS '75 Conference*. IEEE. pp. 174–176.
47 Zhao, F., Li, H., Liu, S., and Chen, J. (May 2011). Preparation and properties of an environment friendly polymer-modified waterproof mortar. *Construction and Building Materials* 25 (5): 2635–2638.
48 Fortin, H., Kandev, N., and Fafard, M. (May 2012). FEM analysis of voltage drop in the anode connector induced by steel stub diameter reduction. *Finite Elements in Analysis and Design* 52: 71–82.
49 Buggy, M. and Conlon, C. (Nov. 2004). Material selection in the design of electrical connectors. *Journal of Materials Processing Technology* 153–154: 213–218.
50 Kojima, J. (2003). Insulation measurement of undersea equipments and failure analysis of the underwater connector on VENUS system. *2003 International Conference Physics and Control. Proceedings (Cat. No.03EX708)*. IEEE. pp. 75–78.
51 Mazzanti, G. (Aug. 2009). The combination of electro-thermal stress, load cycling and thermal transients and its effects on the life of high voltage ac cables. *IEEE Transactions on Dielectrics and Electrical Insulation* 16 (4): 1168–1179.
52 Anders, G.J. and Brakelmann, H. (Jul. 2004). Improvement in cable rating calculations by consideration of dependence of losses on temperature. *IEEE Transactions on Power Delivery* 19 (3): 919–925.
53 Abomailek, C., Capelli, F., Riba, J.-R., and Casals-Torrens, P. (Jan. 2017). Transient thermal modelling of substation connectors by means of dimensionality reduction. *Applied Thermal Engineering* 111: 562–572.
54 Giacometto, F., Capelli, F., Romeral, L. et al. (Aug. 2016). Thermal response estimation in substation connectors using data-driven models. *ResearchGate* 16 (3): 25–30.
55 Carvou, E., El Abdi, R., Razafiarivelo, J. et al. (Jun. 2012). Thermo-mechanical study of a power connector. *Measurement* 45 (5): 889–896.

56 Wang, J.J.A., Chan, J.K., and Graziano, J.A. (Jul. 2011). The lifetime estimate for ACSR single-stage splice connector operating at higher temperatures. *IEEE Transactions on Power Delivery* 26 (3): 1317–1325.

57 Wang J.-A. J., et al. (*2009*). The effective lifetime of ACSR full tension splice connector operated at higher temperature, IEEE Power Engineering Society Joint Technical Committee Meeting, United States.

58 Seacon. EAGE_Workshop_2010. http://seaconworldwide.com/eage_workshop_2010 (accessed Jan. 11, 2017).

59 MIL-C-24217 (1984). *Military specification: connectors, electrical, deep submergence, submarine.* US Dept. of Defense.

60 MIL-C-24231/14 (1987). *Military specification: connectors, receptacle, multi-pin, type V, 14-, 24-, 30-, and 40-conductor*. US Dept. of Defense

61 Jenkins, D., Christiansen, M., and Thumbeck, S. (2013). Essential design and risk management for a next generation ocean dry mate connector. *OCEANS '13, San Diego*. IEEE. pp. 1–4.

62 Pikkert, A., Binnendijk, M., Riet, M. V., and Harmsen, D. (2013). Innovative and safe cable connection in medium voltage switchgear with integrated earthing and testing facilities. *22nd International Conference and Exhibition on Electricity Distribution (CIRED 2013)*. CIRED. pp. 1–4.

63 ORE Catapult, ORE Catapult Reports. https://ore.catapult.org.uk/resources/reports-publications/ore-catapult-reports (accessed Jan. 10, 2017].

64 Teledyne Products, Nautilus WM1.7-60. http://www.teledynemarine.com/nautilus-wm1-7-60?ProductLineID=42 (accessed Nov. 21, 2019).

65 Remouit, F., Lopes, M. F. P., Pires, P., Sebastiao, L., and Rahm, M. (2015). Automation of subsea connection for clusters of wave energy converters, presented at the *25th International Ocean and Polar Engineering Conference.*

66 Beurthey, S. (2011). A new technique for the deep-sea mating of inexpensive connectors. *Nuclear Instruments and Methods in Physics Research, Section A* 626–627 (Supplement): S124–S126.

67 D'Amico, A. (Nov. 2006). Analysis of the cable system layout for the NEMO underwater detector. *Nuclear Instruments and Methods in Physics Research, Section A: Accelerators, Spectrometers, Detectors and Associated Equipment* 567 (2): 535–537.

68 Engström, J., Eriksson, M., Göteman, M. et al. (Nov. 2013). Performance of large arrays of point absorbing direct-driven wave energy converters. *Journal of Applied Physics* 114 (20): 204502.

69 Hou, P., Hu, W., and Chen, Z. (2016). Optimisation for offshore wind farm cable connection layout using adaptive particle swarm optimisation minimum spanning tree method. *IET Renewable Power Generation* 10 (5): 694–702.

70 Painter, H., and Flynn, J. (2006). Current and future wet-mate connector technology developments for scientific seabed observatory applications. *OCEANS 2006*. IEEE. pp. 1–6.

71 Zhao, R., Qiao, C., and Wang, Y. (2009). Optimization of connection design for cable connector in underwater monitoring system. *2009 International Conference on Measuring Technology and Mechatronics Automation*. IEEE. Vol. 2, pp. 861–864.

72 Thumbeck, S., Toth, J., and Barlow, S. (2003). Pioneering subsea fiber optic and electric solutions - communication interconnect advances and solutions for cabled

observatory installation, operation and maintenance (focusing on wet-mate capabilities). *2003 International Conference Physics and Control. Proceedings (Cat. No. 03EX708)*. IEEE. pp. 79–82.

73 Carter, L. et al. (2009). *Submarine Cables and the Oceans: Connecting the World*. UNEP/Earthprint.

74 Sandwith, C., Paradis, J., and Morrison, J. (1977). Underwater electrical cable and connector seals: Some in-house design options, commercial options, and performance/failure analyses. *OCEANS '77 Conference Record*. IEEE. pp. 401–406.

75 Okaro, I.A. and Tao, L. (Dec. 2016). Reliability analysis and optimisation of subsea compression system facing operational covariate stresses. *Reliability Engineering and System Safety* 156: 159–174.

76 Vedachalam, N., Srinivasalu, S., Ramesh, R. et al. (Jul. 2015). Review and reliability modeling of maturing subsea hydrocarbon boosting systems. *Journal of Natural Gas Science and Engineering* 25: 284–296.

17

Examples of Connectors

Lei Su[1], Xiaonan Yu[1], San Kyeong[2] and Michael G. Pecht[2]

[1] *Jiangsu Key Laboratory of Advanced Food Manufacturing Equipment and Technology, Jiangnan University, Wuxi, China*
[2] *Center for Advanced Life Cycle Engineering (CALCE), University of Maryland, College Park, MD, USA*

Various connectors by numerous manufacturers are available in the market. Some of the key manufacturers include Amphenol, Aptiv, Fischer Connector, Hirose, Molex, and TE Connectivity (in alphabetical order). This chapter presents some of their products and focuses on the specifications, types, limitations, advantages, and usage in a particular application. The products are introduced in alphabetical order by manufacturers. For each manufacturer's products, the connectors are listed in board-to-board, board-to-wire, and wire-to-wire order.

17.1 Amphenol ICC M-Series™ 56 Connectors

Amphenol ICC's M-Series 56 connectors are mezzanine-type board-to-board signal connectors. Liquid crystal polyester (LCP) is introduced to form the connector housing, and the contact interface is made of copper alloy plated with gold over nickel. The data rate is 56 Gb/s for the none-return-to-zero (NRZ) application and 112 Gb/s for the four-level pulse amplitude modulation (PAM-4) application. The mating durability of this connector is up to 200 cycles. Figure 17.1 shows the M-Series 56 connectors [1].

17.2 Amphenol ICC Paladin® Connectors

Amphenol ICC's Paladin® connectors are backplane type signal connectors (see Chapter 1 for a discussion of backplane-type connectors). The connectors are available in both type of board-to-board and board-to-wire. The data transfer rate is up to 112 Gb/s to meet the demand for high-performance applications that include the datacenter, server, and supercomputer industries. The connector provides smooth linear transmission beyond 40 GHz and nominal impedance of $95 \pm 5\,\Omega$. The number of differential pairs varies from 8 to 128, and the contact interface is made of copper alloy with gold finish. As a trade-off for the high performance, the operating temperature is limited as $-40\,°C$ to $85\,°C$. Figure 17.2 shows examples of Paladin® backplane connector [2].

Electrical Connectors: Design, Manufacture, Test, and Selection, First Edition.
Edited by San Kyeong and Michael G. Pecht.

Figure 17.1 Amphenol ICC M-Series™ 56 connectors [1].

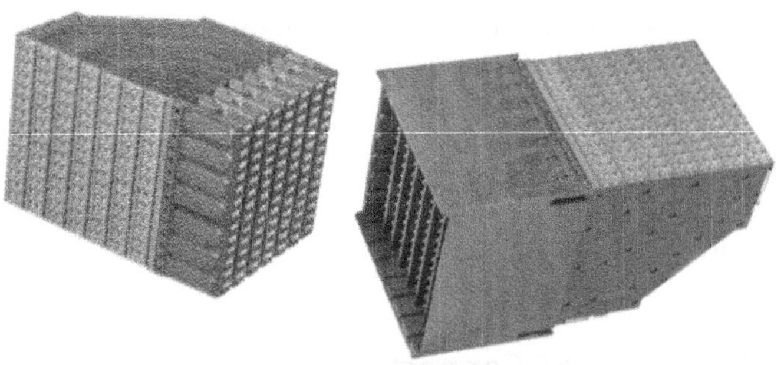

Figure 17.2 Amphenol ICC Paladin® 112G connectors [2].

17.3 Amphenol ICC 3000W EnergyEdge™ X-treme Card Edge Series

Amphenol ICC's 3000W EnergyEdge X-treme card edge connectors are board-to-board power connectors, targeting industrial applications such as networking and datacom equipment. The connector is available as both mezzanine and vertical configuration and delivers 3000 W at 12 V. The current rating is 28 A per contact (5.08 mm pitch) and 42 A per contact (7.62 mm pitch), respectively. The contact interface is made of copper alloy with gold electroplating over nickel. Figure 17.3 shows the 3000W EnergyEdge X-treme card edge connector [3].

17.4 Amphenol ICC FLTStack Connectors

Amphenol ICC's FLTStack connectors are board-to-board type signal connectors. This connector has a floating range of ±0.50 mm in the XY direction and an effective mating length of 0.9 mm in the Z direction, featuring a self-cleaning function and vibration resistance. The data transfer rate is up to 3 Gb/s. The number of contact varies from 20 to 70, and the contact interface is made of copper alloy. The connectors also certify

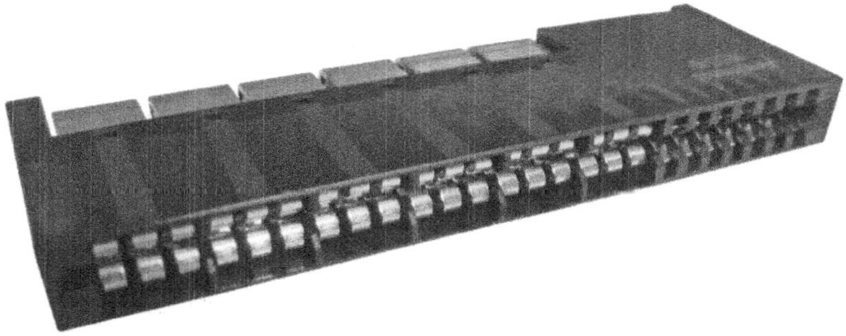

Figure 17.3 Amphenol ICC 3000W EnergyEdge™ X-treme connector [3].

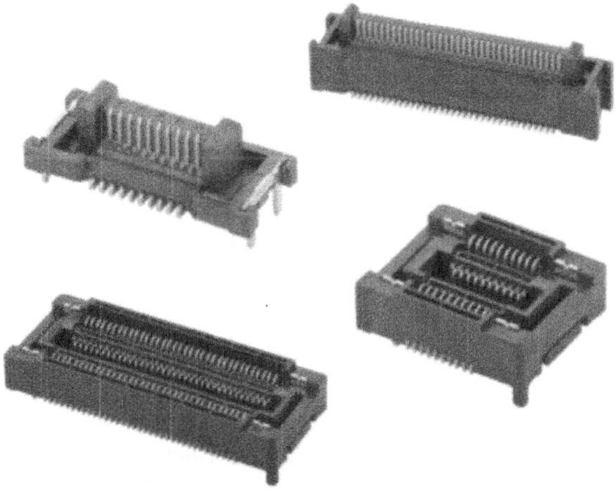

Figure 17.4 Amphenol ICC FLTStack connectors [4].

the US Council for Automotive Research (USCAR) 2 standard. Figure 17.4 shows the FLTStack connectors [4].

17.5 Amphenol ICC HSBridge Connector System

Amphenol ICC's HSBridge connectors are used as signal connectors in the field of automotive electronic systems such as infotainment, telematics, and camera devices. The connectors are compatible with several protocols, including low-voltage differential signaling (LVDS), universal serial bus (USB) 2.0, high-definition multimedia interface (HDMI), and video graphics array (VGA). The connector has five contact points that are made of copper alloy with gold finish. The connector housing is made of LCP and provides thermal resistance of 85 °C for 1008 h. The mating durability of this connector is up to 10 cycles. Figure 17.5 shows the HSBridge connectors [5].

17.6 Amphenol ICC MUSBR Series USB 3.0 Type-A Connectors

Amphenol ICC's MUSBR series USB 3.0 Type-A connectors are designed to provide a standard USB interface under harsh environments with data rates up to 5 Gb/s. The connector housing ensures IP67 dust and waterproof rating under IEC standard 60529.

Figure 17.5 Amphenol ICC HSBridge connectors [5].

Figure 17.6 Amphenol ICC MUSBR series USB 3.0 connector [6].

The mating durability of this connector is up to 1500 cycles. The contact interface is made of phosphor bronze with gold electroplating over nickel. Figure 17.6 shows the MUSBR series of USB 3.0 Type-A connectors [6].

17.7 Amphenol ICC Waterproof USB Type-C™ Connectors

USB Type-C connectors are 24-pin double-sided connectors designed to transfer both signals and power to mobile devices. Their specifications were published and finalized by the USB Implementers Forum (USB-IF) in 2014 and adopted as IEC standard 62680-1-3 in 2016. Generally, the plug parts of USB Type-C are found on cables and adapters, whereas the receptacle parts are found on mobile devices and adapters. Amphenol ICC's waterproof USB Type-C connectors for USB Type C provide a 10 Gb/s data transfer rate that is compatible with USB 3.1 specifications and mating durability of up to 10 000 cycles. An O-ring is added to the connector housing to ensure IPX8 waterproof rating under IEC 60529. The rated current is 5 A in DC for USB power delivery. The contact interface is made of copper alloy. Figure 17.7 shows the USB Type-C™ connectors [7].

17.8 Amphenol ICC NETBridge™ Connectors

Amphenol ICC's NETBridge connectors are board-to-wire type signal connectors designed for Ethernet systems in automotive environments. The connectors have one

Figure 17.7 Amphenol ICC waterproof USB Type-C™ connectors [7].

Figure 17.8 Amphenol ICC NETBridge™ connector [8].

to seven contact pins made of brass. The connectors provide a data transfer rate of 100 Mb/s and 1 Gb/s, according to IEEE standards 100BASE-T1 and 1000BASE-T1. The connectors also certify USCAR 2 standard. Figure 17.8 shows the NETBridge connector [8].

17.9 Amphenol Sine Systems DuraMate™ AHDP Circular Connectors

Amphenol Sine Systems' DuraMate AHDP Circular Connectors are power connectors for heavy-duty applications such as agricultural vehicles, construction vehicles, and marine installations. The number of contacts varies from 7 to 47, and individual contact pins with various diameters from 7.42 to 1.02 mm can be combined to form a connector. The rated current per each contact pin varies from 7.5 to 100 A, depending on the diameter of the contact pin. The contact interface can be plated with nickel or gold. The connector housing provides a strain-relief back shell. The connector is sealed with silicone rubber to prevent lubricating oils or fossil fuels from penetrating inside

Figure 17.9 Amphenol Sine Systems DuraMate™ AHDP circular connectors [9].

and ensures IP69K dust and waterproof rating under IEC standard 60529. Figure 17.9 shows the Amphenol Sine DuraMate AHDP circular plastic connectors [9].

17.10 Amphenol Aerospace MIL-DTL-38999 Series III Connectors

Connectors for military or aerospace applications should be qualified for use in extreme environments, including high/low temperatures and vibrations. Amphenol Aerospace's MIL-DTL-38999 Series III connectors are designed for military applications and are qualified to MIL-DTL-38999, as the name of the product implies. These circular and screw-coupling connectors ensure the mating durability of up to 500 cycles, and the contact pins have a recessed shape to reduce potential contact damage in the mating process. The shell material is available in electroless nickel, stainless steel, zinc-nickel plated, olive drab cadmium, and nickel-polytetrafluoroethylene (PTFE) to meet several specified environments. The connectors provide electromagnetic interference (EMI) shielding capability of 65 dB at 10 GHz. The number of contacts varies from 6 to 128. The current rating is up to 23 A per contact. The operating temperature is from −65 °C to 200 °C. Figure 17.10 shows the Amphenol Aerospace MIL-DTL-38999 Series III connectors [10].

17.11 Fischer Connectors UltiMate™ Series Connectors

Fischer's UltiMate connectors are circular connectors with 2–42 contact points and are designed for reliably transmitting signals or low-voltage power in oil applications or an underwater depth of 100 m. The connectors are protected by an O-ring made of fluoropolymer elastomer to ensure the IPX9 waterproof rating under IEC standard 60529. The operating temperature range is −55 °C to 135 °C, and the mating durability is 10 000 cycles. The connectors ensure discontinuity of less than 1 μs under mechanical shock of 300 g amplitude rating under military standard (MIL-STD) 202 method 213.

Figure 17.10 Amphenol Aerospace MIL-DTL-38999 Series III connectors [10].

Figure 17.11 Fischer Connectors UltiMate™ series 27-pin connector [11].

The rated current of the connectors depends on the number of contacts and varies from 3.1 to 9.2 A under IEC standard 60512. In the case of use as a signal connector with the capability of 10 Gb·s^{-1} data rates, similar levels of HDMI, the connector should contain 19 or more contact points. For the contact interface, the male part is made of brass, and the female part is made of bronze, whereas both parts are finished by gold over nickel. Figure 17.11 shows the 27-pin UltiMate Series connector [11].

17.12 Hirose Electric DF50 Series Connectors

Hirose Electric's DF50 connectors are board-to-wire connectors. The housing has a stainless steel lock in the center of the connector to keep the contact mating in the proper location. The number of contacts varies from 20 to 50, and the contact interface is made of brass plated with gold. The current rating of each contact is 1 A. These connectors are used in office equipment, home appliances, and portable gadgets. Figure 17.12 shows the DF50 connectors [12].

Figure 17.12 Hirose Electric DF50 connectors [12].

Figure 17.13 Hirose Electric MicroSD™ DM3AT connector [13].

17.13 Hirose Electric microSD™ Card Connectors

Since microSD cards are designed for use with mobile phones and other portable devices, the size of connectors for the microSD card is considered carefully. The connectors also provide high mechanical durability to withstand frequent mating and unmating actions. Hirose Electric's microSD card connectors have a height of 1.68 mm, a width of 13.85 mm, and a length of 15.95 mm. The connector housing is made of stainless steel, providing a mating durability of up to 10 000 cycles and 400–600 cycles per hour. Copper alloy plated with gold is used for the four contact interfaces. Figure 17.13 shows a Hirose Electric DM3AT connector [13].

17.14 Molex SAS-3 and U.2 (SFF-8639) Backplane Connectors

Molex's SAS-3 and U.2 connectors are backplane-type board-to-board connectors. The SAS-3 backplane connectors are used in computer-storage devices, providing data transfer rates up to 12 Gb/s via 14 contact pins for signal transfer. The connectors include 15 contact pins for power transmission, and the current rating is up to 1.5 A per contact. The U.2 connectors, formerly known as small form factor (SFF) 8639 standard connectors, have 68 contact pins. They can serve as signal connectors and power connectors. Figure 17.14 shows the SAS-3 backplane receptacle and the U.2 connectors [14].

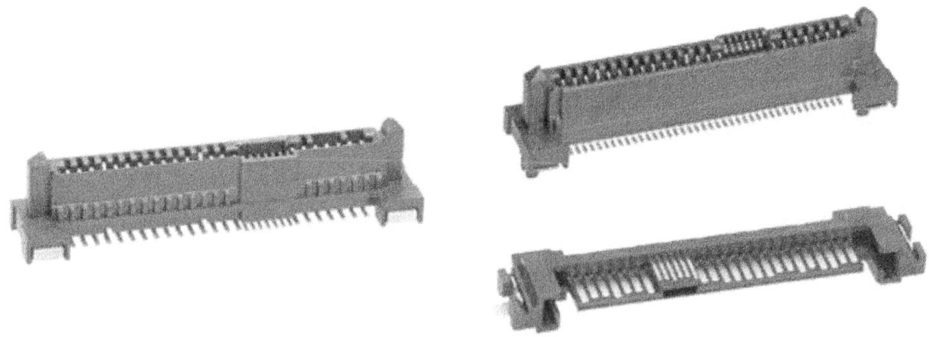

Figure 17.14 Molex SAS-3 backplane receptacle (left) and U.2 connectors (right) [14].

Figure 17.15 Molex NeoPress™ high-speed mezzanine connector [15].

17.15 Molex NeoPress™ Mezzanine Connectors

Molex's Modular NeoPress™ High-Speed Mezzanine connectors provide data transfer rates of up to 28 Gb/s. The connectors provide nominal impedance of 85 Ω or 100 Ω. The stack heights vary from 9.0 to 45.0 mm. LCP is introduced to the connector housing to withstand mechanical stability. The contact interface is made of copper plated with gold. Additional ground plating on the upper and lower housings minimizes crosstalk. Figure 17.15 shows Modular Neopress™ high-speed mezzanine connectors [15].

17.16 Molex Impel™ Plus Backplane Connectors

Molex's Impel™ Plus connectors are backplane type signal connectors. The connectors have a data transfer rate of up to 56 Gb/s with 92 Ω of nominal impedance and are used in mass telecommunications and networking facilities. The connector housing contains grounding tail aligner for lead frames and wafer shields to reduce cross talk. The number

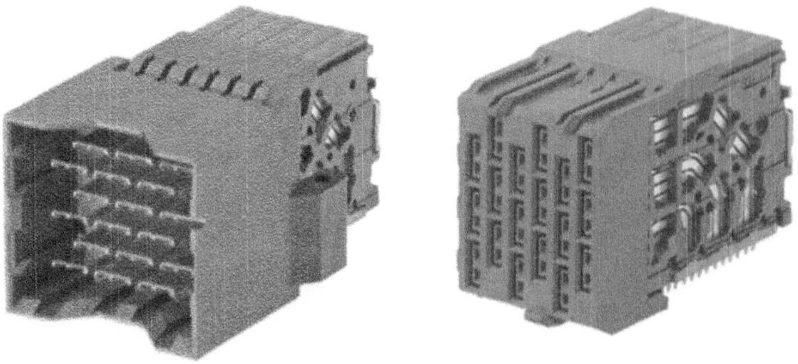

Figure 17.16 Molex Impel™ Plus backplane connectors [16].

of differential pairs is available from 36 to 64, and the pitch can be selected from 1.9 to 2.35 mm. The contact interface is made of copper alloy plated with gold over nickel. The operating temperature is −55 °C to 85 °C. Figure 17.16 shows pin and receptacle parts of Impel™ Plus connectors [16].

17.17 Molex EXTreme Guardian™ Power Connectors

Molex's EXTreme Guardian™ connectors are board-to-wire type power connectors with high current carrying capacity. Each contact has an 11.0 mm pitch and contains six individual points of contact interface made of copper alloy with gold electroplating over nickel. The current rating is up to 80 A per circuit, and a connector can deliver current up to 200 A. The mating durability of this connector is up to 25 cycles. Figure 17.17 shows the EXTreme Guardian™ power connector [17].

Figure 17.17 Molex EXTreme Guardian™ power connectors [17].

Figure 17.18 Molex Imperium™ power connectors.

17.18 Molex Imperium™ High Voltage/High Current Connectors

Molex's Imperium™ connectors are wire-to-wire connectors to transmit large amounts of electrical power. The current rating and voltage rating is up to 250 A and 1000 V per contact, respectively. The contact diameter is available in 8 and 11 mm, and the contact interface is made of copper alloy plated silver over nickel. High-voltage interlock loop (HVIL) is incorporated in the terminal part, reducing high-voltage arcing during disconnects. The termination and contact interface are sealed to ensure an IP6K9K dust and waterproof rating under IEC 60529. Figure 17.18 shows pin and receptacle parts of Molex Imperium connectors [18].

17.19 TE Connectivity Free Height Connectors

TE Connectivity's free height connectors are mezzanine-type board-to-board connectors (see Chapter 1 for a discussion of backplane-type connectors). The connectors are used in applications, including high-end computer industries and communication equipment. These connectors allow for high pin counts in a small footprint to save PC board real estate. A range of position sizes from 40 to 440 positions in contact pitches of 0.5, 0.6, 0.8, and 1.0 mm (IEEE 1386 standard) are available, and 4–20 mm board-to-board stack heights can be achieved in increments of 1 mm by mating various combinations of vertical plug and receptacle housing heights. The contact interface is gold-finished. Figure 17.19 shows free height connectors [19].

17.20 TE Connectivity STRADA Whisper Connectors

TE Connectivity's STRADA Whisper connectors are backplane-type signal connectors that are used in data storage and router applications. The connectors are available in both type of board-to-board and board-to-wire. The data transfer rate is up to 112 Gb/s. The number of differential pairs varies from 48 to 144, and individual differential pair is surrounded by six ground points to increase signal integrity. The contact interface is made of copper alloy plated with gold over palladium-nickel alloy. The insertion loss is less than 1 dB and linear up to 20 GHz. The nominal impedance can be tuned to 85 Ω or 100 Ω. Figure 17.20 shows STRADA Whisper backplane connectors [20].

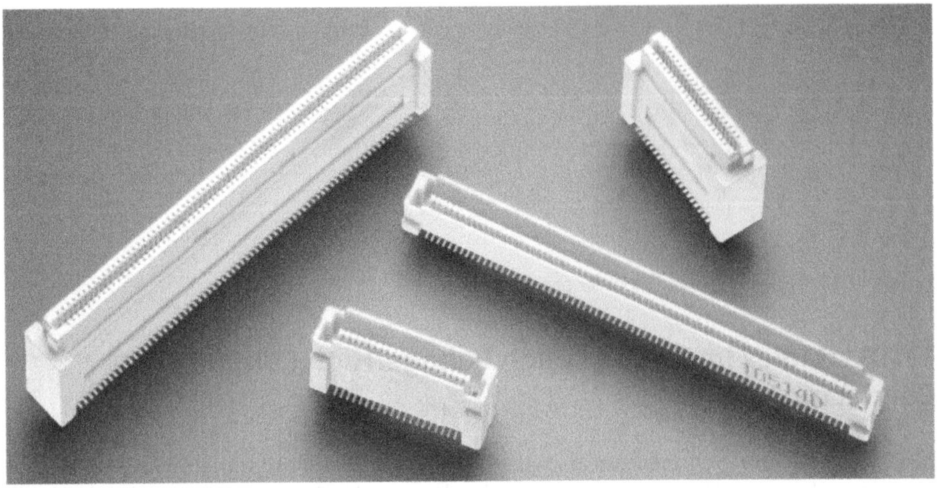

Figure 17.19 TE Connectivity free height connectors [19].

Figure 17.20 TE Connectivity STRADA Whisper connectors [20].

17.21 TE Connectivity MULTI-BEAM High-Density (HD) Connectors

TE Connectivity's MULTI-BEAM HD connectors are a board-to-board connector and can be used as a power connector or signal connector for PCB applications. When used as a power connector, these connectors can carry power up to 135 A. The connectors are mostly applied in data centers, telecommunications, industrial and automotive devices, and power systems. Their mechanical specifications include durability of 500 mating cycles, mechanical shock up to 50 G under EIA 364 standard, and vibration shock up to 4.9 G with a frequency up to 500 Hz. Noble metal electroplating covers the contact interface. Figure 17.21 shows samples of MULTI-BEAM connectors [21].

Figure 17.21 TE Connectivity MULTI-BEAM HD connectors [21].

Figure 17.22 TE Connectivity HDMI connector [22].

17.22 TE Connectivity HDMI™ Connectors

HDMI is an interconnection system to transmit digital audio and video data and was designed by the HDMI forum in 2002. HDMI connectors are used in various home appliances and portable electronic devices, so they are required to withstand frequent mating cycles. TE Connectivity's HDMI connectors provide a transfer data rate of 5 Gb/s and a mating durability of 10 000 cycles. These connectors have 19 contact interfaces that are finished by gold over palladium-nickel alloy. Figure 17.22 shows an example of an HDMI connector [22].

17.23 TE Connectivity AMP CT Connector Series

TE Connectivity's AMP CT connectors are board-to-wire type power connectors for office equipment or home appliances. The number of contacts varies from 2 to 15 for a single row and 8–30 for a double row. Both types of insulation-displacement contact (IDC) and crimps are available for the termination. The connectors have a double row of 0.5 mm pitches, and the number of contacts varies from 50 to 100. The current rating is up to 4 A per contact and is applied both in AC and DC. The contact interface is plated with tin or gold. Figure 17.23 shows the AMP CT connectors [23].

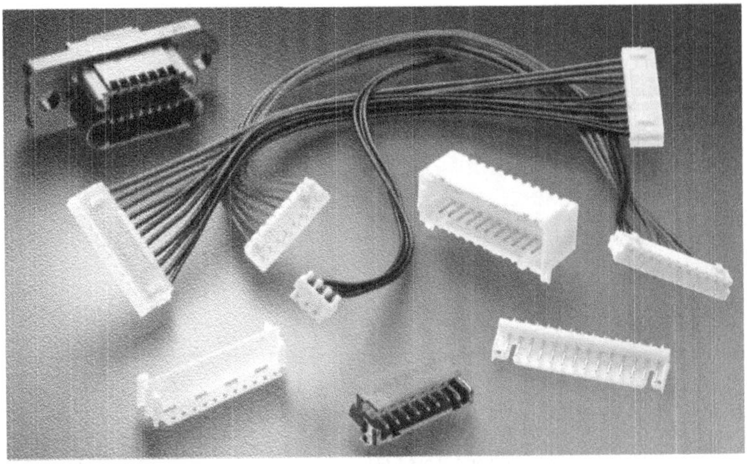

Figure 17.23 TE Connectivity AMP CT connector series [23].

Figure 17.24 TE Connectivity Micro Motor connectors [24].

17.24 TE Connectivity Micro Motor Connectors

TE Connectivity's Micro Motor connectors are used as signal connectors and power connectors for servo motor industries. Four screw-type clips in the housing sustain the contact retention of the connector against vibration and shock during operation. When used as a power connector, the rated current and voltage of the connectors are 5 A and 380 V (AC), respectively. The contact interface is made of nickel with gold electroplating. Figure 17.24 shows samples of the Micro Motor connectors [24].

17.25 TE Connectivity AMPSEAL Connectors

TE Connectivity's AMPSEAL connectors are board-to-wire type power connectors that are used in construction, agricultural, and special-purpose automotive applications. The connectors have a secondary lock in the connector housing to keep the contact retention under the vibrating conditions that automotive applications face. The connectors can withstand vibration shock of 10 G. The current rating is up to 17 A for

Figure 17.25 TE Connectivity AMPSEAL connectors [25].

gold-plated contacts and 8 A for tin-plated contacts. The number of contacts ranges from 8 to 35. The termination and contact interface are sealed with silicone rubber to ensure an IP6K9K dust and waterproof rating under IEC 60529. Figure 17.25 shows the AMPSEAL connectors [25].

17.26 TE Connectivity M12 X-Code Connectors

TE Connectivity's M12 X-Code connectors are circular connectors with a 12-mm locking thread used for transferring electrical signals under the mal-conditions of the railway field. The connectors are used for applications including Wi-Fi networks, onboard Ethernet networks for passenger information, entertainment systems, and surveillance systems. X-code refers to the capability of transferring data at up to 10 Gb/s. There are also other types of coding, such as A-code for 1 Gb/s and D-code for 100 Mb/s. To be used reliably under harsh environments, the connectors are protected by silicone O-rings that can prevent penetration by dust and liquid. The inner housing materials of the connectors are made of copper alloy with nickel electroplating and provide EMI shielding to minimize network disturbance. The contact interface is made of copper alloy with gold electroplating. Figure 17.26 shows the M12 X-code connectors [26].

17.27 TE Connectivity SOLARLOK 2.0 Connectors

TE Connectivity's SOLARLOK 2.0 connectors are electrical power (DC) connectors for applications of photovoltaic systems. The connectors use IDC, which is explained in

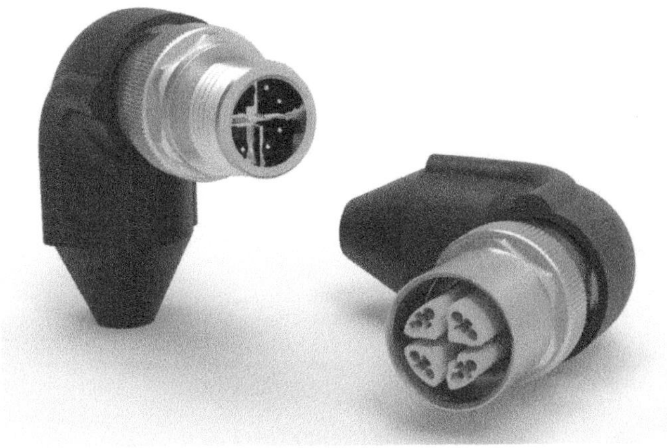

Figure 17.26 TE Connectivity M12 X-code connectors [26].

Figure 17.27 TE Connectivity SOLARLOK 2.0 connectors [27].

Chapter 1. The connector housing ensures IPX8 waterproof rating under IEC standard 60529 to make the connector work reliably in the wet environments. The current rating of the connectors depends on the cross-section of wire: 30 A for 2.5 mm^2, 35 A for 4.0 mm^2, and 40 A for 6.0 mm^2 under the IEC standard 60512 series. Figure 17.27 shows samples of SOLARLOK 2.0 connectors [27].

17.28 TE Connectivity Busbar Connectors

Busbar connectors are needed to transfer or distribute high current power. Major applications include industrial power distribution, power plants, and core network energy systems. TE Connectivity's busbar connectors have a current rating of up to 500 A with a voltage rating of 12 V or 48 V. The contact width varies from 3 to 6 mm. The connector is made of copper alloy with silver electroplating. Figure 17.28 shows samples of busbar connectors [28].

Figure 17.28 TE Connectivity busbar connectors [28].

References

1 *M-Series™ 56 connectors catalog*, Amphenol ICC, Aug 2019.
2 *Paladin® 112G connector catalog*, Amphenol ICC, May 2018.
3 *3000W EnergyEdge™ X-treme connector catalog*, Amphenol ICC, June 2019.
4 *FLTStack 0.50 mm floating board-to-board connector catalog*, Amphenol ICC, Aug 2018.
5 *HSBridge connector system catalog*, Amphenol ICC, Mar 2018.
6 *Rugged MUSBR series USB 3.0 connector catalog*, Amphenol ICC, Dec 2017.
7 *Waterproof USB Type-C™ connectors catalog*, Amphenol ICC, May 2019.
8 *NETBridge™ automotive Ethernet connector catalog*, Amphenol ICC, Aug 2019.
9 *DuraMate™ AHDP connectors catalog*, Amphenol Sine Systems, Feb 2019.
10 *MIL-DTL-38999 connectors catalog*, Amphenol Aerospace, Apr 2018.
11 *Fischer UltiMate™ series catalog*, Fischer Connectors, Sept 2019.
12 *1 mm pitch, board-to-wire lock connectors – DF50 series catalog*, Hirose Electric, May 2019.
13 microSD™ card connectors – *DM3 series catalog*, Hirose Electric, Jan 2017.
14 *SAS-3 and U.2 (SFF-8639) backplane connectors catalog*, Molex, Feb 2016.
15 *NeoPress™ high-speed mezzanine system catalog*, Molex, Sept 2017.
16 *Impel™ Plus backplane connector catalog*, Molex, June 2018.
17 *EXTreme Guardian™ power connector catalog*, Molex, Dec 2017.
18 *Imperium™ high voltage/high current conector catalog*, Molex, Jan 2014.
19 *0.8 mm free height (FH) connectors catalog*, TE Connectivity, July 2018.
20 *STRADA Whisper backplane connector catalog*, TE Connectivity, May 2018.
21 MULTI-BEAM high density *(HD) connectors catalog*, TE Connectivity, Aug 2016.
22 *Connectors and cable assemblies incorporating HDMI™ technology catalog*, TE Connectivity, Aug, 2008.
23 AMP CT connector catalog, TE Connectivity, Apr 2014.
24 *Micro Motor connector catalog*, TE Connectivity, Apr 2018.
25 *AMPSEAL connectors catalog*, TE Connectivity, June 2019.
26 *M12 X-coded right angle connector catalog*, TE Connectivity, May 2019.
27 *SOLARLOK 2.0 connector catalog*, TE Connectivity, Apr 2019.
28 *Busbar connectivity catalog*, TE Connectivity, Feb 2020.

Appendix A

Standards

For reference, this appendix lists the standards and specifications for connectors and related areas. These documents are developed by standards organizations internationally. The letters in each standard denote the organization that is responsible for the document.

AEC	Automotive Electronics Council
ANSI	American National Standards Institute
ASTM	American Society for Testing and Materials
EIA	Electronic Industries Alliance
IEC	International Electrotechnical Commission
IPC	Institute for Interconnecting and Packaging Electronic Circuits
ISO	International Organization of Standards
JEDEC	Joint Electron Devices Engineering Council of EIA
UL	Underwriters Lab, Inc.
MIL	Department of Defense (DoD)
QQ	Department of Defense

Electrical Connectors: Design, Manufacture, Test, and Selection, First Edition.
Edited by San Kyeong and Michael G. Pecht.

A.1 Standard References for Quality Management and Assurance

ISO 9001/9002	Quality program requirements
MIL-STD-109C	Quality assurance terms and definitions (canceled; suggested substitution: ISO 8402 or ASQC-A8402)
MIL-STD-105E	Sampling procedures and tables for inspection by attributes (canceled; substitution: ASQC-Z1.4-1993)
MIL-STD-45662A	Calibration system requirements (canceled; suggested alternative: ISO-10012-1 or ANSI-Z540-1)
MIL-I-45208A	Inspection system requirements (canceled; no replacements)
MIL-I-45607C	Acquisition, maintenance, and disposition of inspection equipment (inactive for new design, and no longer used except for replacement purposes)
IPC-QS-95	General requirements for implementation of ISO 9000 quality systems

A.2 General Specifications for Connectors

MIL-C-39029D	General specifications for contacts of electrical connectors This specification covers the general requirements for removable crimp, solderless wrap, and solder type electrical contacts for use in connectors and other electric and electronic components.
IPC-C-406	Design and application guidelines for surface mount connectors This specification provides guidelines for the design, selection, and application of soldered surface mount connectors for all types of printed boards, rigid, flexible-rigid, and backplanes.

A.3 Safety-Related Standards and Specifications

UL 94 V0	Standard for safety, tests, and flammability of plastic materials and for parts in devices and appliances. Serves as a preliminary indication of acceptability of plastic materials with respect to flammability for a particular application.
UL 746B	Standard for safety of polymeric materials – long-term property evaluation. Measures temperature index of polymers.

A.4 Standard References for Connector Manufacturing

Connector contact	
ASTM-B-16-92	Standard specification for free-cutting brass rod, bar, and shapes for use in screw machines
ASTM-B-194-96	Standard specification for copper-beryllium alloy plate, sheet, strip, and rolled bar
ASTM-B-139-95	Standard specification for phosphor bronze rod, bar, and shapes
Contact plating	
MIL-G-45204	Standard specifications for electrodeposited gold plating (canceled; recommended substitution: AMS 2422 and ASTM B488)
ASTM-B-488	Electrodeposited coatings of gold for engineering use
MIL-P-45209	Standard specifications for electrodeposited palladium nickel (canceled; recommended substitution: ASTM B679)
ASTM-B-679	Electrodeposited coatings of palladium for engineering use
MIL-T-10727	Standard specifications for electrodeposited or hot-dipped tin plating (canceled; recommended substitution: ASTM B545 (type I), ASTM B339 (type II))
ASTM-B-545	Standard specification for electrodeposited coating of tin
ASTM-B-339	Standard specification for pig tin
MIL-P-81728	Standard specification for electrodeposited tin-lead plating (canceled; recommended substitution: SAE-AMS-P-81728)
QQ-N-290	Federal standard specifications for electrodeposited nickel plating
MIL-C-14550	Standard specification for electrodeposited copper plating (canceled; recommended substitution: AMS 2418)
Plating thickness	
ASTM-B-487	Standard test method for measurement of metal and oxide coating thickness by microscopic examination of a cross section
ASTM-B-567	Standard test method for measurement of coating thickness by the Beta backscatter method
ASTM-A-754	Standard test method for coated weight of metallic coatings on steel by X-ray fluorescence

A.5 Standard References for Socket Material Property Characterization

Connector housing	
ASTM-D-150	AC loss characteristics and permittivity of solid electrical insulation
ASTM-D-256	Determining the izod pendulum impact resistance of plastics
ASTM-D-495	High-voltage, low-current, dry arc resistance of solid electrical insulation
ASTM-D-570	Water adsorption of plastics
ASTM-D-638	Tensile properties of plastics
ASTM-D-648	Deflection temperature of plastics under flexural load in the edgewise position
ASTM-D-695	Compressive properties of rigid plastics
ASTM-D-696	Coefficient of linear thermal expansion of plastics between −30 °C and 30 °C with a vitreous silica dilatometer
ASTM-D-732	Test method for shear strength of plastics by punch tool
ASTM-D-785	Rockwell hardness of plastics and electrical insulating materials
ASTM-D-790	Flexural properties of unreinforced and reinforced electrical insulating materials
ASTM-D-792	Density and specific gravity (relative density) of plastics by displacement
Connector Contact	
ASTM-B16	Standard specification for free-cutting brass rod, bar, and shapes for use in screw machines
ASTM-B139	Standard specification for phosphor bronze rod, bar, and shapes
ASTM-B194	Standard specification for copper-beryllium alloy plate, sheet, strip, and rolled bar
ASTM-B-888	Standard specification for copper alloy strip for use in the manufacture of electrical connectors or spring contacts
ASTM-B-740	Standard specification for copper-nickel-tin spinodal alloy strip

A.6 Standard References for Socket Performance Qualification

Dielectric withstanding voltage	
ASTM-D-149	Standard test method for dielectric breakdown voltage and dielectric strength of solid electrical insulating materials at commercial power frequencies
EIA-364-20A	Withstanding voltage test procedure for electrical connectors
IEC-60512-2	Electromechanical components for electronic equipment; Part 1: Basic testing procedures and measuring methods. Part 2: General examination, electrical continuity and contact resistance tests, insulation tests and voltage stress tests
MIL-STD-1344, Method 3001.1	Test methods for electrical connectors: dielectric withstanding voltage
Insulation resistance	
ASTM-D-257	Standard test method for DC resistance or conductance of insulating materials
EIA-364-21B	Insulation resistance test procedure for electrical connectors
MIL-STD-1344, Method 3003	Test methods for electrical connectors: insulation resistance
Contact force	
EIA-364-04	Normal force test procedure for electrical connectors
Mating/unmating force	
EIA-364-13A	Mating and unmating forces test procedures for electrical connectors
MIL-STD-1344, Method 2013	Test methods for electrical connectors: mating and unmating forces
Contact retention	
EIA-364-29B	Contact retention test procedure for electrical connectors
EIA-364-35B	Insert retention test procedure for electrical connectors
MIL-STD-1344, Method 2007	Test methods for electrical connectors: contact retention
Engagement/separation	
EIA-364-37B	Contact engagement and separation force test procedure for electrical connectors
IEC-60512-13-1	Mechanical operating tests – engaging and separating forces
IEC-60512-1-3	General examination: electrical engagement length
MIL-STD-1344, Method 2014	Test methods for electrical connectors: contact engagement and separation force
Contact resistance	
EIA-364-06	Normal force test procedure for electrical connectors
IEC-60512-2	General examination, electrical continuity, and contact resistance tests, insulation tests, and voltage stress tests
MIL-STD-1344, Method 3004.1	Test methods for electrical connectors: contact resistance

Dielectric withstanding voltage	
MIL-STD-1344, Method 3002.1	Test methods for electrical connectors: low signal level contact resistance
ASTM-B-539-96	Standard test methods for measuring contact resistance of electrical connections (static contacts)
Current-carrying capacity	
EIA-364-70A	Temperature rise vs. current test procedures for electrical connectors and sockets
IEC-60512-3	Basic testing procedures and measuring methods: current-carrying capacity test
Capacitance	
EIA-364-30	Capacitance test procedures for electrical connectors
Inductance	
EIA-364-33	Inductance of electrical connectors

A.7 Standard References for Socket Reliability Qualification

Environmental test methods	
MIL-STD-810	Environmental engineering considerations and laboratory tests
EIA-364-C	Electrical connector test procedures, including environmental classifications
MIL-STD-883	Test methods and procedures for microelectronics
Durability	
EIA-364-09B	Durability test procedure for electrical connectors
MIL-STD-1344, Method 2016	Test methods for electrical connectors: durability
Vibration	
EIA-364-28C	Vibration test procedure for electrical connectors and sockets
IEC-60512-6-5	Dynamic stress test – section 5: test 6a: random vibration
MIL-STD-1344, Method 2005.1	Test methods for electrical connectors: vibration
Mechanical shock	
EIA-364-27B	Mechanical shock test procedure for electrical connectors
MIL-STD-1344, Method 2004.1	Test methods for electrical connectors: shock

Environmental test methods	
Temperature life	
EIA-364-17	Temperature life with or without electrical load test procedure for electrical connectors
MIL-STD-1344, Method 1005.1	Test methods for electrical connectors: temperature life
Thermal cycling/shock	
EIA-364-32B	Thermal shock test procedure for electrical connectors
MIL-STD-1344, Method 1003.1	Test methods for electrical connectors: temperature cycling
Humidity	
EIA-364-31	Humidity test procedure for electrical connectors
MIL-STD-1344, Method 1002.2	Test methods for electrical connectors: humidity
Altitude	
EIA-364-03B	Altitude immersion test procedure for electrical connectors
Solder wicking	
EIA-RS-486	Solder wicking test procedure for sockets, plug-in electronic components
Mixed flowing gases	
EIA-364-65A	Mixed flowing gases
IEC-60512-11-7 (or IEC-60512-11-14)	Climatic tests – section 7: test 11g – flowing mixed gas corrosion test
Salt spray	
EIA-364-26A	Salt spray test procedure for electrical connectors
MIL-STD-1344, Method 1001.1	Test method for electrical connectors: salt spray
Sand and dust	
IEC-60512-11-8	Basic testing procedures and measuring methods – Part 11: climatic tests – section 8: test 11 – sand and dust
Gas tight	
EIA-364-36A	Test procedure for determination of gas-tight characteristics for electrical connectors, sockets, and/or contact systems

A.8 Other Standards and Specifications

Reporting	
ASTM-B-868	Standard practice for contact performance classification of electrical connection systems
Preparation for delivery	
MIL-C-55330	Packaging of connectors, electrical and fiber optics

A.9 Telcordia

This generic requirements (GR) document sets forth the Ericsson view of the technical GR for, and characteristics required of, connectors used for joining single-mode optical fibers, and for the jumper assemblies made using such connectors.

Telcordia is a set of GR documents that define the GR and characteristics for joining fibers, cables, and PCBs, and for the jumper assemblies made using such connectors.

GR-1217, "Generic requirements for separable electrical connectors used in telecommunications hardware," contains proposed, generic physical design requirements for separable connector products used in a typical service provider network. The requirements defined in this Network-Equipment Building System (NEBS) document apply to separable electrical connectors used in typical service providers' electronic telecommunications networks. The scope includes printed wiring board (PWB) connectors, coaxial and cable connectors of various types, and integrated circuit (IC) sockets. It includes connectors made with noble and non-noble metal contact finishes. However, these requirements do not apply to non-metallic optical connectors used for mating of fiber optics or to connectors intended for alternating current (AC) power applications. The pin and socket connector, widely used by equipment manufacturers to interface PWB assemblies to backplanes in current telecommunications systems, is the most prominent and the most critical application within this scope. Edge-card connectors with plated contact fingers are also included in the PWB classification.

This GR document includes the Telcordia view of the various requirements, desired features, and characteristics of connectors and jumper assemblies, and the performance tests for comparing such products against the stated criteria. The performance tests for analyzing optical connectors and jumper assemblies are intended to reflect a composite picture of various operating conditions. The GR, objectives, conditional requirements, and test methods are subject to change. GR-326 is regarded as the most complete and rigorous standard for single-mode fiber-optic connectors. Issue 4 defines the latest requirements required of connectors used for joining single-mode optical fibers and for jumper assemblies made using such connectors.

Issue 4 of GR-326 discusses endface geometry of single-mode connectors and contains new and updated criteria for single-mode connector reflectance. It also contains further detail regarding evaluation of samples after salt spray exposure and additional text that addresses similar products and factory considerations as it relates to present-day products. The criteria provide the manufacturer with a basis for quality control and crucial information for quality assurance to guarantee long-term performance in the field.

GR-1217 meets all quality-level standards for electrical connectors, plus provides a comprehensive treatment of electrical connector performance for telecommunications hardware and addresses both noble and non-noble plated connectors. It includes a flexible qualification methodology, which allows for cutting-edge product innovations to be qualified and used in telecommunications products and emphasizes connector performance via qualification tests, as opposed to preferred metallization; connector performance metrics were reviewed and modified where appropriate to account for reduced connector geometrics now typically employed. GR-1217 defines accelerated aging test methods, which can be used to determine reliability of mean time between failures of a particular connector. It aligns with GR-63 and GR-78.

A.10 Society of Cable Telecommunications Engineers (SCTE)

The Society of Cable Telecommunications Engineers (SCTE) is the premier membership organization for technical cable telecommunications professionals. SCTE member benefits, technical training courses, certifications, and information resources focus on providing professionals on every level of the industry with the tools needed to develop and advance their expertise. By connecting members from cable operators and vendors through a wide range of programs and events, SCTE strengthens relationships and promotes technology-focused thought leadership.

SCTE proudly serves as the technical leader by guiding through initiatives and standards aimed at improving the quality of work and advancing technologies as well as sustaining and diversifying the industry.

With the billions of components found across miles of broadband networks, it is important to ensure that all the pieces work together. This has been a major cornerstone in the SCTE Standards Program. Standards for coaxial cables and connectors have been in existence since the program started in the early 1990s. In the future, increased deployment of optical fiber, standards for fiber cables, connectors, and interfaces are what will drive the equipment standards side of the SCTE Standards Program. Standards and recommended practices for this category are developed in the existing and emerging technology areas including: connectors, fiber, coax, related test procedures, amplifiers, power requirements, housings, environmental factors, cable management, home wiring, STB/gateways, RFoG, labeling, and symbology.

A.11 Electronic Industries Alliance/Telecommunications Industry Association (EIA/TIA)

TIA/EIA-568 is a set of telecommunications standards from the Telecommunications Industry Association (TIA), an offshoot of the Electronic Industries Alliance (EIA). The standards address commercial building cabling for telecommunications products and services. As of 2017, the standard is at revision D, replacing the 2009 revision C, the 2001 revision B, the 1995 revision A, and the initial issue of 1991, which are now obsolete [1]. The best-known features of TIA/EIA-568 are the pin/pair assignments for eight-conductor 100-Ω balanced twisted pair cabling. These assignments are named T568A and T568B.

TIA/EIA-568 defines structured cabling system standards for commercial buildings, and between buildings in campus environments. The bulk of the standards define cabling types, distances, connectors, cable system architectures, cable termination standards and performance characteristics, cable installation requirements, and methods of testing installed cables. The main standard, TIA/EIA-568-D.1, defines general requirements, while TIA-568-D.2 focuses on components of balanced twisted-pair cable systems. TIA-568-D.3 addresses components of fiber-optic cable systems, and TIA-568-D.4 addresses coaxial cabling components. The intent of these standards is to provide recommended practices for the design and installation of cabling systems that will support a wide variety of existing and future services. Developers hope the standards will provide a life span for commercial cabling systems in excess of 10 years. This effort has been largely successful, as evidenced by the definition of category 5 cabling in 1991, a cabling standard that (mostly) satisfied cabling requirements for 1000BASE-T, released in 1999. Thus, the standardization process can reasonably be said to have provided at least a nine-year life span for premises cabling, and arguably a longer one [1].

The transmission performance of a cabling system depends upon the characteristics of the horizontal cable, connecting hardware, patch cords, equipment cords, work area cords, cross-connect wiring, the total number of connections, and the care with which they are installed and maintained. The development of high-speed applications requires that cabling systems be characterized by transmission parameters such as insertion loss, PSNEXT loss, return loss, and PSELFEXT. System designers use these performance criteria to develop applications that utilize all four pairs in a cabling system for simultaneous bidirectional transmission. This standard provides minimum cabling component performance criteria as well as procedures for component and cabling performance validation [2].

Purpose

This standard specifies cabling components, transmission performance, system models, and the measurement procedures needed for verification of balanced twisted-pair cabling. Requirements for four-pair balanced cabling systems are provided. This standard also specifies field test instruments and applicable reference measurement procedures for all transmission parameters [1].

A.12 International Electrotechnical Commission (IEC)

The International Electrotechnical Commission (IEC) is a nonprofit standardization organization that provides standards for all electronic, electrical, and related technology. The IEC is made up of members, called "national committees," and each NC represents its nation's electrotechnical interests in the IEC. This includes manufacturers, providers, distributors and vendors, consumers and users, all levels of governmental agencies, professional societies, and trade associations, as well as standards developers from national standards bodies. National committees are constituted in different ways. Some NCs are public sector only, some are a combination of public and private sector, and some are private sector only. About 90% of those who prepare IEC standards work in industry [3].

A.12.1 IEC Standards

EC standards have numbers in the range 60 000–79 999, and their titles take a form such as IEC 60417, "Graphical symbols for use on equipment." The numbers of older IEC standards were converted in 1997 by adding 60 000, for example, IEC 27 became IEC 60027.

The IEC cooperates closely with the International Organization for Standardization (ISO) and the International Telecommunication Union (ITU). In addition, it works with several major standards development organizations, including the IEEE, with which it signed a cooperation agreement in 2002, which was amended in 2008 to include joint development work [3].

A.12.2 IEC Connectors

IEC connectors are electrical power connectors specified by IEC standards. IEC connectors may also refer to the following [4]:

- IEC 60309, connectors primarily used for industrial purposes
- IEC 60320, for use up to 250 V AC for electrical appliances
- IEC 60906-1, 230 V AC connectors for domestic use
- IEC 60906-2, 115 V AC connectors
- IEC 60906-3, safety extra-low voltage connectors for domestic use
- IEC 62196, connectors and charging modes for electric vehicles

A.13 Military Standards (MIL-STD)

These military standards are used to standardize equipment used for military purposes including connectors. The different document types are as follows [5]:

Acronym	Type	Definition
MIL-HDBK	Defense handbook	A guidance document containing standard procedural, technical, engineering, or design information about the materials, processes, practices, and methods covered by the DSP. MIL-STD-967 covers the content and format for defense handbooks.
MIL-SPEC	Defense specification	A document describing the essential technical requirements for purchased materiel that is military unique or substantially modified commercial items. MIL-STD-961 covers the content and format for defense specifications.

Acronym	Type	Definition
MIL-STD	Defense standard	A document establishing uniform engineering and technical requirements for military-unique or substantially modified commercial processes, procedures, practices, and methods. There are five types of defense standards: interface standards, design criteria standards, manufacturing process standards, standard practices, and test method standards. MIL-STD-962 covers the content and format for defense standards.
MIL-PRF	Performance specification	A specification stating requirements in terms of the required results with criteria for verifying compliance, but without stating the methods for achieving the required results. A performance specification defines the functional requirements for the item, the environment in which it must operate, and interface and interchangeability characteristics.
MIL-DTL	Detail specification	A specification specifying design requirements, such as materials to be used, how a requirement is to be achieved, or how an item is to be fabricated or constructed. A specification that contains both performance and detail requirements is still considered a detail specification.

A.14 Standards for Space-Grade Connectors

For space applications, there are requirements that are associated with standards that must be followed. NASA applicable connector standards [6] are given here. ESA applicable connector standards [7] are given in Table A.1.

S-311-P-10D-Connectors – Electrical, Rectangular, Miniature, Polarized Shell, Rack and Panel, for Space Flight Use

S-311-P-4/05C – Connectors, Electrical, Rectangular, Subminiature, Polarized Shell, Detailed Specification for

S-311-P-4/06A – Contacts, Electrical, Coaxial, and High Voltage for Electrical Connectors

S-311-P-4/07C – Connectors, Electrical, Rectangular, Subminiature, Rack, and Panel, Detailed Specification for

S-311-P-4/08C – Contacts, Electrical, Power, Crimp Style, Rear Removable, Size 22D

S-311-P-4/09C – Connectors, Electrical, Polarized Shell, Rack, and Panel, for Space Flight Use

S-311-P-4/10C – Contacts, Electrical, Crimp Style, Size 20

S-311-P-4D – Connectors, Electrical, Subminiature, Rack, and Panel, General Specification for

Table A.1 Specifications of ESA for multicontact connectors.

Doc. or Spec. No.	Specification Title
3401	Generic Specification for Connectors Electrical Non-Filtered Circular and Rectangular
3401/001MS	Connectors Electrical Rectangular Non-Removable Solder Bucket PCB and Wire-Wrap Contacts and Removable Coaxial and Power Contacts, based on type D*M
3401/002MS	Connectors Electrical Rectangular Removable Crimp Contacts, based on type D*MA
3401/004	Contacts Coaxial Crimp-type and Solder-Type for 3401/001 and 3401/002 Connectors
3401/005	Contacts Electrical Crimp for 3401/002 Connectors
3401/007	Connectors Miniature Electrical Circular Bayonet Coupling Crimp-Type Removable Contacts, based on type DFE
3401/008	Connectors Miniature Electrical Circular Push-Pull Coupling Removable Crimp Contacts, based on type DBAS
3401/009	Contacts Electrical Crimp for 3401/007 and 3401/008 Connectors
3401/011	Connectors Subminiature Electrical Rectangular Central Jackscrew Coupling Crimp-Type Removable Contacts, based on type UR
3401/012	Contacts Electrical Subminiature Crimp-Type for 3401/011 Connectors
3401/016	Connectors Electrical for Printed Circuit Boards Removable Contacts Crimp Wire-Wrap Solder and Saver, based on type HE801
3401/017	Contacts Electrical Crimp Wire-Wrap Solder and Saver for 3401/016 Connectors
3401/020	Connector Savers Electrical Rectangular Miniature Removable Contacts, based on type D*BMA
3401/021	Contacts Electrical Male/Female Type for 3401/020 Connector Savers
3401/022MS	Accessories for Rectangular Connectors 3401/001 and 3401/002 and Connector Savers 3401/020
3401/023	Connectors Miniature Electrical Circular Push-pull Coupling Hermetic Receptacle, based on type DBA 7*H
3401/029	Connectors Electrical Rectangular Microminiature, based on type MDM
3401/031	Connectors Electrical Single-in-Line Microminiature, based on type MTB1
3401/032	Accessories for Connectors Microminiature 3401/029 and Connector Savers 3401/041
3401/033	Connectors Savers Electrical Circular Miniature non-removable, based on type DBAS
3401/039	Connectors Electrical for Printed Circuit Boards Non-Removable Solder and Wire-Wrap Contacts and Connector Savers, based on type KMC

Table A.1 (Continued)

Doc. or Spec. No.	Specification Title
3401/040	Contacts Power Crimp-Type and Solder Type for 3401/001 and 3401/002 Connectors
3401/041	Connector Savers Electrical Rectangular Microminiature, based on type MDM
3401/043	Connectors Miniature Electrical Circular Bayonet Coupling Hermetic Receptacle, based on type DBC5*H
3401/044	Connectors Electrical Circular Bayonet Coupling Removable Crimp Contacts, based on MIL-C-38999 series II
3401/045	Contacts Electrical Crimp Type for 3401/044 Connectors
3401/052MS	Connectors Electrical Circular Bayonet Coupling Scoop-Proof Removable Crimp-Type Contacts, based on MIL-C-38999 series I
3401/056	Connectors Miniature Electrical Circular Triple-Start Self-Locking Coupling Scoop-Proof Removable Crimp Contacts, based on MIL-C38999 series III
3401/057	Connectors Miniature Electrical Circular Triple-Start Self-Locking Coupling Scoop-Proof Hermetic Receptacle and Feedthrough, based on MIL-C-38999 Series III
3401/058MS	Contacts Electrical Crimp for 3401/052 and /056 Connectors
3401/060	Connectors Electrical Rectangular Multiple Insert Type Rack and Panel Removable Crimp Contacts, based on MIL-C83527
3401/061	Contacts Electrical Crimp for 3401/060 Connectors
3401/062	Accessories for Circular Connectors 3401/0443401/052 and 3401/056
3401/063	Connector Savers for Electrical Circular Connectors 3401/044, 3401/052, and 3401/056
3401/064	Accessories for Circular Connectors 3401/008 (DBAS)
3401/065	Connectors and Savers Electrical Rectangular Non-Removable PCB Contacts, based on type MHD
3401/066	Contacts Electrical Triax Crimp for 3401/056 Connectors
3401/069	Contacts Electrical Twinax Crimp for 3401/001 Connectors
3401/070	Connector Receptacles Electrical Circular Triple-Start Self-Locking coupling Scoop-proof, P.C.B. Non-Removable Contacts Based on MIL-C38999 series III
3401/072	Lightweight Accessories for Rectangular Connectors 3401/001 and 3401/002

S-311-P-768 – Connectors, Electrical, Miniature, Circular, Low Outgassing [Amphenol (453) Suffix], General Specification for
S-311-P-822A – Connectors, PWB, 2mmc PCITM Style, High Reliability
S-311-P-826A – Triaxial Connector, Ultraminiature
S-311-P-835 – Connector, Electrical, RF, Twinaxial, Microminiature

References

1 TIA's Fiber Optics Tech Consortium, TIA/EIA-568. www.tiafotc.org/ansi-tia-568-d (accessed Mar. 30, 2020).

2 TIA/EIA 568-B.2 (2001). *Commercial building telecommunications cabling standard, Part 2: Balanced twisted-pair cabling components.* TIA/EIA.

3 International Electrotechnical Commission, About the IEC. www.iec.ch/about (accessed Mar. 30, 2020).

4 IEC Webstore, IEC Connector. webstore.iec.ch/searchform&q=iec%20connector (accessed Mar. 30, 2020).

5 Weibull, Military directives, handbooks and standards related to reliability. www.weibull.com/knowledge/milhdbk.htm (accessed Mar. 30, 2020).

6 NASA, Electronic Parts and Packaging (NEPP) Program, Goddard Space Flight Center S-311 Specifications Lookup. nepp.nasa.gov/index.cfm/23192 (accessed Feb. 3, 2020).

7 European Space Components Information Exchange System, ESA/SCC Specifications for Multicontact Connectors. escies.org/webdocument/showArticle?id=575&groupid=6 (accessed Feb. 3, 2020).

Index

Electrical Connectors: Design, Manufacture, Test, and Selection, First Edition.
Edited by San Kyeong and Michael G. Pecht.

d

t